Everest

GEORGE EVEREST
1790-1866

Everest:
The Man and the Mountain

J.R. Smith

Whittles Publishing

Typeset by
Whittles Publishing Services

Published by
Whittles Publishing,
Roseleigh House,
Latheronwheel,
Caithness, KW5 6DW,
Scotland, UK

© 1999 J.R.Smith

All rights reserved.
No part of this publication may be reproduced,
stored in a retrieval system, or transmitted,
in any form or by any means, electronic,
mechanical, recording or otherwise
without prior permission of the publishers.

ISBN 1-870325-72-9

Printed by J.W.Arrowsmith Ltd., Bristol

Dedicated to all those of the Survey of India who have, for over 150 years,
upheld the professional standards of care and integrity
generated by Colonel Sir George Everest.

A SETU HIMACHALAM
(From Cape Comorin to the Himalayas)
Motto of the Survey of India

...for I am by no means disposed to be very humble, or play the courtier, or to kiss the rod which chastises me, or, to borrow from my more lowly countrymen, one of their significant phrases, 'to think etc. etc. of myself'

George Everest,
An Account of the Measurement of an Arc...,
Parbury Allen & Co., London, 1830, p. vii

CONTENTS

Foreword ix
Preface xi

Part I   The life and work of Sir George Everest   **1**

1 Introduction   3
2 Early years: family beginnings and a military education   5
3 India 1806–1820: early survey work, an important appointment, and jungle fever   11
4 The Cape of Good Hope 1820–1821: a time for recovery, and confirmation of historical observations   27
5 India 1822–1825: taking charge of the Survey, and battling with fever again   35
6 England 1825–1830: sick leave at home, arguing for better equipment and conditions of service   53
7 India 1830–1834: a period of intensive field work   67
8 India 1834–1839: better instruments, and remeasurement   91
9 The Jervis affair   105
10 India 1839–1843: final measurements, and the handover of the Survey to Waugh   123
11 England 1843–1866: retirement, a late marriage, and recognition   133

Part II   The Park Estate and family matters   **145**

12 The Park Estate   147
13 Family memories from Sir George's niece, Mary Boole   155
14 Relatives and friends   165
15 Chronology   179

viii   Everest: The Man and the Mountain

Part III   The measuring and naming of Mount Everest   **187**
    16   The mountain: measuring its height   189
    17   The mountain: the issue of its name   211
        Postscript   225

Appendixes   **229**
    1   Instrumentation used in the Survey of India   231
    2   Instrument makers   249
    3   An atlas of India   261
    4   Everest's role in administration   267
    5   The figure of the Earth and geodesy   273

    References   289
Index   **299**

Foreword

Colonel Sir George Everest was a distinguished Survey of India officer and his brilliant career culminated in his becoming the Surveyor General of India from 1830 to 1843. He laid the foundation of modern geodetic surveys in India. His original work in measuring a Meridional Arc from Kanya Kumari (Cape Comorin) in the extreme south to the first Himalayan range (Banog near Mussoorie) is unparalleled in the world. Colonel Sir George Everest was a genius and, to honour him in recognition of his invaluable contribution to science, the highest peak in the world has been named after him as "Mount Everest". The biographical sketch of this great scientist has been brought about by Mr J.R.Smith for the first time in the form of a book and this is a befitting tribute to a great genius.

V.K.Nagar
Surveyor General of India

Preface

A number of people collected information on different aspects of the work of Sir George Everest for his birth bicentenary in 1990 and this gave an opportunity to bring it all together for the first time. In doing so, some earlier statements now appear to be incorrect, and new, fascinating sides to his life have appeared. Some threads have been discovered and followed until a dead end was reached; these remain naggingly unresolved. Assumptions have been made and later proved false. Others have been more fruitful.

So that the picture might be as complete as possible, details of the Great Trigonometrical Survey are included from its inception in 1799, as essential background to the work of Sir George Everest.

In the original documentation and authoritative publications the spelling of any one place name can vary from page to page (for example, Takal K'hera, Tekarkhera, Takalkhera; Mohsin Husain, Mohsin Hussain, Mohsin Hussein). While only one version of any one name is used here (except in verbatim quotations), it may not be the version felt most appropriate by some readers. For this, apologies are proffered but the difficulties must also be obvious.

Generally the references are indicated as [123 p. 4] but with the documents from the Survey of India, housed at the National Archive of India, the form is [263 (266) 192]. Here 263 is the reference number, and indicates the National Archive of India (Survey of India); (266) is the file number; and 192 the page within that file.

Numerous references will be found to the National Archive records [263] since they are such a valuable collection of source material. Much of it is, however, also referred to and quoted within the five volumes by Phillimore [227, 228, 229, 231, 232] and these may well be more easily accessible to the reader interested in investigating the subject in more depth.

After much discussion and thought it was decided to keep Part 1 of the volume in as near chronological order as possible. In addition, a scientific personality, working daily with numbers, cannot adequately be written about without the inclusion of a certain number of formulae and figures, so apologies for the deviations into these, even within the main text. Some more specialised and technical aspects of Everest's work in India are in the Appendices; the non-specialist reader may wish to omit these.

Although Everest never saw the mountain that bears his name and had no direct input to its survey, it would nevertheless seem appropriate to include details of it. In compiling Part 3 (chapters 16 and 17) only two aspects of the mountain (its name and height) have been considered. Those interested in climbing routes, topography, conquerors and the like will have to search elsewhere.

Some publications lend themselves to numerous quotations, others not, and there is perhaps a fine line between the two to get the right balance. Here it will be found that some chapters contain many quotations while others have few. With a person such as Everest, where little is written about him by people who knew him, the flavour of his attitudes, language and ideas can best be obtained by quoting from his own correspondence and publications. Hence the liberal use of this approach where appropriate.

It would be inappropriate in this book to metricate the linear measurements. Since all operations at the relevant period were made in imperial units, these are retained here. Some care is, however, required with the unit 'foot' because a Paris foot (as opposed to an English foot or an Indian foot) occasionally appears. For the metricated reader the conversions are:

1 inch = 0.0254 metre

1 foot = 12 inches = 0.3048 metre

1 fathom = 6 feet (Eng.) = 1.8288 metres

1 mile = 5280 ft (Eng.) = 1.60934 kilometres

1 toise = 6 Paris feet = 6.394596 English feet = 1.9490 m

1 Paris foot = 1.065766 English feet = 0.3248 m

1 Indian foot = 0.99999566 English feet = 0.3048 m

Where the term 'feet' is used on its own the implication is that it refers to Indian feet.

A number of loose ends will be found throughout the story. Obviously with time one could possibly pursue these to a conclusion but it was felt preferable to publish once the bulk of available material had been sifted. This could encourage others to pick up the trail or bring to light otherwise unknown sources. I will be pleased to hear from any readers who can throw further light on George Everest and his relatives.

All in all, this has been a frustrating but fascinating chase through records of the last two hundred years with all the usual sideline distractions for those who are easily led or at least who find it difficult to stay on the beaten path.

Acknowledgements

Many people warrant sincere acknowledgement for the help they have given during the collection of material for this work. Foremost among these are:

Andrew Cook and staff of the India Office library;

Michael Pollock of the Royal Asiatic Society library;

Peter Hingley and staff of the Royal Astronomical Society library;

David Wileman and staff at the Royal Geographical Society library;

Pauline Lane-Gilbert and staff at the Royal Institution of Chartered Surveyors library;

Sheila Edwards and staff of the Royal Society library;

Staff of the Society of Genealogists library;

Miss G. Becker, Librarian, Surveys & Mapping, Cape Town.

In India exceptional assistance was given by:

Dr Perti and staff at the National Archives in Delhi;

Lt Gen S.M. Chadha and Sri V.K. Nagar of the Survey of India in Dehra Dun.

Special mention must also be made of Roy Minnick for getting the publishing work off the ground and for Dr Keith Whittles for bringing it to a successful conclusion. Thanks also to Mairi Sutherland for her editing.

Acknowledgements are due to:

The British Council for their assistance with my visit to the Indian celebrations for the

bicentenary of Sir George Everest's birth;

the Syndics of the University of Cambridge and Director of the Royal Greenwich Observatory for permission to quote from the papers of G.B. Airy;

Frank Christopher of Delhi for his valuable comments at draft stage;

Peter Clark of the Royal Geographical Society;

Corpus Christi College Oxford for permission to use extracts from their Minute Book;

Professor Matthew Edney of the State University of New York, Binghampton, for his very useful text *Mapping an Empire* (1997);

Henry Gray for assistance with the Coat of Arms and family tree;

Jane Insley of the Science Museum, without whose assistance this work would have been very incomplete;

Lt Col J.S.O. Jelly, former officer of the Survey of India, for his valuable comments at draft stage;

Alastair Macdonald of the Ordnance Survey who started me off on the trail of Sir George Everest;

The Education Trust of the Royal Institution of Chartered Surveyors for generous assistance with my investigations on the history of surveying – of which Everest is an important figure in the provision of geodetic control for topographic and other surveys of continental extent.

I would also like to make mention of the late A.P.H. Werner whose *'Calendar of the development of Surveying'* in the *Australian Surveyor* during 1966–1968 first aroused my interest in the historical aspects of surveying.

xiv Everest: The Man and the Mountain

Finally, special thanks go to family and close friends who must be sick and tired of the name Everest and its domination of many years of their life.

Many others, too numerous to list, have played important parts with the supply of information and to all these I extend my heartfelt thanks.

Transcripts from Crown-copyright records in the India Office Records appear by permission of the Controller of Her Majesty's Stationery Office.

Ian Miller of The C.W. Daniel Co. Ltd for permission to use material from *The Collected Works of Mary Everest Boole,* and for the photograph of the same lady used in photograph 23;

The Keeper of Manuscripts, Nottingham University Library for permission to quote from the records held there under reference PwJf;

Simon Hempsell of the Photographic Unit of University of Portsmouth for assistance with the photography;

Bill Johnson of the Geography Department of University of Portsmouth for his work on photographs 1, 2, 3, 4, 9, 10 and 33;

the Royal Geographical Society for permission to use photograph 15;

the Council of the Royal Society for permission to quote the details of the certificate admitting Everest as a Fellow;

the Board of Trustees of the Science Museum, London for assistance with the reproduction of photographs 14 and 17;

A.F.W. Sheffield for his help with the Wing family tree and with photographs 16, 24, 26, 27, 30 and page 2;

Audris Skuja for the photograph of Bradford Washburn's model of Mount Everest used as photograph on page 188;

the Survey of India for supplying the material for photographs 1, 6, 7, 11, 12, 19, 20, 32, 36, 37, 39–45 and for permission to use extensive quotations and photographs 5, 8 and frontispiece from the volumes of Col. R H Phillimore and Survey of India Records;

Dr Bradford Washburn of The Museum of Science, Boston, USA, for the photographs of Mount Everest used in photographs 34 and 35.

A.G.Bradbury for permission to reproduce photograph 31.

Cover photographs of Mount Everest courtesy of Leica Geosystems.

All others are by the author.

J.R. Smith

Part I

The life and work of Sir George Everest

Chapter 1

INTRODUCTION

This is the life story of a single-minded man whose name is a household word although many do not associate the name with a person. He was born over two hundred years ago, yet his name endures and is likely to continue to do so, engraved in the highest peak on earth.

Little if anything is known of the man unless the reader happens to be a geographer or surveyor. Yet surely, if the highest known peak of the world bears someone's name, then the person behind the name must have been exceptional.

Certainly that is so, but not in a glamorous way. The story will not make a box-office success as a film. It does not even rival the exploits of a Livingstone, Cook or Columbus. Perhaps that is why no one has put together before a biographical paper about Everest, let alone a full biography.

(I must qualify this by noting that the five volume *Historical Records of the Survey of India* by Colonel R. H. Phillimore, CIE, DSO, and particularly volume 4, contain a mass of information about both the personal and professional life of George Everest. However, it is scattered throughout the general history of the Survey of India. The vast, minutely detailed, effort that Phillimore put in over many years is itself a monument to his dedication. It has drawn upon every conceivable source of material and must be the first port of call for any query on survey operations in India up to 1861.)

Might it then have been a fool's errand to try to put together the life story of Everest? Certainly it was fraught with difficulties. Some periods had a wealth of information while others were very bare. Nevertheless, I hoped to formulate a reasonable picture of the life of this individual that might appeal to both technical and non-technical reader alike.

Mary Everest Boole's comment did not make the task any easier [44] 'that circumstances into which I cannot now enter, led to the destruction of nearly all written memorials of his life...' No other reference has so far brought to light what the circumstances were, but certainly the sentiment appears to be all too correct. The material referred to must have consisted of only a single copy of everything and none spread around family or friends.

Relatives survived him, but only one, a niece, appears to have put pen to paper on family matters and then only in a cursory way. The direct line died out in 1935 and the other branches are far removed. Perhaps when a scholar, remote descendant or archivist somewhere reads this, up will pop just the information we have sought, but until that time we must accept the available crumbs and make the most of them.

Everest's professional career centred around the measurement of an arc of the meridian, or section of a line of longitude, that stretched from the south tip of India to the foothills of the Himalayas. For someone to spend most of his working life on one project there must have been something very special about it.

If this measurement was of a high order of accuracy the results had a twofold use. Firstly they could combine with other similar measures from around the world to allow better determination of the size and shape of the earth (known as the figure of the earth). Such knowledge has extreme importance in navigation but other scientific fields require it as well.

Secondly, the arc, measured by triangulation (or the observations in a long series of connected triangles) together with other triangulation, would form the skeleton on which the surveying for detailed maps of India could be based. As with the human body or a building, the flesh or fittings cannot be added until there is a firm and reliable foundation.

In 1834 Everest received a letter that referred to him as a Kumpass Wala[*] – a term to which he took considerable offence. In response he said 'I am the Surveyor General and Superintendent of the Great Trigonometrical Survey of India... I objected to a low, familiar, appellative which, though it may be in common use in a bazaar, I cannot allow to be applied to me as my official designation...' He requested to be addressed by the same title as that used by the Commissioner: Surveyor General Kishewar Hind '...which is a literal translation of that assigned to me by my masters". [263 (289) 175–7]

A surveyor, scientist, linguist, engineer, astronomer and religious philosopher: Everest had many sides, but through them all he had a purposeful goal, went for it, and did not suffer fools gladly on the way. He 'was so indefatigable, that his contemporaries...were accustomed to speak of him as *Neverrest*' [18 p. xiii] His determination to see the Trigonometrical Survey through to its conclusion must be second to none in the annals of persistence under extreme conditions, yet up to now no one has chronicled his efforts. I hope that this volume will go some way to redress the balance and spread an understanding of him that is justly warranted.

---

[*] a derogatory term literally meaning 'user of a compass' or lower class of surveyor.

Chapter 2

EARLY YEARS: Family beginnings and a military education

*P*icture, if you can, Greenwich, England, in the late eighteenth century. Greenwich Observatory, built over 100 years earlier, looked down over the river from its lofty perch. The Everest family (said to have been pronounced Eve-rest) had resided in Greenwich for most of the century, if not longer. Perhaps the two were of similar vintage.

Ancestors

At the end of the 1600s Tristram Everest was a butcher in Church Street, the same street where one finds St Alphege's Church containing memorials to various members of the family. He married Anne Fisher in 1682 and they had 14 children: Elizabeth, John, Ann, Tristram, Elizabeth, Ann, Mary, Margaret, Anthony, Dorothy, Fisher, Elizabeth, Thomas and William. Their years of birth ranged from 1683 to 1710/11 at a steady interval of two years. For many of them only the names and birthdays are known.

The multiplicity of the names Elizabeth and Ann could imply that the early ones died young. Three married: Tristram, Margaret and the third Elizabeth. Of the 14, the one of interest to us is Tristram, born in 1688. Like his father, Tristram was also a butcher in Greenwich and similarly went in for a large family. He married Mary Hack around 1710 and they had eight children: Charles, Mary, Mary, John, John, Mary, Ann and Tristram Hack. Their years of birth ranged from 1711 to 1728. Again, the multiplicity of names probably represents early deaths of two Marys and a John. Tristram died in 1752.

Charles did marry and have three children, but the second John, born on 20 July 1718, is the offspring of interest here. He married Susannah Cole by licence on 5 April 1741 at St Mary Magdalene, Old Fish Street. John did not carry on the family tradition of butchering but instead practised law and was articled in 1736 to Wm Radley of Greenwich. Keeping up the family tradition, John had 10 children: Susannah, Mary, Mary, Elizabeth, John Cole, Tristram, William, George, Robert and Kitty with birth dates ranging from 1742 to 1756. He died in 1769.

Of the children, three are known to have married and had families: John Cole, Tristram and George. As before, it is the Tristram (actually William Tristram) who interests us. He was born in 1747 (baptised 22 November) in Greenwich and, like his father before him, became a solicitor. Among his posts Tristram was solicitor to the Board of Greenwich Hospital and to Chelsea Hospital. In 1766, records show him offering a reward on behalf of the Board of Greenwich Hospital for the recovery of iron bars stolen from Crane Wharf in Greenwich. He married Lucetta Mary Smith on 2 August 1786. Here we have reached the father and mother of the future Sir George Everest. There were six children: Lucetta Mary, John, George, Robert, Thomas Roupell and Charles. Their dates of birth ranged from 1787 to 1802. Tristram died in 1825.

George Everest is Born

George Everest was born on 4 July 1790 but the location remains in doubt. Some references place his birth at Gwernvale, an estate near Crickhowell, Brecon, whilst others quote Greenwich. His father did have an estate at Gwernvale, but George's baptism took place on 27 January 1791 at St Alphege's Church, Greenwich. Whilst the baptism certificate records the date of birth it makes no mention of the place of birth. As recently as the 1930s the College of Heralds was approached for its opinion and the outcome favoured Greenwich although no written proof exists. The *Richmond Herald* could only find a statement by Sir George on the 1861 census that he was born in Greenwich. This, of course, may or may not have been true.

Gwernvale still exists today as the Gwernvale Manor Hotel. The site reportedly has been inhabited for 800 years, and it seems to have passed into the Everest family

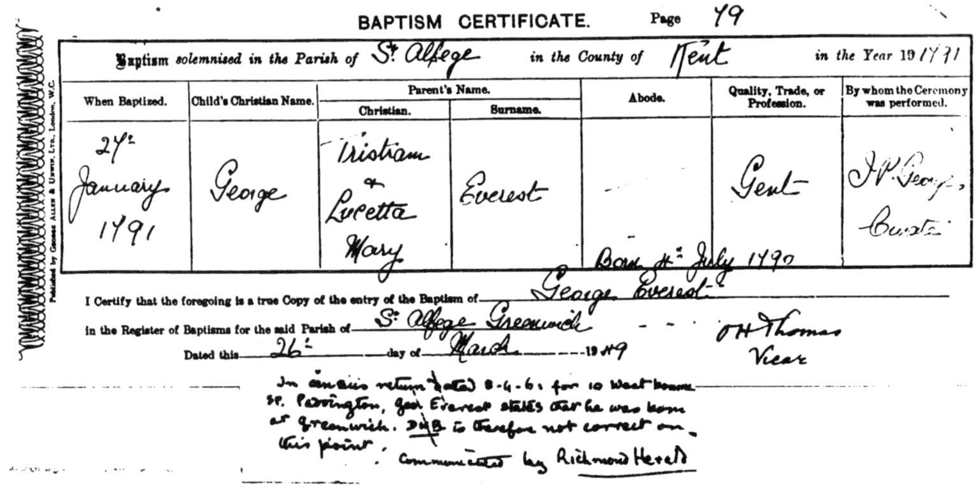
BAPTISM CERTIFICATE. Page 79

Baptism solemnised in the Parish of St Alfege in the County of Kent in the Year 1791

When Baptized.	Child's Christian Name.	Parent's Name. Christian.	Parent's Name. Surname.	Abode.	Quality, Trade, or Profession.	By whom the Ceremony was performed.
27th January 1791	George	Tristram & Lucetta Mary	Everest	Born 4th July 1790	Gent	J. P. George, Curate

I Certify that the foregoing is a true Copy of the entry of the Baptism of George Everest in the Register of Baptisms for the said Parish of St Alfege Greenwich

Dated this 26th day of March 1949

O H Thomas, Vicar

In census return dated 8-4-61 for 10 Westbourne St. Paddington, Geo. Everest states that he was born at Greenwich. DNB is therefore not correct on this point. Communicated by Richmond Herald

Figure 1 *Baptismal certificate showing the date but not the place of birth.*

in the 1780s. This makes it a little uncertain as to whether George's father or another relative originally purchased the estate. The Everests did some rebuilding and Gwernvale then remained unchanged until this century, although it seems that the estate passed out of the family on the death of George's father in 1825. A local history of 1898 suggests that Mr Everest built the mansion as it looked then and that he used to reside there for the greater part of the year [190 p. 400]. The older property was possibly somewhat removed from the site of the present building.

In his will, proved in 1825, Tristram Everest left 'all my messuage farm lands and hereditaments situate in the Parish of Crickhowell, Co Brecon and all other my messuages lands tenements hereditaments and real estate whatsoever and wheresoever... to the said Anthony M. Hawkins, William Church and George Smith ... These gentlemen, 'his dear and highly valued friends' from either Middlesex or London; together with George Everest they were the executors of the Estate. From the same will George inherited his father's '... large telescope and also my gold watch with the Chain and Seals and all my written books and books of account, maps and papers.' In addition some monies and shares passed to George. [132]

Brothers and sisters

Among George's siblings, one died young and three died unmarried. John, a Fellow of Corpus Christi College, Oxford, died in 1820 at the young age of 32. He is buried in the Cloisters of the College with a simple pavement stone inscribed 'J E 1820'. Lucetta Mary died in 1857 aged 69, and Robert died in 1874 aged 75. Robert had an MA from Oxford where he had studied geology, a particular interest of his later in India, and where he wrote many papers on the topic. He had been a Chaplain in the service of the East India Company (Bengal) and after retirement returned to live the rest of his life in Ascot. On one occasion George referred to Robert as 'my own much loved brother' and they certainly met up several times in India. His nephew Lancelot Feilding Everest described his Uncle Robert as 'always kind and generous to us children'.

Before discussing George in any detail it is worth saying a little about Thomas Roupell, born in 1800, who did marry. (Somewhere in the Everest ancestry one might expect to find the name Roupell again since it is so uncommon as a christian name as to surely have some hidden significance.) He became Rector of Wickwar in Gloucestershire – some five miles north of Chipping Sodbury – and married Mary Ryall, a sister to Dr John Ryall, Professor of Greek and Vice Principal of Queen's College, Cork. In 1837 a 'flu epidemic left Thomas a wreck, and he was advised to go to France for a while. He left a Curate in charge at Wickwar and was actually away some ten years until, as his daughter Mary put it [67], he discovered the true cause of his health problem and put it right.

While in France, Thomas became closely involved in the work of Samuel Hahnemann who practised homoeopathy. Although somewhat eccentric he was a man of learning and, as with George, was a close friend of the eminent scientists

Babbage and Herschel. It is likely that Thomas was not only a patient of Hahnemann but might even have acted as a guinea pig. Hahnemann imposed his philosophies, involving rigid rules about cold baths and walks before breakfast, on the Everest family. This left Thomas' wife and the nanny in the unenviable position of having to treat the two children cruelly. Mary's brother George used to scream at the bath – from which Nurse had just cleared the ice, sometimes with the help of a stout stick – but Mary said that she was too proud to scream. The family returned to England in 1843 and Thomas died in 1855.

His death hastened his daughter Mary's marriage to George Boole, the Professor of Mathematics at Queen College, Cork. George Boole is well-known for his mastery of logic but had a tragically short life. He was the son of a shoemaker from Lincoln with no more than a basic school education. Yet at 20 Boole opened his own school and spent all his spare time studying languages. (His fascination with mathematics and logic came later.) In 1849 he became Professor and a Fellow of the Royal Society. He died at his home in Ballintemple of a feverish cold and congestion of the lungs in 1864 aged 49.

A military education

George, and probably his brothers and sister also, received only an ordinary, non-scientific education, customary for young gentlemen of that time. Certainly he would have had a good helping of Latin and some theology but also probably a large amount of flogging, fagging and fisticutting. [67 p. 1095]

In his early years, George lived on Lime Kiln Lane in Greenwich, which runs down past the Town Hall to near the National Maritime Museum and Royal Naval College and the River Thames. With that and the Greenwich Park nearby, no doubt youngsters met a wide variety of characters and had plenty of scope for mischief. Considering the rough naval presence in that part of London at the time one can hardly wonder that he and other boys got into lots of scrapes. Apparently George was a very troublesome boy but not of the deceitful or really wrong kind. The ladies of the household did not welcome his escapades. On one occasion he was found in a quarry firing at a rock with no care for where the bullets might ricochet.

Perhaps this type of behaviour prompted the idea to send George to the Royal Military Academy at Woolwich. On 11 July 1804 he was nominated by the Earl of Chatham, Master General of the Ordnance, and proposed by John Manship, a Director of the East India Company, to be a gentleman cadet in the Royal Regiment of Artillery. He first went to the Royal Military College at Marlow for a short while and then moved on to Woolwich where he completed the course. [130], [228]

From 1798 the Royal Military Academy provided officers for the East India Company as well as for the King's Service. This resulted in an increase in cadet numbers to 100, with 40 earmarked for the East India Company. As this proved insufficient other 'extra' cadets were drawn from neighbouring schools. [156] (A Royal Charter on 31 December 1600 had incorporated the East India Company, an English company formed for the exploitation of trade with the Far East and India. Later, as it gradually became involved in politics it acted as an agent of the British Government in India from early 1700s until the mid 1800s. An annually elected Court of Directors ran the company, but it ceased to exist in a legal capacity in 1873.)

From 1803 – the year before Everest started – the extra cadets transferred to the Royal Military College Marlow where in 1803 there were 20 cadets destined for the East India Company and 60 for the King's Service. As vacancies occurred at the Academy, transfers were effected from the College and this was the route taken by Everest.

Among the staff of the Academy at that time were Dr Hutton, Everest's Professor of Mathematics, and Sir James Ivory. Hutton saw the mathematical potential of Everest and forecast a distinction, which young Everest duly obtained. In fact George qualified for a commission before the age at which the law admitted such a position.

The syllabus

The syllabus studied by Everest included the elements of arithmetic, vulgar and decimal fractions, square and cube roots, principles of algebra up to quadratic equations, classics, and French, progressing to the elements of Euclid, trigonometry and mensuration of solids, conic sections, applied mechanics, laws of motion and fluxions, fortification and drawing. These were supplemented with various days devoted to science and use of globes. One should not read much significance into the fact that George passed highly on the course since, at the time, the course was criticised for its low standards. The course lasted for little more than a year in Everest's case.

In 1802 a new arrangement was laid down for the conduct of mathematical studies throughout the Institution itemising six points:

1. Every student must take his turn to come up to the master and to do as much work as he was able.
2. Since this process is time consuming any student found unprepared should be immediately sent back to his place.
3. Any students who persist in not working will be brought to the notice of the Lieutenant Governor or Inspector.
4. All the rules and definitions are to be repeated by heart. Only the first example in each rule is to be entered in the written book; the rest are to be on a slate and erased in the presence of the Master.
5. Every rule, example, definition, theorem and problem is to be learnt and distinctly said and explained to the Master.
6. The geometrical part of the course to be commenced after going through the algebra. All theorems and problems to be entered in a book with figures and by these figures only are the demonstrations to be admitted. [13]

George's training would probably have been among the first influenced by the ideas of Robert Dawson. In 1803 Dawson put forward suggestions for better instruction of prospective survey personnel.

> First the mathematical principles of survey were taught, and applied to Practice, by performing a Survey of a small quantity of Ground, ... when the Survey is completed, the way of plotting or projecting it on Paper is taught, and when the Ground Plan is laid down, Methods of determining the Areas of enclosures etc. on it, will be given.
>
> Secondly the principles of trigonometrical survey were inculcated. The piece of country ... may be consider'd as a military position which is to be taken up and fortified, ...considerable precision being essential in the determination of general distances and heights – these are first ascertained Trigonometrically, which gives an opportunity of explaining the Process, and affording necessary practice, in Trigonometrical Surveying, Roads, Rivers and particular boundaries, will then be survey'd instrumentally.
>
> Thirdly instruction in reconnoitering was given, and fourthly the delineation and relief of surfaces, practising from Models as well as Ground. Finally ...He [the student] makes fair drawings, of the plans produced from his own Surveys, which concludes the Instructions. [73]

This six-month course after Academy training was instructed by people such as Major William Mudge. The period of Everest's education coincided with the early years of the Ordnance Survey. As a result it could well have been that the surveying content of his course was influenced by developments for the national survey.

On 30 November 1805 Everest received a certificate from Woolwich allowing him to leave the establishment on his being recommended for promotion. He had attained the qualifications required of gentlemen cadets destined for the King's Service, and a recommendation, through the Master General of Ordnance, to the Director of the East India Company.

Departure for India

George sailed for India on the *Sir W. Bensley* in 1806 as an Artillery cadet in the Bengal Presidency. His departure from England hit his mother very hard. Despite his escapades he was a favourite among the family, and his absence probably hastened her death in 1809.

Chapter 3

INDIA 1806–1820: Early survey work, an important appointment, and jungle fever

*O*ur young cadet arrived in India on 11 July 1806. He had entered the Bengal Artillery with the rank of Lieutenant on 4 April. This initial rank of Lieutenant is interesting. It happened that all the cadets of his season were promoted at once to that rank to fill existing vacancies. [261 p. 251] Everest referred in passing that from the age of 16 (i.e. after arriving in India) he was in garrison, cantonment or camp, but he gave no further details, and records do not indicate his activities. [96 p. vii]

Survey work in Java

Records suggest that Everest was at the siege of Kalinjar (Callinger) near Allahabad in 1812 when that fort was captured, but no corroborative proof exists. Then in 1813 records place him in Java. In fact one source [261] suggests that he went to Java much earlier – 'after about a year's service' (i.e. 1807) – but again there is nothing to substantiate this. So sketchy is this period of his life that one can but assume he was on the mundane and routine duties that might be expected of an occupying army. He recorded in his memo on the Great Trigonometrical Survey [94] that early in his career he had a considerable interest in astronomy. Then in 1812–13 a friend, Robert Smith of the Engineers, requested that Everest go with him on a survey task. This was refused since it was considered that no officer should be employed out of his own line.

Why Java? In the Napoleonic Wars the French not only overran the Netherlands but in 1811 also occupied Batavia, the capital of Java, which was the chief Dutch possession in the East. [227] Such a close proximity of the French worried the Directors of the East India Company which had scattered settlements in the eastern islands. Hence the Governor General of India, Lord Minto, organised a large force, mainly of Madras troops, to recapture Batavia, which they did on 26 August 1811. Sir Stamford Raffles was then appointed Lieutenant Governor of Java and it was held until 1816 when it was restored to the Dutch. Such a delay was not without its reasons since in 1812 William Macbean George Colebrooke started a reconnaissance survey of the eastern end of the main island and of neighbouring islands. In July of the same year he started a survey of the Solo river, a task he continued until October.

Before Everest joined the Great Trigonometrical Survey (GTS), India was organised under three Surveyor Generals: one for each of the Bengal, Madras and Bombay Presidencies. Then in June 1814 the Directors of the EIC decided that economies must be made, and the Survey work of the three Presidencies would be brought under one man. The order for this did not reach Calcutta until November 1814, and it was 17 April 1815 before the Governor General, Lord Moira, communicated the name of Colin Mackenzie as the first holder of the new post of Surveyor General, operative from 1 May 1815.

In 1815 William Colebrooke applied for Lieutenant Everest to be appointed to complete the Java survey. Because of Everest's knowledge of mathematics and astronomy, the Lieutenant Governor of Java agreed. He also added to Everest's duties the determination of positions and estimated heights of certain mountains in the longitudinal mountain range of Java. Where and how Everest gained the knowledge to perform this task is not immediately apparent since up to that time, other than his one year at Woolwich, there is no indication of further education. Quite possibly Everest was self-taught.

By June of 1815 Everest was ordered to Samarang to complete the survey of the Solo river, originally started by Colebrooke, by the most accurate methods possible, preferably from a series of astronomical observations. He probably had little choice of method since the only instrument mentioned was a 'telescope with micrometer' (perhaps a form of simple theodolite), which he purchased for 100 Spanish dollars.

In particular he was required to determine information about the mountains of Merappie and Mer Baboo, and this he supplemented with a survey of the bay of Pachitan complete with soundings. He took from July 1815 to January 1816 to do this. Then orders came to evacuate Java and restore it to the Dutch. Everest was immediately deputed to make a rapid survey of the harbours and communications of the south coast before evacuation. It was during this operation that he had his first recorded bout of fever, due probably to fatigue, and was out of action for some days.

During this time other surveys were being carried out in Java by Captain Godfrey Baker, who reported that Everest had fixed the positions in latitude and longitude of 10 mountains in Java. The notebook of the reconnaissance survey sometimes attributed to Everest was perhaps more likely the work of Baker.

After the restoration of Java to the Dutch, Everest sailed from Batavia on 28 September 1816 in the *Lady Barlow*, reaching Malacca on 8 October and was back in Calcutta on 20 November 1816 to continue his military duties. [228 pp. 135–8]

River clearance

Everest's next recorded assignment was in 1817 when he was deputed to clear the rivers Ichamati and Matabhanga (connecting the Ganges and Hoogly to the north of

Calcutta) of obstructions to navigation. Although this was not a survey operation it did require considerable ingenuity, and Everest proved adept at the task. His suggestions as to how the obstructions could be removed were initially 100 per cent successful but, because of the nature of the problem, this was not a permanent cure, and a resident engineer was required. His ability, however, did not go unnoticed.

Figure 2 *India circa 1830 showing the principal locations mentioned in the text.*

Work on visual telegraph lines

In the nineteenth century and even much of the twentieth, mail was carried across India by runners. These would cover stages of some eight miles each and the mail might thus cover 70 miles a day. In 1813 a scheme was suggested for visual telegraph lines, as had already been successfully established in Europe during the previous 20 years. The telegraph lines allowed the rapid passing of urgent messages by semaphore signalling. Everest was employed on setting out one of the first such lines in India. [229 pp. 269–70]

By order of 21 October 1817 he was detailed to proceed to Calcutta and start on a survey of the proposed telegraph line from Calcutta to Chunar, a distance of some 400 miles. In this he was assisted by Lieutenant R.B. Fergusson of the Ramgarh battalion. Before they had reached Sherghati, Everest had received notification of his appointment to the post of chief assistant to Colonel William Lambton (Superintendent) on the Great Trigonometrical Survey, but he remained in Hindoostan to see the telegraph line through to completion before moving to Hyderabad.

By mid February 1818 Everest had reached Sherghati (latitude 24° 40' longitude 84° 50') with the telegraph line. This was somewhat over halfway, with distances between the sites averaging only 9½ miles because of the thick forest. As they reached more open country this rose to 12 or 13 miles.

One particular problem in addition to the terrain was that the telescopes with which they had to work were of limited range. They are referred to as '…the common achromatic telescopes generally used by military men in the trenches or in reconnoitring parties…' Certainly theodolites were then available, but this was essentially a military engineering rather than national survey task. Whether the military had equipment of similar standing to that of the survey is not known. Where feasible the lines of sight between tower positions were required to clear intermediate ground by a minimum of 30 feet but this was often not possible in the wooded areas. In fact, on many occasions the observer had to be raised 70 feet to see a target eight miles away that was 90 feet above the ground.

(The distances between towers were only required very approximately so complicated techniques were not necessary. It sufficed to use a perambulator (measuring wheel) or to measure selected angles between the telegraph line and conspicuous points. One peak in particular served for six different sites. Such a technique suggests that the conspicuous points had been previously surveyed and that running coordinates were maintained along the telegraph line. It would then be simple to calculate each successive distance.)

It was May 1818 before they reached Chunar, and Everest remained there during the rains. In June 1818 he sent a detailed report to Lambton.

> The land between Calcutta and Burdwan is flat and swampy; highly fertile and populous, and studded everywhere with villages in which there are almost always found trees rising to a great height and obstructing the view. An observer standing

on the ground would…never be able to extend his prospect beyond 5 miles, and very seldom beyond three miles in any direction. …After reaching Bankoorah, which is about 100 miles from Calcutta, the face of the country is divided into uplands and valleys, and about 20 miles farther westward we enter upon a wide extended plain, in which there are several isolated hillocks and hills almost approaching to the size of mountains… 2000 feet is far too great for telegraphic purposes because the telegraph being a dark coloured object requires a light background; in this part of the line therefore, the eminences of 300 and 400 feet have generally been chosen…

There is a peculiar vapour in the hot weather which affects the atmosphere at a less height than 100 feet… and causes so great a divergency in the rays of light, that telescopes of large magnifying powers are of little use, and in such situations the telegraph distances have seldom been greater than 7½ miles, whilst in the hilly tracts 18 miles has not been too great. [263 (91) 363–70]

On 16 October 1818 Everest left Chunar for the march to Hyderabad and the beginning of the task that was to occupy the remainder of his working life. By the end of 1818 Everest had joined Lambton on the Great Trigonometrical Survey.

Early surveys of India

The Great Trigonometrical Survey was not the first major survey work in India. As early as 1764 Hirst is said to have measured a base near Jellinghi. Then in May 1788 Michael Topping is recorded as having done some triangulation* together with a measured base of 6½ miles at Porto Novo.

That base had been measured with 2 rods of 25 feet each laid end to end. [227] In addition Reuben Burrow had measured a degree of longitude near Cawksally during 1790–1 with a 50 feet steel chain and in 1791 measured a meridian degree with bamboo rods of nearly 200 feet length. In 1791 Michael Topping said:

> …Were these Triangles carried throughout India… the Geography of the Company's Territories might soon be rendered complete… One good Geometrical survey of a Province, or line of coast, is sufficient, whereas after ten bad ones the work requires to be gone over again. [228 p. iv]

Then in 1799 Thomas Sydenham measured a base of about three miles length near Ardenelle by means of a steel chain. All of these surveys were, however, somewhat isolated and unrelated.

Among the first propositions for a regular survey of India was the suggestion in 1778 by Captain Robert Kelly to produce a general map of the Deccan and Carnatic based on actual surveys and corrected with astronomical observations. The whole

---

* The technique of triangulation had its first practical use in 1615 when Willibrord Snellius introduced it in Holland. Its essential features were the measurement of one or more lines to the best possible accuracy and the observation of all the angles in a series of triangles spread across the country. The base or measured line gave scale to the whole figure, and the corners of each observed triangle became well co-ordinated positions. Each side of a triangle could be many miles in length, controlled particularly by the requirement for the ends to be intervisible.

map was to be divided into parallelograms each of one degree in both latitude and longitude. Each sheet he intended to illustrate with an appropriate sketch. The whole idea, however, was turned down. [250 p. 189]

As early as 1784 there was a proposal for an arc measurement in India. It was as a result of suggestions from Alexander Dalrymple and General Roy – the founder of the survey of Great Britain as we know it today – that Major William Lambton, 15 years later, was able to obtain permission to carry out one of the most ambitious survey schemes ever devised. General Roy had written in 1787:

> The British Dominions in the East Indies offer a scene particularly favourable for the measurement of five degrees of latitude on the coast of Choromandel... Two degrees of longitude, at each extremity of this arc, should likewise be measured.
>
> The plains of Bengal,...afford another situation where it would be of great consequence to determine the lengths of a degree or two of latitude, and as many of longitude. These two operations could not fail to be patronized by the East India Company who should defray the expense; ... [247]

In February 1800 Lambton wrote:

> ...the utility of such work, and the advantage and information which the nation would derive therefrom, are so clearly understood that no argument is necessary to demonstrate its advantages. The Surveyors of particular districts will be spared much labour when they know the position of some leading points to which they can refer, because when these points are laid down in the exact situations in which they are upon the globe, all other objects of whatever denominations, such as towns, forts, rivers etc which have a relation to those points, will also have their situations true in Latitude and Longitude. [265 (1) p. xv]

Lambton's idea was to have a mathematical and geographical survey that would cover the whole sub-continent with a network of triangulation and in so doing have a meridian arc right up the centre of India from the southern tip as far as the foothills of the Himalayas. It was to be capable of extension in any direction and in fact at one stage it was suggested that the chain of triangulation should go over the Himalayas but that was not achieved.

Later we read that:

> Colonel Everest had speculated on carrying it [the arc] further, beyond the Himalayas and across the wild regions to the north, until it struck the Russian triangulation within the dominions of the Czar. An arc stretching from the Indian Ocean to the Polar Sea would, indeed, as he himself describes it, have been a 'vast project'. [18 p. xiii]

The Great Trigonometrical Survey

Lambton's proposal was submitted to the Madras Governor, Lord Clive, with the support of such influential people as his commanding officer Arthur Wellesley (later the Duke of Wellington) and Wellesley's brother Richard who was the then Governor General. With backing of this standard Lambton could not fail to get permission to proceed. [81], [196] In particular the timing was opportune since, with the fall of Tipu Sultan of Mysore in 1799, the East India Company had acquired vast territories in southern India. Phillimore commented that:

> The debt that Indian geography owes to William Lambton can hardly be adequately expressed, for without him it is difficult to see how the boon of a great trigonometrical survey would have reached India… The presence in India of a man of Lambton's genius and character, knowledge of mathematics and interest in geodesy, was entirely fortuitous… Lambton was geographer as well as geodesist, and from the commencement of his work tried to combine his geodetic work with a general survey of the country. [228 p. 9]

Not only would such a scheme provide a sound foundation for future mapping of the country but arc measurement also had scientific value. From knowledge of the linear distance apart of two points of similar longitude and their angular separation found from astronomical observations, it is possible to determine the figure of the earth, in particular, its size but, in combination with other similar arcs around the world, also the shape of the earth.

The formal orders for the start of the survey were issued on 6 February 1800. Among Lambton's early assistants were Lt John Warren of HM 33rd Foot and Ensign (later Captain) Henry Kater of HM 12th Foot. In October 1800 a baseline (see footnote, p.15) was measured at Bangalore, and the zenith sector (Figure 29) was used for the first time at Dodagoonta for the accurate determination of latitude by star observations.

In April and May 1802 a base was measured at Madras (known as the St Thomas' Mount baseline), and Lambton set out on the task of connecting this base by triangulation (see footnote, p. 15) to that at Bangalore, continuing on to the coast at Mangalore. In the process of doing so he determined the direction of the meridian at six locations relative to Madras observatory. These gave the meridian lines of Balroyndroog, Mullapunmabelta, Savendroog, Yerracondah, Kylasghur and Carangooly. By late September 1802 Lambton, together with Warren, was in a position to use the 36 inch theodolite (which had arrived in India that month) on a meridian arc from Cuddalore to Madras (Trivandeporum to Paudree) and to use the results for an evaluation of the length of 1°.

Lambton initially assumed that the figure of the earth was such that $e^1 = 1/150$ for the purposes of his preliminary computations. This was so far out that it gave meridional arcs in latitude 13° too small by 0.485% and arcs of parallel too long by about 0.143%. As a result of his longitude arc the width of the Indian Peninsula at 13° north was found to be only 360 miles. This compared with Rennells' map of 1793 which gave 404 miles and the calculations by Colebrooke in 1800 of 386 miles – a 'loss' to the country and to the Company of some 40 miles!

---

[1] a ratio of the axes defining the ellipsoidal shape

Lambton reached the Malabar Coast near Mangalore in 1805 and on the way Warren measured a base at Bangalore. In February 1806 Lambton returned to Bangalore from Mangalore and extended the meridian arc south to a base at Pachapalaiyam (latitude 11° 00', longitude 77° 30'). From 1807 the coastal chain was extended south from Cuddalore (latitude 11° 43', longitude 79° 45') to close on a new baseline at Tanjore (latitude 10° 45', longitude 79° 00'). The following season the Great Arc was continued to Cape Comorin. The longitude arc through Tanjore was later joined to the Great Arc near the base at Coimbatore (latitude 11° 00' longitude 77° 30').

The period from 1809 was fraught first with military operations and then with shortage of staff, but by 1811 Lambton was able to add another baseline near Gooty in latitude 15° 00'. Gradually he extended the arc northwards and by 1815 had yet another baseline, this time at Bidar near Damargidda at latitude 17° 55', longitude 77° 40'.

Thus had been the progress of the Great Arc by the time Everest was privileged to join Lambton. In July 1824, when writing to Colin Mackenzie, the Surveyor General, Everest described the Great Arc in the following terms:

> the objects embraced by the Great Trigonometrical Survey of India are both geographical and geodesical. The survey was commenced in 1799 solely with the view of connecting the East and West Coasts of the Peninsula in order to determine the exact difference of Longitude; it was afterwards carried from Dodagoonta, Cape Comorin, along the meridian, and on arriving at Punnae it was again carried Northward by which means certain... distances in a direction due North and South were determined, these distances were arranged into portions averaging nearly 900000 feet each at both extremities of which observations were taken with an instrument called a Zenith Sector whereby the Arc of Amplitude or difference of latitude as it is called was determined [263 (171) 216]

The meridian arc took until 1841 to complete but in parallel with it a number of other meridional and longitudinal arcs were observed across the country. By 1893 the whole country was criss-crossed by a grid of arcs with perhaps the principal ones being:

Longitude arcs
1. From Peshawar along the foothills of the Himalayas to the Eastern frontier
2. From Kurrachre (Karachi) through Sironj to Calcutta
3. From Bombay to Waltair
4. From Mangalore to Madras

Meridian arcs
5. From Kurrachre (Karachi) to Peshawar
6. From Cape Comorin to Dehra Dun
7. From Cape Comorin via Mangalore and Bombay to arc 2.
8. From Cape Comorin via Madras and Waltair to arc 2.

There are many other arcs forming the overall grid iron pattern developed by Everest, and all are infilled with lower order triangulation. The grid arrangement of triangulation chains was useful for providing a strong framework in an economical manner. In any country such a skeleton of fixed points was the basis of all subsequent mapping activity, and India was no exception. After all the effort put in during the nineteenth century India now has an excellent framework and one that other countries have since emulated.

In 1807 Lambton had suggested that his work on the Survey should come directly under Government control but this came to nothing. By 1817, when he was in need of assistants Lambton again raised the issue. This time the idea was accepted by Governor General Lord Moira (Francis Rawdon Hastings, Lord Hastings from 1817). By a decree of 25 October 1817 the work being executed by Lambton was to be brought under the control of the Supreme Government in Calcutta and from 1 January 1818 to become known as the Great Trigonometrical Survey of India. (In 1878 it was to become simply The Trigonometrical Survey.)

It was a lucky situation since Lord Hastings had knowledge of the workings of the Ordnance Survey of Great Britain and as a result appreciated the importance of geodetic survey and the sound foundation that a network of accurately located triangulation points provided for mapping.[81] It was also Lord Hastings who was instrumental at the same time in appointing the young artillery officer, Captain George Everest, as assistant to Lambton when the lynchpin of the GTS was 65 years old and obviously getting towards the time when the reins would have to be handed on. The post was to be effective from 1 January 1818 (before Everest started on the telegraph line).

Lieutenant Colonel Young (Secretary to the Governor General) wrote to Major Craigie (Secretary to the Government Military Department) on 25 October 1817 with reference to this appointment of an assistant to Lambton.

> The Mathematical qualifications for conducting such labours are of a very high order and possessed by few in India, they require to have been kept up by habitual exercise, and moreover the extreme accuracy indispensable in Trigonometrical calculations on the scale of Col Lambton's undertaking, demand a dexterity in the use of the Instruments and a scrupulous degree of attention in what may be termed the practical part of the labour which can scarcely be conceived by persons unaccustomed to it...
>
> The Governor General therefore has selected for this office Capt Everest of the Artillery whose eminent degree of Science as a Mathematician he is assured and whose talents are known to the Vice President in Council both by his surveys in Java... and by his successful exertions as an Engineer in recently clearing the navigation of the Matabhauga and other Rivers... [263 (171) 248–9].

For a second time it was mathematical ability that had got him a posting that he wanted. When one bears in mind that his ability in this area must have been mostly self-taught it was all the more credit worthy.

Neither Lambton as Superintendent of the GTS nor Mackenzie as Surveyor General had any say in the appointment, which could so easily have gone wrong since Lambton's choice would have been James Garling. Apparently Lambton had even written to Everest to say that his application for the post was unsuccessful as he had already engaged another officer. [229 pp. 342–3] (Or did he really mean earmarked, since the appointment was not within his favour to give?) The appointment carried a salary of Rs 600 per month in addition to which there were various allowances according to circumstances, a salary which was about half that of Lambton's. [299 p. 252]

In getting to his new location with Lambton, Everest recalled:

...On my way from Chunar I have employed myself in taking a route survey of the journey lying through Mirzopoor, Rewah, Myheer, Jubbulpoor and Magpoor to Hyderabad... 759 miles ...at rate of 12⅓ miles per day ...occupied whole time and left me no leisure to protract or plot. [263 (154) 57]

He took the latitudes – probably by sextant – of about 35 places mostly by three or four observations to the north and south of the zenith. He also took barometer and thermometer readings at 60 places with the mean of morning and evening observations and distances were recorded by perambulator. The results were not plotted until 1840 – over 20 years later! [213]

Everest joins Lambton

Once Everest had joined the GTS, Lambton naturally wished to satisfy himself with the ability of his chief assistant. The two of them went out on 8 January 1819 so that Everest could gain instruction at the hand of the master. They began by observing three triangles of the Great Arc near Bidar. Although they stayed together until April, Everest was in full control by the end of January. After returning to Hyderabad to organise his party, Everest set out on the triangulation to the east from the Ceded Districts to the area between the River Kistna and the Godavari (Goodavery). This runs more or less along the 19° N line between 78° and 80° E. The area had no allegiance to the East India Company since at that time it was still under the Nabob of Golconda or the Nizam as he was commonly known.

The work was started in June 1819 and depended on the meridian chain at Kylasghur and then that at Karangoolee. It was a wild area where each village had its mud fort defended by *jinjals* (*gingal*: a large musket fired generally from a rest; or a light gun mounted on a swivel) and men armed with swords, matchlocks and spears. Many of these were in a state of open resistance to the native Government. The area was teeming with malaria and all manner of unpleasant manifestations. [96 p. 6] It was obvious that it was not the best of places to go to carry out operations which, to the uninitiated, could appear to be very suspicious. Literally everything – wildlife, humans, vegetation and climate – was against a safe completion of this survey.

Everest started by having trouble with the guard of State police and this he described in his Report of 1830. [96]

These people most heartily disliked the expedition … and seized every pretext of skulking away from the camp and back to the city. My representations …were answered by Colonel Lambton with a desire, both on his part and that of the British Resident, that I should seize the first feasible grounds for making an example…

The infliction of corporal punishment is an odious task; but in this case there was really no choice between that and giving up the operations. Urged, therefore, on one side by my superiors, and irritated on the other by the total disregard shewn … I took an opportunity about a month after leaving Hyderabad to chastise one of these defaulters with some severity; in consequence of which the whole body, about forty in number, burst into open mutiny, seized the native gentleman whom the minister had deputed as their chief, and declared they would quit my camp and carry him back with them.

It was in a grove of mangoe trees surrounded by a ditch and bank that they had selected their spot of encampment. There sat the Daroga [headman], surrounded by the mutineers, some with their swords drawn, others looking on. It … became my duty to assert my authority, or give the matter up entirely as hopeless.

With the Great Trigonometrical Survey of India there has always been an escort of regular sepoys … not belonging to the standing army. Colonel Lambton had detached twelve of these under my orders. I drew up a small party of eight men with loaded muskets in front of the grove where the rebellious Juwans (young soldiers) were lording over their superior, and declared my intention of firing a volley into the midst of them unless they immediately laid down their arms. Their resolution quailed before this decisive step, and they now became as meanly humble as they had been audaciously insolent. So, having deprived them of their weapons, and placed them under the surveillance of my sepoys, I made a severe example of three of the principal offenders by publicly flogging them and turning them out of camp with ignominy…

Threats of vengeance buzzed around me for some weeks after this occurrence, and it was necessary to be armed and well prepared to resist assault on my person. But the natives of India are not a malice-bearing race and, finding when they knew me better that good behaviour was a perfect security against all unkindness, they became at last willing, obedient, and obliging as I could desire.

Everest described in graphic terms his first season there in the height of the monsoon, in thick forest, with swollen rivers and finally succumbing to sickness.

In June 1819 he set out with Henry Voysey

… to carry a series east to meet the meridian of Karangoolee, then south to meet the series of that meridian where it had been left at Polichintah and Sarangapullee on the south bank of the Kistna river; then north to meet the Godavery and lastly to run down the Kylasghur meridian to meet the former points on the Kistna… [96 p. 6–8]

The joys of travel

Even as Everest got to his first station, the violent rainy season set in. Previously harmless rivulets became raging torrents and all cross-river communication was out of the question. [229 p. 229]

Everest described it thus:

> There is a stream near Hyderabad called the Moosee, which falls into the Kistna below the ferry ... by which I had intended to pass to the station of Sarangapullee (2 miles south of Kistna, 79° 46' E)... the Moosee being at ordinary times barely ankle deep... I had ordered ... the supplies for my camp to be prepared at a village on the southern bank; but when ... I reached the crossing I found this rivulet, so insignificant at Hyderabad, now filled to overflowing, carrying away trees and other floating objects in its foaming current.
>
> Thus cut off from all communication with the provisions which had been prepared for my followers ... I learned ... that there was about fifteen miles distant a place called Kompullee, below the confluence of the Moosee and Kistna, where there used to be a ferryboat ...By the following evening the camp was transferred to Kompullee, where we once again had abundance to eat; and, having turned the flank of the Moosee, had at last attained the north bank of the Kistna which, pouring down over a bed of rocks shelving and dipping at all angles, was really a formidable obstacle.
>
> As it was of great importance that my carriage-cattle should be conveyed to the opposite side, I had my elephants brought to the water's edge; but neither caresses nor menaces could induce them to try the passage. Probably it was fortunate that they did not make the attempt; for these powerful animals, though more at home in water, perhaps, than any other quadrupeds, are from the size of their limbs ... in need of what sailors term sea room, and in a river which, like the Kistna, abounds with rocks ... were very liable to receive some serious injury, of which their natural sagacity rendered them peculiarly apprehensive.
>
> The boat which was to convey me and my party across the roaring and angry flood was put into the hands of the cobblers to... undergo the necessary repairs, for it was an old, and crazy, and very leaky vessel, which had for sometime been laid up high and dry; but now, when no alternative was left, but either to await the subsidence of the flood, or to trust ourselves to this frail craft, I found that there was no ... reluctance on the part of my people to risk their persons...
>
> The boat, or leather basket, contained about six persons, with a proportion of dead weight; so, having reduced the baggage and followers to the smallest possible quantity sufficient to carry the instrument [an eighteen inch theodolite], my little party embarked, and in three journeys which, as it required to undergo repairs after each, occupied till nightfall, the vessel had conveyed to the southern bank all whom I intended should accompany me...
>
> I left the camp, with tents, cattle etc, under charge of Mr Voysey, with directions to proceed onwards to Polichintah along the northern bank, and await my arrival there; and, as the station flag of Sarangapullee was in sight about twelve miles off, and in appearance hardly two... attended by one of the sub-assistants [Olliver and Rossenrode], and after some hours toiling over rocks and through jungle, I reached it just as the setting sun was shedding its last rays on the horizon.
>
> Thus separated from my baggage, and without shelter against inclemencies of the weather, I learned to know what an Indian climate must be to the houseless

European. The sky had during the day been bright and cloudless beyond compare; but shortly before sunset black threatening clouds began to grow together into a frowning mass; and at last, when all their batteries were in order, a tremendous crash of thunder burst forth and, as if all heaven was converted into one vast shower bath, the vertical rain poured down in large round drops upon the devoted spot of Sarangapullee.

I had procured a charpaee [a rude bedstead or litter] from a village about five miles off and, having bent down the branches of a young tree and covered them with rice straw, I had hoped by the assistance of an umbrella to protect myself against the effects of the storm; but, on awaking in the morning, I found that I had been lying all night with my clothes soaked through; and yet, so sound had been my sleep from fatigue, that I had been totally unconscious of the circumstance. [96 p. 9–12]

One of the problems on the GTS was that the climate in India was such that the best daytime observing conditions occurred during the worst of the rainy season. Lambton had as a result always insisted on the field season coinciding with the rainy season. Thus not only were there delays due to rain, cloud and mist but it was also the worst time of year for malaria and other sicknesses and the surveyors were continually struck down with severe bouts. [229 p. 442]

It is easy to conceive what a reckless waste of life and health was caused by this exposure to the pitiless pelting of the tropical rains, in forest tracts teeming with miasma [infectious matter, pestilential vapours hanging in the air], no constitution, European or Asiatic, could bear up for any length of time against such a complication of hardships as thence arose, – eternal watchings by day to the prevention of all regular exercise-tents decomposing into their original elements – servants – cattle – baggage – clothes – bedding –all daily dripping with rain – every comfort which the indwellers of cities and leaders of regular lives deem essential to happiness and even to existence, remorsely sacrificed.

The introduction of lamps and heliotropes has totally changed the face of things, and by rendering the rainy season the least fitting period for observing luminous objects... [265 Vol. 1 p. 19]

Jungle and yet more jungle

Everest commented:

In those gloomy days when the mists descend and obscure the horizon, it was the chief relaxation of Mr Voysey and myself ... to employ our followers with handspikes and ropes in tearing off the loose masses of granite, and letting them find their way to the bottom of the hill. Certain it was a magnificent spectacle to see an enormous mass, seven or eight feet high, descending along the slippery side of the spheroid, and striking fire in the progress – yet cautiously at first, and as if afraid to venture – suddenly, when it met with some hindrance, it would bound up and roll over like a planet in free space, and, lastly, when it attained the limits of the jungul, it would tear down large trees, and make the welkin roar again as it tumbled into the abyss below.

> Doubtless all this may be very childish, but ... the French academician, De la Condamine ... and his companions resorted to precisely the same methods of amusing themselves... We did not continue the pastime during the night for fear of injury to my followers; but if our amusement was by accident prolonged a little beyond twilight it is inconceivable how grand the sight became, for wherever the rock slid along the bare side of the hill it was accompanied by a dense train of such enduring sparks as we see emitted from the impact of the hoofs of the pampered coachhorse on the London pavement; and the light emitted when it struck any obstacle was sufficient to enable us to trace its progress, and make it resemble a whirling mass of phosphoric matter. [96 p. 13]

From Kundagutt he looked for a station to the east of Hydershahipett and the only high point he could see was a long, black-coloured mountain resembling an elephant, some 60 miles away. He thus sent four of his most skilful flagmen, with an efficient guard, to seek and occupy two likely locations on that mountain.

> It took me about three weeks to run southward along one side of the series, and to return northward by the other side to Hydershahipett. Nothing whatever having been heard of my detached parties, great apprehensions were entertained by me for their safety; but at last a gap began to break open in the black mountain ... and after a fortnight's further waiting I had sufficient daylight behind to distinguish the colours of the Great Trigonometrical Survey flying on one spot, and a signal mark on the other. The secret of the delay now came out. The station of Hydershahipett was on the verge of the great forests of teak and ebony, far into the depths of which was situated this elephant mountain, called Punch Pandol. The access to it was by a circuitous route, unknown to any but the few struggling natives who lived in those forests, in a state closely bordering on savage life.

> The nearest village was Poomrarum, about five miles from the summit, from which it was necessary to cut a road for the instruments and tents;... and how my unfortunate flagmen could have had perseverance enough to go through with such a task, how they could have coaxed any uninterested persons to accompany them; how, after having pierced through a forest of teak trees, seventy, eighty and even ninety feet high, thickly set with underwood, and infested with ... tigers and boa constrictors, without water or provisions and with jungle fever staring them in the face ... utterly passes my comprehension.

> I had, indeed, been warned months before, that these junguls were the seat of the most deadly fever, which attacked men's mental as well as corporeal faculties ... and how by means of conciliating treatment and prompt payment, my people had managed to collect a sufficient body of hatchet men to clear away every tree that in the least obstructed the horizon over a surface of nearly a square mile...

> I was now far advanced into this terra incognita; ... to the eastward and northward no sign of humanity could be seen. Yet it was necessary to pierce far deeper into the forest to meet the Godavery, and, having fixed on a station which I judged to occupy it. Day after day having elapsed without hearing of them, I detached a second party, and some days afterwards a third under one of my sub-assistants [Rossenrode], but still no progress was made. At last came a melancholy letter from my sub-assistant,

telling me that he was ill and going to die; and then as a last resort, I despatched my principal sub-assistant Mr Joseph Olliver, …and to my great delight I at last saw my flag flying on the selected hill, and received written intelligence of the name of the nearest hamlet Yellapooram … and of my former parties, many of whom began to suffer from the effects of the climate…

… towards the end of October I marched from Punch Pandol towards Yellapooram, through the wildest and thickest forests I had ever invaded …The distance… is about thirty one miles in a straight line; along the route we took it is little short of sixty four.

The eminence was most fortunately situated, and seemed to have been placed there on purpose to accommodate me, for, had it been a hundred yards to the north, the ray to my western station of Kotaajpoor must have been obstructed … Three parties were immediately detached to occupy the three peaks,… and I hoped in, …a few days to complete the observations… in which, had success attended me, I should, to use Colonel Lambton's words "have performed a very magnificent work indeed to start with!"…

From Yellapooram hill … to the north, south and west the eye wandered over one uninterrupted mass of foliage, ornamented with all the different shades of green and yellow. Not a vestige of a human habitation or of cultivated land was anywhere to be seen, and the whole expanse seemed to be marked out for eternal solitude. [96 p. 9–21]

Fever

Buoyed up hitherto with the full vigour of youth and a strong constitution I had spurned at the thoughts of being attacked by sickness, against which I foolishly deemed myself impregnable; but my last day's ride through a powerful sun, and over a soil teeming with vapour and malaria, had exposed me to all the fatal influences of these formidable forests. [96 p. 19]

By the evening of 2 October 1819 Everest was seized with typhus fever (or possibly malaria) and within five days nearly 150 of his camp followers were similarly laid down without the strength to carry the theodolite. They had to make all haste back to Hyderabad for which 'A litter was made for me; Mr Voysey had a palanquin… but the jungul fever pursued my party like a nest of irritated bees, long after we had quitted the precincts of the forest…' Unfortunately 15 of Everest's followers perished by the time the party reached headquarters. Some were eaten by the tigers.

Everest had reached Hyderabad a few days before the others and the British Resident immediately sent the public elephants, doolies (litters) and camels with a strong escort to hasten the return of the others, who were little more than corpses when they arrived. For some while both Everest and Voysey had bouts of delirium.

Not to be beaten, Everest set out again in June 1820 with the same team to complete the interrupted work to the north of Yellapooram. However after a few weeks he had another violent attack of jungul fever (a remittent tropical fever, a form of malaria which, while getting much better at recurring periods, never entirely leaves the patient) and rather than sacrifice himself he went on extended sick leave and left Mr Olliver to fill in the remaining blanks of the scheme.

In a statement to the East India Company in 1821 it was said:

> It appears that severe sickness had seized the whole of the party which was sent out on survey in the year 1819. Captain Everest, assistant to Lieutenant Colonel Lambton had suffered so seriously from illness as to be under the necessity of proceeding to sea for the recovery of his health.

On 1 October 1820 Everest sailed from Madras for the Cape of Good Hope and a better climate in which to recuperate.

Chapter 4

THE CAPE OF GOOD HOPE 1820–1821: A time for recovery, and confirmation of historical observations

*I*t was 25 November, 1820 when Everest reached Table Bay and had a chance to recover from the fevers to which he had succumbed. Little is known of the year he spent there, other than the work he did on the meridian arc that had been observed by M l'Abbé de LaCaille between 1751 and 1752. Everest wrote to Lambton, a day or two after arriving at the Cape, but no other word was received from him until nearly a year later. In fact by 9 August 1821 Lambton commented that he was at a loss to know what had happened. [233] Everest did, however, indicate that '…instead of seeking my own ease and amusement I employed myself in objects connected with my professional pursuits …' [263 (171) 240–1]. What could that have referred to? There were no major survey schemes in operation that he could take an interest in nor any instrumental developments to attract his attention. The Cape Observatory had yet to be constructed, but it is just conceivable that he was able to discuss the design, layout and furnishing of it since construction began soon afterwards.

Everest's one oblique reference is to the compilation of a pocket catalogue of some 600 stars which he had extracted from Piazzi's collection lent to him by 'my excellent friend Mr. Fellowes'. [96 p. 24] Fellowes (or Fallows) was the Rev. Fearon Fallows, the first astronomer at the Cape from 1820 to 1831. This does suggest that Everest could have had a hand in the site selection and design for the Observatory if he was so friendly with Fallows. However, Fallows only arrived at the Cape on 12 August 1821, three months before Everest left. While the two could well have spent much of that time together, it does not explain what Everest was doing during his first nine months there.

Soon after arriving in the Cape, Everest wished to carry out the task suggested to him by Colonel Lambton, of verifying, or at least investigating, the work of LaCaille. Whilst he '…immediately … commenced my inquiries…' [87 p. 258] he could not make any progress, since there was no copy of LaCaille's *Journal* to be found in the Cape, and Everest had to send to London for one. That could not be achieved in a few days; it was July 1821 before he had any documentation to aid his research. Unfortunately Everest had no access to, or perhaps even knowledge of, LaCaille's papers in the *Memoires* of the Paris Academy, and hence had little numerical data to go on.

R.S. Webb, when describing LaCaille's *Journal* said 'In this little volume ... very few pages contain anything directly concerned with the measurement of the arc of meridian... [it] cannot be recommended as an historic account of the geodetic operations.' [290 v 6 pp. 89–90] This is all that Everest had to work with.

The arc of LaCaille

We know little of Everest's investigation into the arc of LaCaille. It would appear that he wrote upon the subject only once.[87] Maclear devoted a paper to it [207] and referred to it when he reobserved the arc.[208] Other than that there is nothing to draw upon.

L'Abbé Nicolas-Louis de LaCaille (1713–62) conceived a plan to make a trip to the Cape and had arrived there on 19 April 1751 specifically to make astronomical observations. He lodged at No. 7 Strand Street (now part of the Old Mutual Centre), Cape Town, the house of a Jan Laurens Bestbier, a cavalry captain, and found there a place suitable for the construction of a small observatory*. It was only a 12 ft × 12 ft observatory, and contained but two 6-foot radius sectors, a pendulum clock, quadrant and various telescopes. Its construction had been completed by 17 May 1751. His initial ideas bore no mention of an arc measurement although he had gained experience in this in France where he had worked with the Cassini family.

Although LaCaille initially made no mention of an arc, by 6 September 1751 he wrote that he had found an area near Groene-Clof (Kloof), 12 leagues north of the Cape, that would be suitable for baseline measure. Thus he either had the idea for an arc soon after arriving in the Cape, or he had it in mind all along but did not say so before his arrival for fear of upsetting the Dutch. He felt that it was desirable to measure a meridian arc to discover whether the southern hemisphere was similarly formed to the northern one, and it was required for the accurate determination of lunar parallax.

Thus it was that, with the assistance of Governor Tulbagh, LaCaille surveyed by precise triangulation, a meridian arc of just over 1° in extent and included in this a baseline of eight miles length. Mention of the work of LaCaille is recorded in his *Journal* [195], but he specifically omitted details on calculations or investigations of an astronomical character. For such technical information recourse had to be made to various of the Paris *Memoires*, particularly that for 1751 (pub. 1755), and the work *Grandeur et figure de la terre* by Delambre of 1912.

It would appear that LaCaille did considerable reconnaissance during August 1752 and made his observations during September and October with the baseline measure

---

* The present Cape Observatory was not founded until the 1820s – just after Everest was there. It was established by Order of His Majesty King George IV on 20 October 1820 through Governor Tulbagh. The architect was John Rennie, chief engineer to the Admiralty, although it was not built until after his death, when the work was under the superintendence of M De Ruyter.

taking place from 17 to 29 October 1752. The scheme consisted of two large triangles and a base extension figure. The base was measured by wooden rods to be 6467.25 toises (= 38 803.50 Paris ft) and the arc length determined with a 6 foot radius zenith sector computed to 69 669.1 toises for an angular difference of 1° 13' 17.3" derived as the mean from 16 values of the amplitude.

These observations resulted in a length for 1° of 57 037 toises. After consideration of the length of the toise standard used by LaCaille, Lalande later amended this to 57 040 toises. In his *Fundamenta Astronomiae* LaCaille gave the arc as 1° 13' 17.5", a value which would modify the equivalent for 1° to 57 034.4 toises at the mean latitude of 33° 18.5' S. This suggested, when compared with other arcs, that the earth must be more flattened towards the south pole than towards the north.

Surprised by these results, LaCaille repeated the base measurement, recomputed and confirmed the whole scheme. By 8 March 1753 LaCaille was leaving the Cape and making his way to the Ile de France (Mauritius).

Locating the stations

As early as 1768 Cavendish considered that the degree at the Cape might be affected by local attraction. Now it would seem that Lambton, or maybe even Everest, had got wind of this idea, since Lambton instructed Everest, if his health permitted, to look into the problem.

As already seen, Everest endeavoured to obtain the appropriate documents as soon as he arrived in the Cape but had to wait many months. Receipt of the *Journal* of LaCaille must have both initially pleased him after such a long wait but also disappointed him since the calculations were missing. It is certain that with his meticulous manner, if the observations, calculations and results had been included, Everest would have at least repeated all the reductions to his own satisfaction.

That was not to be, so he set out to recover as many of LaCaille's station positions as possible. 'I was enabled by the 26th of July to visit the locations described as the sites of his stations, and in fact to traverse the whole theatre of his labours.' [87 p. 258]

In addition to the two terminals of the base there were four major stations connected in two triangles. The celestial observations had been made at the two vertices of these triangles – namely Klip Fontein and Cap, the location of LaCaille's observatory at Cape Town. The two central stations were at Capok Berg and Riebek's Castle. Everest used LaCaille's descriptions of, and observations about, the stations to identify them.

Capok Berg: 'a rounded hill, of easy ascent, belonging to a range of granite estimated to be about 600 feet high'. Everest considered that he was able to correctly locate this point although no artificial mark remained.

Riebek's Castle: 'a mountain of some 1500 feet, about 2½ miles in extent and

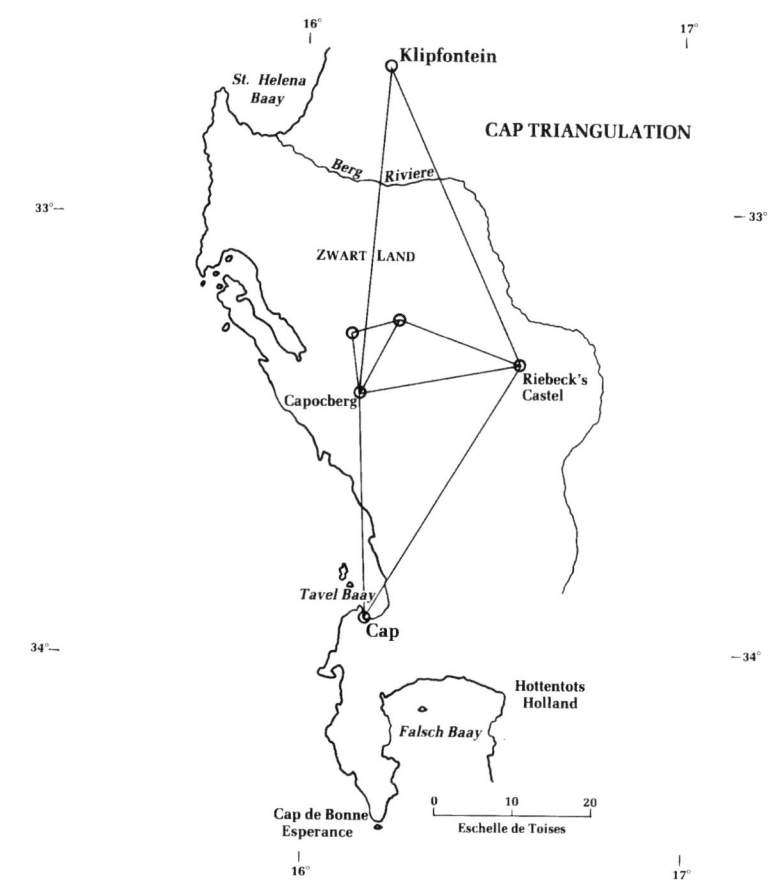

Figure 3 *The observations at the Cape of Good Hope by L'Abbé de LaCaille in 1752.*

divided into several different peaks'. Again Everest was sure that he correctly identified the point used by LaCaille.

The base: by using knowledge of the angles as indicated on the plan, and working backwards from the previous two stations, Everest found a position near Coggera and a point at Klip Berg which he was satisfied formed the terminals of the original base. Note that any angles he used had to be scaled from the small-scale plan. That plan, reproduced with his paper, was at a scale of approximately 1:1 350 000 – not a scale at which angles can be scaled very accurately, although of course he could have had a much larger-scale version.

Klip Fontein: the northern terminal. Particularly in relation to this point, Everest was helped by the memories of an aged lady who pointed out where the old granary, that had housed LaCaille's zenith sector, had stood and also described the fires used as signals. This led Everest to question both the use of large fires as targets and that

LaCaille must, apparently, have positioned his instruments off centre of the target. Such conditions immediately led him to question further the base measure because of the roughness of the terrain (admittedly some 70 years on) and the need to have used coffers, inaccurate targets and no correction for off-centre targets.

Cap: this was the most difficult to tie down. Everest traced it to a particular house, No. 7 Strand Street, and described it as the one next to that in which he was staying. Fallows stated that he had 'ascertained, with the assistance of Captain Everest, who had touched at the Cape, that the house in which LaCaille's observations were made was undoubtedly the same as that occupied in 1821 by Mr D Witt'. [146 p. lx]

Where within that house the observatory was placed was unknown to Everest but he surmised that LaCaille may have taken advantage of the flat roof and had a, not very substantial, structure built upon it. This would obviously have been susceptible to large errors with an instrument set on a frail basis. The particular house was corroborated by the existence of a brass plate perforated with a small hole, fixed in a wall 'for the obvious purpose of determining the sun's passage over the meridian'. [87 p. 262] It was said to have been placed there by LaCaille.

The position of this station was verified by Donald McIntyre who reported his findings in 1951 in his paper for the Bicentenary of LaCaille's visit to the Cape.

Deflection of the plumb line

In addition, Everest questioned the actual site because of its proximity to the formidable Table Mountain. Such a vast quantity of matter must have had a considerable effect on the plumb line (large mountain masses deflect the plumb line away from hanging vertical). LaCaille would certainly have been aware of the effect of attraction since it was in 1738 that Bouguer, whilst he was on the Peruvian arc expedition, had attempted to determine the effect of Chimborazo on the plumb line. However, there is no indication that a correction of any sort was applied, and Everest followed this assumption.

Studying the topography in the vicinity of both Klip Fontein and the Cap, Everest considered that, rather than any effect at these stations cancelling out, the reverse would have occurred; in which case the arc would have been too great by the combined effects. So that he might quantify this, Everest compared the arcs of Bouguer and La Condamine in Peru 1738; of Cassini in 1740 and LaCaille's results. The values he used were:

			Mean latitude	Value of 1°
1 Peru	1738		1° 30' N	56 749 toise
2 Cassini	1740		49° 22' N	57 074
3 LaCaille	1752		33° 18' 30" S	57 037

From knowledge of the earth's semi-axes a and b Everest determined the compression of the earth's elliptical form as $\varepsilon = (a-b)/a$. Comparing 1 and 3 gave $\varepsilon = 1/172$; 2 and 3 gave 1/1250 and 1 and 2 gave 1/300.

Since the last one agreed closely with other calculations of the time, Everest then worked backwards from the value of 1/300 to determine by how much the Cap parameters would need to be changed in order to agree. Taking arcs 1 and 3 he found the Cap arc should be 56 918.4 toise, and taking arcs 2 and 3 he found the Cap arc should be 56 918.7 toise.

Perpetuating a mistake

In taking this to its final stage Everest quoted the LaCaille arc value from Hutton's *Philosophical Dictionary* of 1795 and would appear to have perpetuated a transcription error. The figure of 410 814 feet (note that here it is Paris feet that are used, not English) quoted by Hutton – and perpetuated in Grant's *History of Physical Astronomy* of 1852 – should read 418 014 feet to agree with the measure quoted by LaCaille of 69 669.1 toise. The value is then erroneously turned back to toise as 68 469 – the value used by Everest. How easy it is to perpetuate mistakes!

It would seem that Hutton had the misprint value of the arc in feet and the length of a degree as 57 037 toise. Combining these gave a value (again erroneous) for the arc of 1° 12' 01.55", a result that was queried by R.S. Webb but not unravelled. [290 Vol. VI p. 118] As Everest pointed out, he had to use the values given in an old edition of Hutton's *Dictionary*, because all his reference works were in India, and he had to make do with what he could find in Cape Town.

Luckily the effect of the error hardly changes Everest's conclusion. Where the figures he used resulted in a difference of 8.99" to attribute to attraction, the figure should have been 9.15". So particular was Everest that, surely, if he had had a hint that the figure he was using was incorrect he would have moved all obstacles to obtain the correct value. After considering the value of 5.8" obtained at Schehallien and that of 7.5" at Chimborazo, Everest felt that it was quite feasible that most of the 9.15" (he used 8.99") could be attributed to the combined attraction at the two terminal stations since this would only be about 4.5" at each.

Airy, in the *Encyclopaedia Metropolitana* of 1845, stated that had Everest had access to his reference works he would have found some of his objections were without foundation. LaCaille stated that the ground was cleared and levelled for the base measure; he also indicated that the instruments were not placed at the centre of the stations and gave the numbers necessary for the reductions. The magnitude of fires at 45 miles need not be so great as to throw any great uncertainty on the terrestrial measure. The instability of the observatory was of little consequence so long as the zenith sector observations were taken with care. In the same publication, Airy gave the length of the arc as 445 506 English feet, a figure which was derived by use of the toise/foot conversion value of 1 toise = 6.394 596 English feet and thus tallied with the other figures quoted in the same work, of 69 669.1 toises, 1° 13' 17.5" and 57 034.4 toises. (Note that the toise/Paris foot relation was 1 to 6.000.)

A task for Maclear

In concluding his report Everest suggested that rather than reobserving LaCaille's arc it would be preferable to have a new triangulation of greater extent. This would incorporate the new Observatory, then under construction, and connect with the old stations of LaCaille. When Everest met with Fallows he outlined his feeling that the arc should be remeasured. The suggestion impressed Fallows, and he wrote to both the Admiralty and Herschel that it was a project that should be pursued. The Admiralty did not co-operate at that time but, when Maclear arrived at the Cape in 1834 to succeed Fallows, part of his brief was to seek an appropriate site for a baseline and for related geodetic work. [284 pp. 50–5]

As Everest had done, so Maclear set out to determine the exact location used by LaCaille in Cape Town and the likely positions of the other points. Maclear was advantaged to have access to more than just LaCaille's *Journal* so was able to be more precise in his search for the original station positions. In particular he referred to the Paris *Memoire* 1751. Thus, while Everest had talked in terms of a flimsy structure for the Observatory, Maclear was able to state that it was built by order of the Governor Tulbagh at the bottom, or furthest end, of the court of No. 7 Strand Street, where he lodged, and that the pillars to carry the larger instruments were constructed with all possible solidarity. At that time there were also the remains of marks indicating the meridian direction from the Observatory.

Maclear was able to find a person who declared that the position suggested by Everest as the granary was in fact the foundations of an old house and that the correct position of the granary was some distance away, but no trace of foundations could be found. Maclear in this paper raised points similar to those of Airy, in refuting some of Everest's conclusions, albeit due to lack of available information when Everest was there. It could be that the comments attributed above to Airy were really taken from Maclear's paper of 1839.

Maclear made extensive investigations in the Klip Fontein area for verbal and other evidence and was able to excavate two further foundations in the locality, one of which he believed to be the true site of the granary. Full details of his investigations are given in [207].

In 1836 Maclear began preparing for his measurement of a new scheme and by March 1838 was observing at Klip Fontein with a Bradley 13½ foot zenith sector hopefully over the spot used by LaCaille. He was able to conclude that half of LaCaille's error was a function of the base line measurement and half due to the deflection of the plumb line.

In July 1990 two amateur astronomers from Cape Town successfully used resection[*] to relocate Maclear's station at Klip Fontein and uncovered the quart bottle that had been buried at the time. It was left undisturbed. [212]

---

[*] Resection is a method of determining an unknown position by observing angles at that point between several known positions.

Thus, whilst Everest drew appropriate conclusions from the scant information available to him, several of these conclusions were unjustified in the light of fuller details from LaCaille's writings. His deductions on the effect of attraction were, at least partially, borne out by the subsequent work of Maclear.

Everest dated his report to Lambton on the Cape Arc as 31 August 1821, a copy of which he sent to Joseph Dart of the East India Company, dated 3 September 1821. While the date on which he sailed back to India is unknown, it must have been by late October or early November that year as he reached Madras on 31 December 1821, ready to renew his efforts on behalf of the Survey of India.

He summed up his stay in South Africa saying 'the fine climate … had most thoroughly renovated my health'. [96 p. 23]

Chapter 5

INDIA 1822–1825: Taking charge of the Survey, and battling with fever again

*R*ather neatly, Everest arrived back in India right at the end of 1821, and by 6 February 1822 he was at Hyderabad. From there he had to trek some 300 miles to Takalhera to meet up with Lambton.

Work begins on the Bombay longitudinal series

After the year's break it was Lambton's thought that Everest should now start from the meridian arc and begin a longitudinal arc towards Poona and Bombay, later known as the Bombay longitudinal series. Such an arc would provide control for topographic maps as well as assist in the determination of the precise relationship between the Bombay and Madras Observatories.

It is interesting to note here the following comment made about Lambton. Besides the computation of triangles, co-ordinates and heights, he was continuously occupied with calculations for 'a desideratum still more sublime', the figure of the earth and abstruse phenomena affecting terrestrial and astronomical measurements. Abstruse and sublime perhaps, but Everest was to follow closely the footsteps of his predecessor in this field. Let it not be forgotten that Lambton, in 1818, produced parameters for the figure of the earth using the three sections of arc that he had measured up to that time, combined with the English, French and Swedish values. He derived a value for the compression (particular ratio of the axes) of 1 : 310.28 with the semi-axes of 20 918 747 and 20 851 326 feet. [228 p. 262]

The countryside had scattered hills of a few hundred feet elevation across the plain, which hyenas found to be convenient homes. No jungle, no fever, no mosquitoes, no bandits, no raging torrents, combined with a kind and well-natured population, made this a much more favourable scheme for Everest to observe. [229 pp. 134–5]

It was here that he had to resort to the use of towers. The lay of the land was such that sight lengths were naturally restricted to some 20 miles or so. There was a particular problem with a station to the north of D'haroor and its view towards

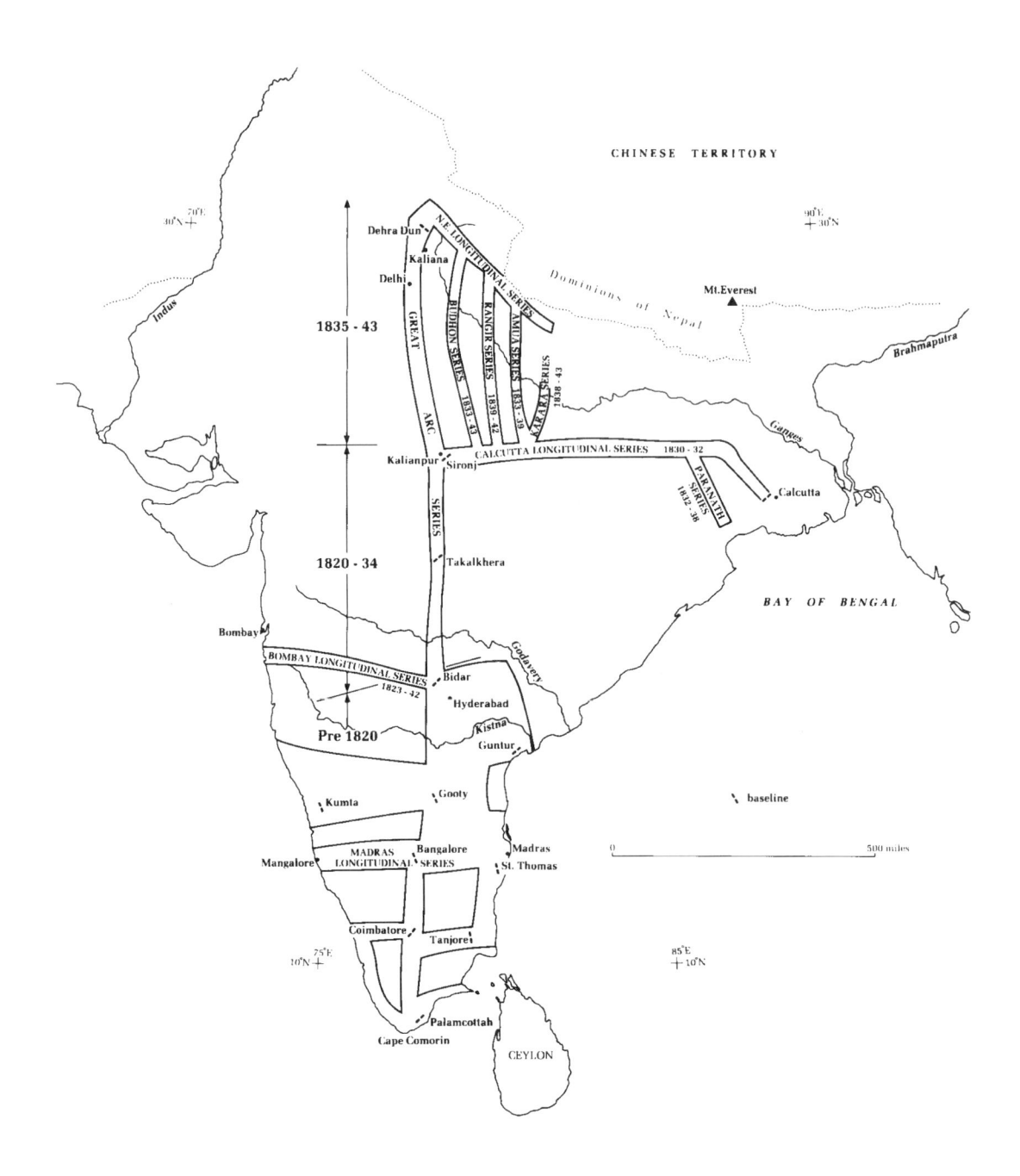

Figure 4 *The arcs of triangulation in India observed prior to 1843.*

Chorakullee, so he commanded that stone towers be built at each end until the points became intervisible. When both were of the order of 20 feet high, a particularly clear morning allowed sight not only of the other tower but the whole range of hills as well: such was the dual effect of a clear atmosphere and the phenomena of refraction. [96 pp. 25–32]

This atmospheric quirk was ably demonstrated at the same stations. Everest had a mast and a lantern hoisted on the top of one tower. By keeping watch on the lantern through the telescope at the far station, the image of the lantern was seen to rise up the cross-hair in the telescope until it had moved by 4 minutes of arc, equivalent to an apparent rise of the lantern of 21.5 feet. Given the right combination of circumstances, nature could be of considerable assistance to the observer in such a country.

Using lamps

Up to this period Lambton had always insisted on observing during the rainy season, when visibility of opaque signals – flags, masts, piles of stones – tended to be best. Unfortunately, this coincided with the most suitable conditions for a multitude of fevers to thrive. By changing to night observations, Everest was able to take advantage of the increased effects of refraction as well as the healthier dry season. [229 p. 246] After this experiment, Everest was set on a course to reverse completely the timing of field operations. He replaced the observation of opaque signals by day with the use lamps at night.

This is not to suggest that lamps had not been used before. So-called blue lights were used by Lambton when there were long distances and thick atmospheric conditions to overcome. Each station where they were used required 22 blue lights. Each light weighed 1½ seers (three pounds) and a camel could manage to carry 160 – or 7 stations' worth. They were reserved for lines in excess of 25 miles, often allowing the field work to be completed in a fraction of the time that would have been needed by the previous methods.

The lights were lit for short periods of three of four minutes at a time with a set interval of about the same length in between. This allowed the instrument to be read when the light was not required. With preplanning of the observation programme, longer intervals could be introduced where they were thought necessary, for changing zero (changing the setting of the graduated circle in the theodolite) or other such activities. When such lights were used, the method gradually adopted was for each light to be observed separately in relation to a reference light instead of reading all stations sequentially round the horizon. No doubt this particularly made it easier to programme the sequences for having the distant light lit and unlit, and was probably also more economical on the number of lights required.

It was at this time that Everest invented a new form of lamp that became known as a vase lamp; it was what might be called a cheap and cheerful but effective alternative to expensive, and not readily obtainable, reverberatory lamps (see Appendix 1).

It was found possible to use these at distances of up to 40 miles. Lights that were not all-round visible presented the problem of making sure that the slit was facing in the appropriate direction, otherwise all observations would be abortive. This was overcome by the initial placing of marks round about each station point in the directions of each ray. Then, as a particular station was observing to a lamp, so the slit could be correctly aligned. The timing required for each light to be visible was achieved by use of a chronometer and a detailed schedule of lighting and rest times.

One point that had to be considered was that not only the light could be seen but also the cross-wire against it. One without the other would be of no use. At that time it was possible for the reticule to be of fine spider web – for which the local jungle spiders were ideal – or fine silver wire. Everest tended to prefer the latter, particularly in the micrometers. Whilst a blue light might burn for five minutes, it was found, on a particular experiment over 39 miles, that the wires were visible for only three minutes when the powerful great theodolite was used and only one minute with the smaller instruments.

Lambton passes on the reins

In 1822 Lambton measured his last baseline at Takalkhera. Everest commented:

> This method of measurement (by chain on the ground) is remarkable for its simplicity but is, I think, objectionable on two accounts: that it is not in nature to present a perfect flat, other than in stagnant water, and that the tension of the chain cannot always be the same when drawn by the force of the human applied to a capstan. But Colonel Lambton called these objections absurd and pedantic... at the same time... the Lieutenant Colonel had put up the zenith sector at a station in the very alignment (of the baseline), which was 196 chains from the south end. He was in a constant state of exertion and fatigue by day, in superintending the measurement, which he performed exposed to all the effects of a tropical sun, and unaided, except by Mr Voysey and a few natives, for all of those on whom he might have relied in this hour of necessity, were disabled by sickness and the reckless exposure to which he had subjected them. By night instead of reposing himself from his labours, he continued to take zenith distances ... but here he had no person whatever to relieve him from his toil ... The consequences... his constitution received a death-blow, from which it never afterwards recovered. [265 (12)]

It would appear from these quotations that Everest was of a very like mind to Lambton both in dire lack of acceptable support and in driving himself well past the limits that might reasonably be expected.

It was 18 September 1822 when Lambton wrote to Everest that he was handing the large theodolite over to him. With it were five signal flags, an observing tent and 26 coolies for their carriage. Lambton's instructions were that Everest should proceed to the great meridional series somewhere between 18° and 19° and find the most suitable position for a baseline that would allow westerly progress towards Poona and Bombay.

Lambton expressed his confidence in Everest's ability, so much so that he declined to issue detailed instructions.

> To carry these operations from the meridional series entirely through to Bombay, will be a work of great extent and delicacy; but as I have full reliance on your judgement and abilities in the execution of it, I forbear giving you any specific instruction with respect to the minutiae of the performance… [204], [263 (171) 257–8]

These he left for Everest to formulate other than stipulating line lengths of 60–70 miles and the observation of the Pole Star at each extremity until the latitude and longitude of Bombay was fixed in relation to Madras Observatory and also it gave the opportunity to compare their height values against sea level. Little spirit levelling was taking place at that time, and heights were generally transferred by vertical angle observations. Ideally these should be reciprocal and simultaneous, but that was not often possible. Vertical angle observations taken around midday can be quite reliable, but if taken at night or early morning or evening they are affected by the uncertainty of the refraction effects.

Everest parted from Lambton for the last time on 15 October 1822. He commented later that 'Certain trivial circumstances had combined to ruffle that perfect cordiality which had existed between us in 1818, attributable, doubtless, to faults on either side; but we entertained the most thorough mutual esteem and respect for each other…' [96 p. 25]

(After representations in 1842 by Everest and in 1849 by his successor, Andrew Waugh, the Royal Society supported the recommendation that Lambton's triangulation should be reobserved. All the triangulation south of Bidar was completely revised during 1866–74. [229 p. 239])

In 21 years on the Trigonometric Survey, Lambton had completed 165 342 square miles of triangulation at a cost of £83 537. In 1848 Waugh, in commenting on Lambton, said that 'no man could have achieved more with the means at his disposal. His name is reverenced … as the Father of Indian Geodesy, and anecdotes of the talents and energy of this great man have been handed to his successors'. [229 p. 239]*

Everest was to become the very distinguished 'Son' of Indian Geodesy but one who probably contributed at least as much, and possibly more, than the 'Father'.

The death of Lambton

Everest started from the meridian arc stations of Damargidda and Boorgapilly (18° N 77° 45' E) – stations at which Lambton and Everest had first worked together in the field during January 1819. This was a much more amenable area in terms of climate,

---

* Similarly James Rennell was referred to as the Father of Indian Geography. Sandes [250 p. 191], when referring to the same title for Lambton suggested that the term geodesy was invented by A.R. Clarke in 1880 with his textbook of that name, but this cannot be so. As indicated, Waugh used the term in 1848 and Everest used it from time to time in referring to geodists.(*sic*)

wildlife and terrain than that which Everest had endured to the west of Hyderabad before going on sick leave.

When Everest and his team had reached about 76° E at the side Chorakullee to Sawurgaon, around halfway to Bombay, he received the news on 3 February that Lambton had died at Hinganghat on 20 January 1823. At the time Lambton had been accompanied by Dr Morton, an assistant surgeon. For the last month of his life he had been declining rapidly, although he had persisted in pushing forward.

As soon as the sad communication reached Everest, he immediately ceased his triangulation just short of Sholapur and returned to Hyderabad. Unfortunately the delay was sufficient to allow Dr Morton, as an executor, to dispose of Lambton's property as well as a number of Government items before Everest arrived on the scene. In addition he had moved a number of public documents to Hyderabad. [229 pp. 236, 443]

Everest was furious but was too late to stop the hurried sale of the effects. His opinion was that it realised only an eighth of what it would have fetched at a properly conducted disposal. In particular among the public papers were apparently the plans, records and manuscripts covering 23 years of field work. Everest made all haste to meet Morton and salvage what he could from the mess that had been created.

As was his style, Everest opened an acrimonious correspondence regarding the surrender of the official papers and instruments that Morton stated had been handed over to him as the executor. He wrote on 11 March 1823 to Morton from Hyderabad

> ... In consequence of receipt of a letter from Mr Joshua de Penning, Senior Sub-Assistant of the Great Trigonometrical Survey of India, of which an Extract is enclosed, I have arrived in Hyderabad with the view of requiring at your hands all public documents in any way connected with the department which has fallen under my charge by the demise of the late lamented Lt Col Lambton.
>
> I have to request at the same time that you will be so good as to afford me for transmission to the Supreme Government a statement of the reasons which have induced you to take possession of documents avowedly of a public nature and will further beg the favour of you to explain how it has come to pass that such of the Superintendent's Trigonometrical instruments as have been for several years employed in the public service and repaired and carried at the public expense have been disposed of without reference being made to me on whom the charge of them necessarily devolved until it could be ascertained what claim the Government might have upon them. You will perceive that my situation fully authorises me to call for this explanation when I state that I have every reason to believe that certain parts of particular instruments rather costly in their nature have been made in India by the late Superintendent and duly charged on the contingent monthly accounts of this Department with the Government. [263 (171) 70], [88]

Morton replied to Everest:

> In reply to the second requisition contained in your letter I beg to state that I took possession of the above mentioned instruments for their security until called for by

due authority and in reply to the third requisition of your letter I beg to state that the instruments which were sold were pointed out by Mr Josua de Penning to the committee as private property of the late Lieutenant Colonel Wm Lambton and are included in their inventory of the late Superintendent's private effects a Duplicate of which is in my possession. And I beg further to state that the above instruments with the other effects having been delivered over to me by the Committee as an Executor were sold for the benefit of the Estate not supposing that Government could have any claims upon private property. [263 (171) 72]

By the end of March Everest was getting somewhat irritated and impatient.

It is I imagine unnecessary for me to remind you that all papers relating to this department whether manuscripts or not are public property and that no person without the sanction of Government can retain the smallest memorandum or item relating to any survey...

There were in the possession of the late Lt Col Lambton a great number or articles of this kind in manuscript ... I think it will be found on examination that more than two thirds of the papers you have retained relate to observations and calculations connected with the Great Trigonometrical Survey.

With this impression on my mind I conclude you will have no objection to allow me to inspect the papers in question ... [263 (171) 85]

Morton responded 'I am not authorised to allow you to inspect the papers and manuscripts...without the concurrence of Mr Stuart the other Executor...' [263 (171) 87] Mr Stuart was with Lambton's Agents in Calcutta.

After some further correspondence Morton agreed to let the documents be inspected by a member of the Resident's staff and that those items deemed official could be surrendered.

The matter rumbled on into June. Although Everest considered that he had managed to recover all the official material, he was still unhappy about the disposal of various personal items that he considered should not have passed to outsiders.

During this acid correspondence, another who came in for criticism was Josua de Penning, one of Everest's assistants. He had been given instructions by Everest about the public property but was somewhat weak in his dealing with Dr Morton over the matter. This criticism was probably enhanced by the rather poor opinion that Everest held of de Penning's survey results.

Everest takes over

Everest assumed immediate control of the Great Trigonometrical Survey although at that time he had yet to be officially appointed to the post. This formality was to come on 7 March 1823 when he became the second Superintendent of the GTS with the stipulation that he worked under the Surveyor General.

Just before Lambton's death, the Directors of the East India Company had en-

Figure 5 *One of the versions of the Survey of India logo.*

quired as to how long it was likely to be before the Survey was completed because it was costing £6000 a year. Then, after Lambton had died, they discussed the idea of a complete Map of India. Valentine Blacker, the Surveyor General at that time, strongly advocated the trigonometrical methods and was against any idea of watering them down with a mixture of astronomical observations and rapid exploration although that might just be necessary in flat wooded country. He was also very much against any limit being placed on the observation of the Great Trigonometrical Survey. [263 (204) 87–9]

In the same vein, Blacker went on to extol the necessity of achieving good values for the figure of the earth because of the great reliance placed upon this when compiling navigation tables – an area in which the Company had considerable interest. To enhance his case he also mentioned those other areas of science that similarly relied to a greater or lesser extent upon accurate values of the figure.

It was about the time when Lambton died that Henry Voysey renewed his earlier assertion that more interest should be taken in recording the wet and dry bulb thermometers. Some five years previously he had suggested an ingenious method of determining the evaporation by placing a piece of wet muslin over the bulb of a thermometer to coincide with the observation of vertical angles – the inference being that refraction affected vertical observations far more than horizontal ones. Everest records that the first use of the wet and dry bulb in the field was 13 January 1823 at Netoli and from then onwards the practice continued. [229 p. 249]

Fever again

From the time of taking over until August 1823, Everest was occupied with the mass of accompanying administration. At the same time he had yet another bout of fever, this time accompanied by rheumatism and partial paralysis, so that he was a semi-cripple for the next two years. [229 p. 444]

He described one particular period:

> ... about the 20th of August, I had a smart attack of bilious fever, owing to too much labour of computation, which rendered the use of mercury necessary. I got better of this in a few days; but mercurial pills were given me as a constant dose, and one morning, having been overtaken some miles from home by a violent shower which wet me through, I found myself on my return again rather feverish.
>
> The evening of the following day [3 September 1823] is one of which I shall carry the remembrance with me to the grave. I was seized suddenly with an uneasy sensation in my loins; and on the following morning a very violent pain in all my bones, accompanied by typhus fever, showed that the embers of my Yellapooram illness had only been smothered for a time, to burst out more formidably ... For six months after this I was never able to lie in any other position than on my back, and even then, if my sleep exceeded the period of three hours I was awakened by one of these convulsive paroxysms, attended with agonising pain such that it reminded me of the iron boot of torture described by the author of Waverley, or as would have been caused by driving a wedge forcibly in between the bones of the leg, and in that position, turning it round. [96 pp. 35–6]

In a similarly graphic manner, Everest described the attack again in a letter of a year later to the Surveyor General, thus

> My original illness was a fever caused by too much attention to business, in consequence of which I was obliged to take mercury. I was recommended by the gentleman of the faculty to ride every morning to perfect the cure, and on the 3rd September last year [1823], in one of my morning excursions, I was wet through, and my left hip and loins, as well as my left shoulder, were immediately seized with the most violent pains accompanied by typhus fever. This illness has continued to torment me ... without intermission, and within the last four months has arrived at a crisis by the formation of an abscess at my hip, and another at my neck, from both of which fragments of decayed bone have repeatedly been extracted, sundry incisions and other surgical operations of rather an unpleasant kind have often also ... performed... [229 p. 404]
>
> ... Finding by 18th October that I could bear the motion of a palanquin, I quitted Hyderabad ... in company with Mr Voysey, and marched along the high road to Karinjah ... [96 p. 41]

Although advised by the medical men to go on leave immediately if he were to avoid his assured demise, he was determined or, one might say, obstinate to the extreme in wishing to carry on with the programme he had set himself.

> But it was a desperate resolution; for my limbs being in a great measure paralysed I was in the unpleasant necessity of being lowered into my seat at the zenith sector, and raised out of it again, by two men, during the whole of the observations with that instrument. At the great theodolite, in order that I might reach the screw of the vertical circle … frequently … I have been under the necessity of having my left arm supported by one of my followers; and on some occasions my state of weakness and exhaustion has been such that without being held up I could not have stood to the instrument… [96 pp. 35–41]

By then he had been appraised by Mr Voysey of the area to the north – an exercise he had been engaged upon when Lambton died. The valley itself was about 1000 feet above sea level and then the countryside was generally open. The problem areas would be the valleys and dried river beds which were choked with jungle and, according to Voysey: [229 p. 243]

> it is in these places that the miasma is generated. Here also reside an abundance of tigers, the terror of travellers, and so great is the alarm that if you cannot find … Goands for placing your flags, you will possibly find some difficulty in engaging others.. unless attended by at least two sepoys [European trained, Indian soldier]. I had the misfortune in March last to see one of my servants perish miserably before my eyes, without being able to afford any aid, under the fangs of a ferocious animal which had carried off 5 human beings in 3 preceding months. If I had had a Goand for my guide this accident would not have occurred, as these men are perfectly acquainted with the haunts of these animals, and give warning of approach of them. [229 p. 243]

After leaving Hyderabad with Voysey, the expedition met up with de Penning at Takalkhera in the valley of Berar. The value derived by Lambton for the base there was accepted, but Everest made a meticulous connection to the adjacent stations and added fresh zenith distance observations.

What necessitated this was that Lambton had overlooked a problem that had not arisen before. The distribution of stars suitable for observation was uneven to the extent that some were not readily observable with the zenith sector. Thus Everest was forced to adopt a technique that was far more open to errors than the method previously used. Even after making 372 observations at Takalkhera for latitude he eventually discarded them. [229 p.253]

The length of the Takalkhera base was 37 912.561 31 feet reduced to sea level – or around 7 miles, at an average elevation of some 1250 feet. It was begun on 6 January 1822 and was measured along the ground by means of a 'chain stretched between two small wooden capstans placed one at each end. The register heads were fixed on to plates of lead … imbedded into the earth, and the vertical angles were determined as usual, by a transit instrument … exactly in alignment'. [96 pp. 22, 128–9] The chain was compared with the standard chain both before and after the measurement. This was the standard chain that had been compared against Cary's 3 foot brass scale at Hyderabad in June 1821 and was to be later used at Sironj by Everest in 1825. [229 p. 249] (Any measuring instrument, e.g. chain, had to be calibrated for its length against a chain kept in the field for that purpose. This second chain was itself calibrated against a scale of verified length, e.g. Cary's scale, which was kept at a secure location).

While for the GTS baselines there was a standard for comparison there was a problem for the revenue surveyors (who surveyed property for tax purposes). The Surveyor General (Hodgson) was asked to supply a metal rod defining the English yard, feet and inches to each Collector (the official who assessed the worth of each property). He was unable to comply since he had neither the materials nor the required skilled technicians and he replied that they could only be properly manufactured in England. [229 p. 163]

Staff problems

Unfortunately, at this time, Everest lost both de Penning and Voysey. The former had wanted to resign immediately after the death of Lambton, but was persuaded to stay for a while longer. He left on 1 February 1824, after some 23 years service with the Survey, and retired to the coast at Madras. He was succeeded by Joseph Olliver. Voysey simply admitted that he could not manage on the salary he had and would have to look elsewhere. He had originally been appointed by Lambton to oversee the welfare of the staff as well as carry out geological work. In the event he suffered as much from fevers as did the others.

In addition to the problems of fever and shortage of staff, there was the problem of money. With a large party ever on the move, and in such remote territory, the acquisition of ready money at regular intervals to cover pay and provisions was a constant headache. In fact Everest was often carrying with him funds from his own resources equivalent to the total monthly cost of the party. This was around the same as three times his own monthly salary. In addition there were often charges for extra coolie hire, instrument transport and instrument purchase that were not met out of government funds. The supervising surveyor of 170 years ago had really to be a person of considerable private means.

In 1823, when Everest succeeded Lambton, his party consisted of six sub-assistants, a surgeon and dresser, 24 coolies, 30 sepoys and around six other bodies: up to 70 persons in total, to be kept and paid initially from one's own pocket – a situation that it would be difficult to envisage today. At one stage Everest was even committed to the equivalent of some five times the total monthly bill, or over a year's salary, effectively on loan to the government, and on which he lost money in various encashment costs in remote areas.

Somewhat in character this resulted in a continuous stream of heated correspondence until the situation was partly alleviated with the appointment of a special commissariat official to control all financial matters. Control was at times so tight that the Surveyor General had to apply to the government before he could authorise the engagement of 15 pack bullocks for the carriage of consumable supplies as they were additional to the seven authorised camels! [229 pp. 326–7]

Survey problems

In February 1824 Everest expressed some concern over the Survey situation as it was left by Lambton when he died. He wrote to the Surveyor General:

> Even omitting this consideration how were the triangles of the series to be connected? The station of Unjengaon is not seen from Nair or Inlka and the connection must have been carried on either by two sides and the contained angles or by the miserably acute angled triangles about Unjengaon and Badali, both of which methods were totally inadequate for such delicate operations.
>
> This country is in truth one of very great difficulty and the only method of surmounting the obstacles which nature here offers is by use of night lights, for the night refraction is so immense that I dare not specify its extent until my calculations are all prepared. [263 (171) 160–2]

An additional problem at this time was illustrated by Olliver when he wrote to Everest:

> I do not doubt that I should have succeeded in clearing the station and preparing it entirely to your satisfaction by the 17th at furthest if I had it in my power to procure sufficient assistance but I could not even get a guide, and what is worse, the Killader of a neighbouring fortress called Ghunnoor, about 6 miles distant threatened to confine all my people.
>
> There was a great scarcity of water, and no means of bringing it to the top of the mountain, and provisions were not to be had nearer than eight miles so that I could not with due regard to the safety of your people, keep them upon the hill for 24 hours together. [263 (171) 207]

Justifying his position

At the end of May 1824, Everest had another feverish attack when he was near Hoossungabad. It was about the same time that he was called upon by the Directors to justify his fitness to be in charge of the Survey. This was a challenge that he took up readily. He seemed to thrive in such circumstances that could be committed to paper, as will be seen later in his dealings with the Duke of Sussex.

He stressed how there were few aspects of mathematics that he had not studied and how he had spent six years immersed in the intricacies, both theoretical and practical, of geodesy. He described the privations of his journey the previous year from Hyderabad through 340 miles of jungle, when he almost lost hope of getting to cross the River Godavery. He discussed the excruciating pain he was in when he sat up for 30 nights at the zenith sector and 20 at the transit instrument to obtain his position. All these obstacles, which at one time felt like insurmountable barriers to progress, he had overcome. Everest was particularly proud of the symmetry that he had been able to maintain in his triangles and proudly invited the Surveyor General to visit him in the field near Sironj, but Blacker was unable to accept.

On several occasions Everest had bemoaned the fact that among his assistants he had none who 'had a particle of mathematical knowledge beyond decimals, the use of Taylor's Logarithms, and the square and cube root'. [96 p. 33] In addition, he was not at all happy with the field work of de Penning and others in the section Bidar to Takalkhera and referred to the situation:

> Though it might be perfectly unobjectionable to entrust the conduct of a series ... along one of the subordinate meridians, or even part of the principal series under certain limitations, to... a person who from long trial had been found ... skilful,... yet it is rather too much... to leave the entire and almost uncontrolled management of so delicate a work to any person whatsoever. Expert ... as Mr de Penning unquestionably was, yet he was a mere practical man, without any knowledge of the common principles of mathematics; and as to Mr Lawrence ... he was notoriously given to intoxication... Hence ... the fifth section [that to within 60 miles of Ellichpoor near Takalkhera] was a very vulnerable performance, and highly open to objections'. [263 (171) 350]

In July 1824, when making out a case to the Surveyor General for sick leave, Everest included an insight into his background.

> In respect to written testimonials or certificates of my qualifications I beg explicitly to avow that I have none to produce but I would most respectfully observe that none were ever required from my late venerable predecessor, and that at least I stand precisely in the same light as the late Lt Col Lambton did on commencing his career. I entered the service of the Hon'ble Company at the early age of 14, passed under their patronage through the Royal Military College of Marlow in the space of 5 months and the usual course of education at the Royal Military Academy at Woolwich in the space of 7 months and arrived in India before I was 16 years old [in fact he arrived a week after his 16th birthday] since which period I have never passed the Cape of Good Hope and consequently have had no means of procuring written testimonials of the nature alluded to in the letter of the Hon'ble The Court of Directors...
>
> ...
>
> ...that there are few subjects in Mathematics which I have not studied and that my attention has for the last 6 years been increasingly devoted to an attainment of both the theoretical and practical parts of my profession as a Geodist [Geodesist]...
>
> ...
>
> ...had I been an incapable man I should naturally have sought some pretext for delaying the decisive hour which was to expose me to the gaze of the whole scientific world. I should have trembled at the thoughts of succeeding to a man of Lt Col Lambton's high fame and pretension and should at least have availed myself with avidity of the common privilege allowed sick officers of proceeding on leave of absence... at least no want of zeal or energy can be imputed to me... [263 (171) 242-5]

Closing the gap to Sironj

During the next 18 months Everest closed the gap of those 60 miles to the south of Takalkhera from Pilkher, which had been left uncompleted two years previously. He then pushed the Survey forward across the Gawilgarh hills to Sironj. While in this area he had considerable assistance from an old friend, Captain Robert Low, who was Assistant to the Agent to the Governor General for the Nerbudda districts and operated from Betul. Unconfirmed comments suggest that Everest had been at the siege of Callinger (Kallinjar) in the same battalion with Low. [229 p. 444]

It was just before commencing the Sironj base measure that Everest's short temper, probably a result of his illness, came to the surface again and this time Olliver was the target. The latest bout of fever had him demanding that Olliver come to Sironj immediately. When this did not materialise, poor Olliver, observing triangulation a few miles away, was accused of subordination and put under arrest. So tactless was the reaction that the usually most co-operative Olliver became himself somewhat obstinate, and it took the Surveyor General to smooth the troubled waters.

Everest wrote:

With the view of satisfying myself respecting the urgency of the case which might have induced you to set up on your own judgement in opposition to mine I came along the whole base line this morning and I must avow that I saw nothing which could not very safely have been left for 24 hours…

Since therefore upon the examination of the whole features of the case it appears to me that your conduct is quite indefensible upon any grounds which I can discover or you have urged, am compelled to adopt the measure which I have stated. I have to desire that you will consider yourself under personal arrest until the pleasure of the Supreme Government is known.

To Major General Arnold, General Officer Commanding at Saugur, Everest wrote 'In consequence of extreme misconduct on the part of my sub-assistant I have been compelled to make a representation regarding him to the Supreme Government and in the meantime remove him from all duty and responsibility…' [263(91) 489–501]

Sironj baseline

The baseline at Sironj was measured during November and December 1824, and then the latitude was determined at Kalianpur – about 10 miles from Sironj. For this Everest constructed a small observatory and made 388 observations to 17 stars to obtain a latitude value of 24° 07' 11.837". Here he was able to use the same stars as at Takalkhera and so use the preferred method of amplitude determination. [229 p. 253] It seems likely that he took meridian transit observations to both north and south stars rather than observations to positions near transit followed by corrections.

The line was measured, as the previous one at Bidar, by chain, and before use this was carefully calibrated against a Cary 3 foot brass standard scale. It is not easy to compare a 100 foot steel chain accurately against a 3 foot scale so Everest began by constructing a suitable structure. This required a 120 feet long polished stone mounted on 15 pillars each 3 feet high, with brass marks soldered in at 5 feet intervals for the comparison marks. [96 pp. 51–2]

The subsequent base measurement itself was another of those occasions when Everest considered that he had to do everything himself because of the shortage of staff upon whom he could rely. He even supervised the driving of every picket, the reading of the ten thermometers and checking the free movement of the weight. This base was measured on coffers and gave a result of 38 411.899 12 feet reduced to sea level.

By Christmas Day 1824 Everest was able to write to Blacker:

> Baseline completed extending to 7.273 miles. A rough basaltic column 3½ feet long sunk at NE limit. Similar one with a circle at centre sunk at SW end ... under my own eye with the greatest possible accuracy below the arrow of the chain and thus these two limits will probably last for ages. [263 (171) 306]

Olliver was again the target of an angry missive in February 1825, although this time he was not the real culprit. Apparently from early December 1823 Everest had insisted that when in camp, his nights were not to be disturbed by man or beast. A horse, which Everest thought belonged to Olliver, had kept him awake for several hours.

> I must desire that you will ... prevent such an annoyance again occurring, for it is not only in direct disobedience of my standing orders that neither men nor cattle should make any sort of disturbance within my hearing, but it is absolutely impossible that I should continue to perform my duty in my present state of health if my natural rest is thus ... broken. [229 p. 445]

Later he found out that the offender had not been Olliver but Rossenrode and the wrath was to turn that way some months later sparked by a different, quite minor offence.

> ... when some neighing horses were fastened so near my tent as to prevent my getting sleep. This nuisance continued for three nights in succession and, notwithstanding all my endeavours, I was unable to discover who were the promoters of the riot. On the 4th day I had a proclamation made ... that the sentries had orders to turn all neighing horses and other noisy beasts out of camp, and a Naik and 4 sepoys were directed to go the rounds and see this proclamation was carried into effect.

It now appeared who the real offender was that had set himself up in opposition to all authority

> ... You were that person; ... it was a horse of yours which created the nuisance, and... when the Naik went to execute the orders given him, your horse keepers violently

resisted him in the performance of his duty, and said that your neighing horses should not be removed without "cutting their throats". Disgraceful and insubordinate as this was in a civilised camp, yet it was trifling compared to your own conduct on the occasion. You were in the observatory at the time … and you burst forth in a torrent of insolent railing towards me, which would have more befitted a lewd scold in the purlieus of Billingsgate or Wapping, than a person who had been accustomed to the decencies of life. [263 (172) 447–54]

Overcome by fever

The observations were complete by 26 March 1825 by which time Everest decided that he had to succumb to the will of his doctor and apply for sick leave. He had earlier applied in January 1824 for such leave, stating that since the previous August (1823) he had often been dangerously ill. At that stage, by the time permission had arrived he was feeling sufficiently better to carry on. However he was soon reporting that Mr Griffiths, the medical successor to Voysey stationed at Hoossungabad (Hoshangabad), had to go some 40 miles into the heart of the forest to treat him, and he then became a perpetual patient. Everest expressed considerable praise for Griffiths who, probably appreciating the stubbornness of this patient, warned him of the consequence but did not expressly forbid him to carry on observing at night, or sitting long hours calculating. [263 (171) 277–81]

By 30 March a recurrence of his disorder was so much more violent than any he had previously endured that he was deprived of nearly all power of motion. On this occasion he was taken to the home of Major Feilding at Goonah, where he stayed. He had in any case intended going there to carry on his computations in more leisurely fashion than at the cantonment of Saugor, but this new bout put much of that out of his mind.

On 2 May 1825 he wrote to Blacker,

the state of my health during the last month was such as totally to incapacitate me from either mental or bodily exertion but since the last operation performed on me I am a little better and hope to acquire sufficient strength to undertake the journey to Cawnpore before the rains set in. [263 (171) 341–3]

He stayed with Major Feilding almost two months, and then on 25 May he began the journey first to the Ganges at Cawnpore and then on to Calcutta, which he reached on 12 August 1825. [96 p. 44]

His leave was granted 'on account of the bad state of his health, and the valuable services rendered by him are brought to the Court's favourable notice.' [263 (217) 153] In yet again requesting leave he still maintained his devotion to the task he had set himself and requested that he might take 'full and complete copies of my whole work, in order that the scientific results may be calculated by myself, and submitted when entire to the Hon'ble Court of Directors'. He was granted permission to take copies with him so long as the originals remained in the Surveyor General's Office.

He was to receive help with the computations from Richardson and Taylor of the Royal Greenwich Observatory.

Life in the wild

Mention should be made of Everest's relation with the wild creatures that inhabited most of the areas he had to work in over the proceeding years. He made the following comments.

> I had a small tent pitched at the top of the hill, and a larger one at the foot with my camp... One of my followers brought up one morning in a large jungle leaf a heap of these detestable insects (scorpions), which he and others had killed... in my lower tent. Upon counting them... there were, young and old, in number twenty-six...

> The tigers, too, were very large and ferocious. The inhabitants of a hamlet near my station... were preparing, whilst I was there, to abandon their homes in consequence of the perpetual prowling of these animals around them. From that station I intended to take azimuths of verification; but, when I sent out a party with a reference lamp for that purpose, it was necessary to surround them all night long with shouts and revelry, and the blaze of fires, and discharges from musketry, so that the observations, which should be made in peace and tranquillity, became useless...

> I never saw a tiger in the wild state in India. Not a man of mine had ever been carried off, though I had in my excursions with the telegraph line, and in the jungles of the Godavery, invaded ... the forests in which they chiefly abound. And to this lucky cause, probably, may be attributed the belief which the natives generally entertained of my being possessed, by means of astrology, of some necromantic powers, so that tigers had no power to harm me or those who were under my immediate protection...

> The faith placed in the healing powers of the great theodolite and other instruments employed at any time in observing stars were such that I have had people come many miles to entreat permission to bow down before the lower telescope of this imposing instrument; and strange as it may appear, it is no less true that men and women who had been lame or blind for years, others who had the palsy, and others again who were swollen with dropsy, were among my applicants.

Before going on leave he wrote a report of 69 paragraphs plus appendix to Blacker. A few extracts from this serve to illustrate aspects of his life at that time.

> ... on the banks of the Godavery is a tract of country altogether unexplored, and covered with some of the most tremendous forests in the world, it is hardly possible to imagine any part of the Earth more dreary and desolate or more fatal to human existence, yet over this deadly tract I had to throw an extensive net of triangles and though I succeeded most satisfactorily, yet I suffered very seriously from two successive attacks of the fever peculiar to that climate, and my constitution received a shock from which it has never effectually recovered.

This fever returned … in 1823 … became so violent about September of that year that I was deprived of the use of my limbs and my life was for a long time in considerable danger... [263 (171) 360]

To the North of the valley of Berar and within a few miles of Ellichpoor rises a vast chain of Basaltic Mountains which extend as far as the eye can reach in an East and West direction, and appear to be covered with forests quite impenetrable. These wildernesses are almost destitute of inhabitants and the few human beings who dwell there, a wretched set of wild Goands, are engaged in perpetual conflicts with tigers and other wild beasts by which and the barrenness of the soil they are frequently driven from their miserable hamlets. Water is hardly to be met with and provisions unless brought from a distance are nowhere procurable, the Mountains viewed from the Valley of Berar appear altogether interminable, and from the features they present seem to rise ridge beyond ridge so nearly equal in height as to preclude the possibility of selecting a series of suitable geometrical points, besides which they are really the seat of the most deadly fevers, and in the Dekkan are reported to be much worse than is actually the case. [263 (171) 365–6]

He sailed from Calcutta on 11 November 1825 bound for England. Other than the year at the Cape of Good Hope, Everest had not been away from the sub-continent since first arriving there in 1806.

Chapter 6

ENGLAND 1825–1830: Sick leave at home, arguing for better equipment and conditions of service

*E*verest left India in November 1825 for a spell of sick leave – in England – where he must have arrived in late February or early March 1826. Whilst on that leave he several times sought extensions of the period until finally he was told by the Court of Directors that, if he was not back in India within five years of leaving there in the first place, then he would be deemed to be no longer an employee of the East India Company. Whilst he was absent from India, work on the Great Trigonometrical Survey was in abeyance. Certainly the Company had no wish for this to be so, but there was no suitable person to step into the work that Everest had been masterminding.

As Everest stated, he was not the sort of person who would have tried courting the favour of the Governor General, Lord Amherst (formerly William Pitt), to keep his post open. On the contrary, he preferred to gain whatever advancement he could by demonstrating through his achievements that he deserved it. [96 pp. 116–17]

This period of five years is one about which least is known. Obviously from the time of his return to home shores in early 1826 his first problem was to recover his health. During May–June 1826 he was in Margate and commented that he was still not ready to start work in a regular fashion but would gradually take up his computations.

Ready or not, it was during this same period that he wrote a memoir of 130 paragraphs on the Great Trigonometrical Survey. [94] Occupying some 48 pages, it ranged over many aspects of survey in India. He discussed route surveys and how they tended to be flat earth surveys that required periodic astronomical control to contain the accumulation of errors. He explained how a baseline was measured by chain using wooden troughs or coffers supported on tripods, and the subsequent angular measurements. After discussion of some of the instrumentation, Everest commented on the possible streamlining of the administrative aspects of the GTS. Brief mention is made of some aspects of his career and how he was thwarted in 1812–13 when a colleague wanted Everest to join him on a survey operation.

This substantial memorandum brought the reader up-to-date with the activities and requirements of the GTS and how Everest saw its needs. Among his particular regrets was that the Company would not sanction the return of various instruments to England for repair although the sequel to that was the purchase of more modern items.

Fellow of the Royal Society

In 1826 he was proposed for fellowship of the Royal Society. As was the normal procedure, the nomination was read on a number of separate occasions from 16 November 1826 to 1 March 1827. Then at the next meeting on 8 March 1827 he was duly elected. The document read:

> *Captain George Everest of the Bengal Artillery, a gentleman well versed in mathematics and various branches of natural philosophy and who is superintendent of the Great Trigonometrical Survey of India, being desirous of becoming a Fellow of the Royal Society – we whose names are hereunto subscribed do of our personal knowledge declare that he is worthy of that honour and likely to prove a valuable and useful member.*

Henry Kater

Francis Baily

J. South

J.F.W. Herschel

Benjm. Gompertz

Davies Gilbert

Henry Foster

Charles Wilkins

Balloted for and elected March 8 1827 [11]

Henry Kater had left India through ill health shortly after Everest had first arrived there.

It is interesting to note that all of Everest's notable achievements came after he was elected a Fellow. It was 1828–9 when the Everest theodolite and other instruments were constructed to his design for transport back to India. His first Report that included details of his figure of the earth did not appear until 1830 and his second in 1847. It might then be asked on what grounds he was elected.

In earlier years some harsh words had been exchanged over the apparent lack of interest that the Royal Society had shown in William Lambton. In fact not until the French Academy of Sciences had him as a corresponding member did the Royal Society consider him worthy of a Fellowship. Perhaps this persuaded eminent Fellows to ensure that the same criticism could not be levelled at them over his successor. [229 p. 468]

Support from India House

Additionally in early 1827 Everest received a considerable boost for his activities from N.B. Edmonstone of India House. In a long memo to the East India Company, Edmonstone extolled the virtues of Everest and everything he was trying to achieve. He began by suggesting that the Court of Directors had a less than adequate appreciation of the complexity and importance of the whole idea of an arc measurement. Stress, he said, should be laid on the scientific aspects as a precursor to any general mapping activities rather than that the two should run parallel.

To emphasise this he quoted from Lambton's report of 1819.

> This arc is already greater in amplitude than any single arc that has yet been measured on the surface of our Globe, and if it be prosecuted to the above extent, it will be the grandest monument of practical Science that has ever been or ever can be exhibited in any age or nation.

> The happy results that have hitherto succeeded one another at the different stations of observation for determining the lengths of degrees on the meridians hold out strong inducements for such an extension and if it continue to be free from anomalies arising from local attraction it will in itself be sufficient to determine the magnitude and figure of the earth.

Europe was the birthplace of arc measurement, whether one goes back as far as the initial development of the concept of triangulation, to the early arc by Picard or to the inspiration that led to the famous arcs of Peru and Lapland in the 1730s. European countries were not only the model for arc measurement but also for experiments in related fields such as the determination of the variation of gravity with latitude, the problems of refraction and the standardisation of units of measurements. The last two of these were extensively pursued in India but neither Lambton nor Everest had the time or the personnel to investigate the problems of gravity. This did not, however, stop the Academy of Science in Paris from awarding a Diploma to Lambton and accepting him as a corresponding member. Lambton particularly corresponded with Delambre who, in one of his letters of 1818 wrote,

> let us not however too much regret that which is still wanting but let us rather congratulate ourselves on the astonishing precision to which we have arrived. Let us redouble our efforts to diminish the slight anomaly by new researches and let us multiply as much as possible our observations and those scientific enterprises which like yours will confirm the Glory of the Philosophers of the 19th century.

Thus the centre for geodetic knowledge at that time was still the French Academy of Science. Although there were stirrings within England, geodesy was not pursued with such vigour by the Royal Society as it was in France. To glean the latest ideas on theory, technique and instrumentation it was essential for both Lambton, and Everest after him, to keep in touch with the scientific bodies of Europe. Hence the honour bestowed upon Lambton was also of considerable assistance in the furtherance of the interchange of ideas and knowledge.

The special scientific nature of the work on the arc prompted Edmonstone to question whether it should be under the overall control of the Surveyor General or whether the separate Superintendent should be his own master. After all, the qualities necessary to comprehend all aspects of work on the figure of the earth were somewhat special and different from the needs for the control of general mapping.

> Surely it is no more than becoming the character, situation and dignity of the Court of Directors to show to the world that they duly appreciate the magnitude and importance of these objects of Science and are disposed to afford to the prosecution of them their liberal patronage and support unclogged by a strict adherence to official Rules and Form when they interfere with that which ought to be the sole object in view, the successful prosecution of this great National undertaking.

> It is deeply to be regretted for the interests of Science and the Credit of the Court therefore, that the luminous and interesting representation of this subject contained in Capt Everest's able and elaborate Memoir have apparently failed to attract the consideration of the Court and that all his suggestions with regard to the measures necessary for the prosecution of the work have, with one exception, been passed over in silence. The unavoidable effect of this system of procedure if unremedied must be to arrest its progress and in all probability frustrate its accomplishment.

> Capt Everest has represented in strong terms the defective state of the instruments which have hitherto been employed and suggested their being sent home for repair improvement or substitution... He has minutely described those defects, the consequences of age and long use and adverted to the improvements which the progress of science has introduced.

As luck would have it, because the existing instrumentation was not returned to England for repair, Everest's hand was strengthened when it came to seeking funds for new instruments. In fact it could be said that the Bengal Government did him a favour in this respect.

Edmonstone stressed that only the best instrument makers should be used because of the extremes of accuracy required in the construction if the results expected in the field were to be consistently achieved. After indicating the makers he had in mind Edmonstone then decided that some comments on Everest himself were appropriate as a way of adding further strength to the case he was making. He concluded his memo thus:

> The astronomical and mathematical attainments of this officer are of the highest order, he engaged in the great work undertaken by Col Lambton many years ago with enthusiastic ardour and after long practical experience under that most able man succeeded to the Superintendence of the work. I state on good authority that at this moment there is not an individual in India competent to the prosecutions of that great object the continuation of the Measurement of the Arc of the Meridian and the calculations appertaining to it (it is considered that there are not more than three officers in England who are capable of parallel operations in Europe).

> His constitution has completely sunk under the mental and bodily exertions and his

return to England became absolutely necessary for the preservation of his life. [Assistants Garling, Connor and Young had lost their lives.] The govt. abroad justly considered that he might be beneficially employed in this country for purposes connected with the progress of the Trigonometrical Survey in India and it indeed admitted by the Court that his scientific acceptance is indispensable to work up the materials which he has brought home, but not withstanding the peculiarity of the circumstances of this case (to which it may be expected that there is not any other analogous, the Court refuse to consider him in any other light than as an officer on Furlough under the Regulations of the Service.)

A case so peculiar as this in which the most important interests of Science are concerned and when a service connected therewith is to be performed which this officer alone is capable of performing and whom it is resolved to a certain extent to employ for that purpose ought surely be made an exception without the hazard of an inconvenient precedent.

It may perhaps have been suspected that in his Representations and propositions Capt Everest is actuated by motives of a personal nature, but I must do him the justice to declare my conviction that he is far above the meanness of being influenced by motives of that character. He is enthusiastically devoted to the scientific objects in which he has so long been engaged and in the success of which his fame and reputation as a man of science are deeply concerned. He is proudly conscious of having (it may be said) gratuitously employed in that undertaking all the energies of mind and body at the sacrifice of his health and the hazard of his life he naturally feels deeply grieved and mortified at finding his scientific labour and the great national object of them held in the low estimation which is indicated by the proceedings of the Court. I humbly conceive that every consideration of justice, policy and public credit requires the manifestation of liberal encouragement to an officer so circumstanced and so characterised as is Capt Everest and the adoption of whatever measures are best calculated to facilitate and promote the success of the scientific operations which he was appointed to Superintend.

This indeed would only be consistent with that more just and elevated view of the magnitude and importance of the work in question on which it would be so becoming a great body such as the Court of Directors superintending and guiding the vast concerns of our Indian Empire, to Act.

… I beg it may be understood that what I have urged with respect to Captain Everest is the result of a conviction founded on much enquiry, that without his scientific aid, or that of another equally qualified (if such there be available) the great work in hand cannot go on successfully nor what has already been done, be turned to account – my acquaintance with him is slight – my motive is purely public – he has entered into communication with me on this subject of the work to which he is so devoted, merely because he knows that I was in correspondence on it with the late Col Lambton and took a lively interest in it although not pretending myself to any scientific knowledge. [79]

In any endeavour to obtain support surely no one could want more detailed and complimentary backing than this. Whilst it had been a decision of extreme vision that the East India Company so readily took up Lambton's original idea for an arc, here they were being pushed into recognising that they were now controlling the most extensive such arc in the world.

Even so it was a continuing brave move on their part to go along with all the requests, trials and tribulations of 40 years and the associated costs. The rewards of their foresight and the perseverance of Everest have now been visible for 150 years with the renowned foundation of all mapping within the vast country of India being a model for all to see and for some countries to copy. Unlike the construction industry, for example, where designs have subsequent buildings, highways and the like as proof of the success or otherwise of the enterprise, with arc measurement, or triangulation in general, there is little initially except for a few scattered pillars to see for all the efforts. Years later the value becomes apparent, but in the interim all has to be taken on trust.

When so much relies on the people who hold the purse strings, it really requires strong scientific presence in the board room when it comes to putting faith and money into apparently nebulous projects. Thus it was that Edmonstone with his memo did considerable service both to Everest and, in the longer term, to the profession as a whole. There have been many instances particularly in scientific history when such faith was required and found wanting but in this instance the right person was in the right place at the right time to push the case in the appropriate quarters.

New instrumentation

Before leaving India, Everest had requested that while on leave he should be kept on duty so that he could make investigations into new instrumentation for the Survey and he also asked to be allowed to take detailed records of observations with him for carrying on the computations. The Directors went so far as to accept all his advice about instrumental requirements but did not place him officially on duty. This led to an argument over his pay.

Everest commented that 'For the last 16 months, and for the first 6 months of 1827, his time has been so much occupied that he has given up all society, and has taken one expensive journey to Ireland, and frequent minor journeys on … the business of the Company'. [229 p. 446]

Everest felt that he should be on full pay as if he were in India but the Directors would not grant this. Apparently if he had only gone to the Cape to recuperate he would have been entitled to full pay, but not so in England. After extended arguments he was permitted regimental allowances for 'the first six months of 1827 and the regimental allowances together with the pay of … rank from November 1828 up to the date when you shall cease to be employed'. But why was he to receive no pay from July 1827 to October 1828? Other than that he was on the continent for part of this period visiting Rome, Venice, Vienna and Milan, there seems no good reason for the change in circumstances.

The event that happened in November 1828 was the start of the computations of the GTS and possibly this could have triggered the added remuneration. The computations for his 1830 Report were carried out by William Richardson and Thomas Taylor of the Royal Greenwich Observatory. It was Richardson who introduced Everest to Henry Barrow – a co-operation that was to go sour within a few years. [229 p. 257]

Pendulum apparatus

In December 1828 Everest showed considerable interest in some new pendulum apparatus. He wrote to Joseph Dart, Secretary to the East India Company about it.

> I would chiefly allude to the pendulum and astronomical clocks by Jones, Charing Cross, which I learnt from Colonel Forrest have been made for the Engineer Institution at Bombay. I have seen these instruments and their apparatus in the warehouse, and they appear to be constructed in such a manner as to do justice to the liberal patronage of the Honourable Company and credit to the maker, but it would in my humble opinion more materially lend to establish the latter claim and at the same time give to the former that honourable celebrity which is so highly merited, if previous to the despatch of these instruments to their destination they were erected in the Observatory at Greenwich and there subjected to a full course of experiments.
>
> ...but besides this, the principle of the pendulum in question would render such a measure highly advisable not to say indispensable, it is not a convertible but an invariable pendulum; it does not measure absolute length like that which was invented by Captain Kater and erected in Portland Place in 1817 but leaves those lengths in different parts of the earth to be deduced from a comparison of the number of vibrations in twenty four hours with the corresponding vibrations made in some other place where the length had previously been determined and thus, data are furnished for ascertaining the increments of gravity and the figure of the earth. ...In the event of this measure being adopted it may be a question for the Honourable Court to consider how far it may be advisable to allow a certain number of the young gentlemen from Addiscombe to attend during the experiment in order that the methods of adjustment and use may be duly impressed on their minds ... [92 pp. 243–5]

Both the Court and the Astronomer Royal agreed to assist and Lieutenant Colonel Pasley at Chatham agreed to let some of his students take part. Everest suggested three groups of six students each and for each group to have four days of instruction. It was not, said Everest, his intention to turn all engineers into astronomers. However, he was particularly impressed by the interest shown by Mr Western, and he proposed to keep him extra time and said that he should be instructed in astronomy.

Board was arranged at the Green Man Inn on Blackheath for March and April 1829. The total bill there came to £64 for 11 days and this was queried (but paid) as rather high by the Military Committee. [94]

It would appear that it was not only the pendulums that were experimented with, since Everest referred to the Alt-azimuth instrument, in effect a theodolite, requiring some adjustment. He discussed this with Simms who was apparently at the Blackheath exercise. [92 p. 251] In addition they had a transit instrument, clocks and set of compensating bars. The group was also joined by Lieutenants Hastings and Murphy who had been with Colonel Colby in Ireland on the measurement of the Lough Foyle baseline of 41 641 ft in 1827–8.

The Irish Survey

In the summer of 1829 Everest was in Ireland for a second time. In particular he visited W.R. Hamilton at the Dublin Observatory. His visit was recalled by Hamilton in a letter to the Rev Dr Robinson on 3 August 1829.

> Captain Everest ... lately paid me a visit here. He was much delighted with the eight-feet circle, which he assisted me in putting to some more severe tests than I had myself done before. The little repeating circle did not please him equally on examination, since he found, what I also had remarked, that the altitude screws communicate a motion in azimuth ... [152 pp. 335–6]

After his return from Ireland, Everest wrote an extensive report on the Irish Survey [90] in which he highlighted various aspects of the operation there as worthy of introduction into the Indian survey. Each district had its own headquarters and staff of officers, non-commissioned officers (NCOs) and soldiers to do the minor triangulation and detail work. Engraving at six inches to the mile was done centrally in Dublin. The total establishment of the Survey of Ireland was nearly 900 bodies ranging from the Superintendent and four captains down to 391 assistants and 26 engravers. As an incentive there was a system of payments possible to those thought to warrant special consideration. These could be awarded directly by the Superintendent without recourse to higher authority.

For the major triangulation the Survey had one large theodolite of 36 inch diameter and one of 24 inch, the former by Ramsden and the latter by Troughton. In total the department had over 150 theodolites of various sorts.

Observing was possible either by day or night. In the latter case reverberatory lamps with argand burners or Drummond lights were used. As an alternative for daylight use, Colby had a variety of heliotropes. Of particular interest was the use of compensating bars and their transportation from place to place by horse-drawn spring carts.

If similar operations to those in Ireland were to be adopted in India then a body of officers would require training. At the same time, if such officers were to stay with the Survey, their conditions of employment would need to be made attractive. As the situation pertained in India at the time, the post of field surveyor was no more than a temporary position and not a settled one in which one could take pride. For example, if the field surveyor were to fall sick then, unlike all other staff officers, he would lose his salary and his appointment. Everest also commented on the even worse situation of the labourers in the Survey. Here it is evident that he had great sympathy with the family and recreational needs of his staff but it was a matter of persuading his superiors of the advantages of better conditions of employment.

Everest aired his views on the evils of prospective surveyors who were seldom or never selected on their talents as appropriate to the work. He then started to offer suggestions as to how the various defects might be best overcome. To begin with he was much in favour of military control because of the discipline required in the execution of the work and he stressed his desire for a staffing system akin to that of Colby.

The surveyor generally had to supply his own instruments since only the major items were provided by government. This, Everest suggested, was not in the interests of the Company because of the frequent inability of the individual to afford the best instruments. The variety of terrain made it necessary to have different instruments for different circumstances.

In particular Everest felt there was a need for an 18 inch theodolite with three microscopes on the horizontal circle. He recommended that each of the three Presidencies should have seven theodolites at 12 inches, the same number at 8 inches and 24 at 7 inches. These last would be for minor triangulation to control topographic surveys. With regard to the old 3 foot instrument, he suggested that it should be brought to England for repair and overhaul.

He then reminded the Company that they had recently agreed to his ideas of trying simultaneous observations for amplitudes on the Great Arc and that this would require two zenith circles to be obtained. [90]

Surveyor General of India

On 12 August 1829 Everest was appointed Surveyor General of India in addition to the post he already held of Superintendent of the Great Trigonometrical Survey. This new position was to take effect from the time he returned to India on 8 October 1830. On making this appointment the Directors stated that they 'entertained a high opinion of your services as Superintendent of the Grand Trigonometrical Survey, and of your scientific requirements and general qualifications'. [229 p. 301] Note that in different places one finds the GTS defined variously as the Great or the Grand Trigonometrical Survey. Previous holders of high honours were as follows.

Surveyor General of Bengal:

1767–1777	James Rennell	1777–1786	Thomas Call
1786–1788	Mark Wood	1788–1794	Alexander Kidd
1794–1808	Robert Colebrooke	1808–1813	John Garstin
1813–1815	Charles Crawford		

Surveyor General of Bombay:

| 1796–1807 Charles Reynolds | 1807–1815 Monier Williams |

Surveyor General of Madras:

1810–1815 Colin Mackenzie

In 1815 the three Presidential Surveyors General were brought together under one head as *Surveyor General of India*. Those who held this extended post were:

1815–1821 Colin Mackenzie	1821–1823 John Hodgson
1823–1826 Valentine Blacker	1826–1829 John Hodgson
1829–1830 Henry Walpole	1830–1843 George Everest

Superintendents of the Grand Trigonometrical Survey were:

| 1800–1823 William Lambton | 1823–1843 George Everest |

The mode of appointment of a Surveyor General (and some other posts) was strange. After Hodgson left on 24 January 1829 the Governor General Lord Bentinck took several months to select Henry Walpole as successor. However, the Directors, regardless of the Governor General, made their own selection and in August 1829 they told Everest that he was to be Surveyor General. So, even before Walpole took over on 30 October 1829, he knew that as soon as Everest returned from England he was to take over. This suggests there was scope for a little more liaison between the Directors and the Governor General.

Damage to the great theodolite

The equipment that had served the GTS well ever since it was instigated by Lambton had seen hard service and was becoming worn and unserviceable. Indeed the 3 foot diameter theodolite had been severely damaged on two occasions and, although successfully repaired, the time was ripe to study newer equipment currently in use in Europe. The first damage had been in 1808 near Tanjore when, while raising the theodolite up the side of a pagoda, a rope snapped and the instrument hit the side of the tower. Besides a broken screw and clamp, the limb was so distorted as to make it unusable without extensive repair. The second of these occasions was on 10 February 1825 when a severe storm blew a tent down on the instrument. (This was despite ten men holding the fastening ropes.) One of the screws was broken, but the damage was not as severe as during the first incident. Everest did, however, lament the destruction of one of his barometers. [263, (316)]

The baselines up to this time had been measured by chain, and perhaps by now developments had produced a better replacement. Everest was encouraged to travel throughout Europe, visiting the leading instrument makers and studying some of the potential items in use either in observatories or in the field. He probably visited Rome, Vienna, Venice, Milan and Naples, and possibly other centres. One of the interests on such a journey was to see the various transit circles and zenith instruments constructed by the notable German maker, Reichenbach and others. It would be likely that he also visited centres further north in Germany and the Baltic countries.

Everest's first Report

On 25 February 1830 Everest wrote to W.R. Hamilton in Dublin about his Report on the Great Meridional Arc. [96]

> I have just delivered to them (The Hon E I Company) the manuscript copy; and if they print it, you may be quite assured that somehow or other you shall have a copy at your disposal. Now the subject of this letter is to beg that you will read it through, and if, when you have done so, you think it merits such a favour, that you will write a full, fair, and thoroughly impartial review of it, such as you, of all men I know, are most able to write. In asking this favour I do not by any means intend to avail myself of our private friendship and the mutual esteem we bear each other, to shelter myself from criticism; but I confess I shrink from seeing some scribbling charlatan, who cannot comprehend the subject sufficiently to enable him to detect its merits or expose its errors, interfering to assail me with absurdity and draw me into an unavailing correspondence. [152]

Hamilton replied on 5 March 1830:

> ...I gladly accept your promise to send me a copy of your Report, when printed, of the Arc already measured, and am sure that in reading it I shall derive much pleasure and instruction. But I cannot so far mistake the state of my own attainments as to imagine that I could usefully perform the task which your partiality would assign to me, of writing a review of that report. A young person may possess natural talent, and aptitude for scientific speculations; but it is almost impossible that a young man should have the degree of experience requisite for deciding well on the merits of an extensive national work; and I feel sure that I am no exception to this great practical theorem... [152]

On 4 June 1830, the day before he sailed for India, Everest sent a copy of the volume to Hamilton.

> I sail tomorrow for India on board the *Cornwall,* and shall feel extremely obliged if you will do what in you lies to get my bantling well served out, for if he sleeps he will assuredly die a premature death.

> Mr Airy has promised to take it in hand, as far as his multifarious occupations will permit, and he said that you and he together might perhaps be able to concoct a review to be inserted in the Quarterly... [152 p. 374–6]

Only 500 copies of the Report were printed, and of these Everest had but three to take back to India. Of the others, he sent 11 copies to Baron Humbolt for distribution among scientific personalities in Europe and 27 were presented to various scientists and acquaintances in England. Of the three that he took to India one was for Lord Bentinck, one for Colonel Feilding and one for himself.

Testing the compensating bars

Everest referred to the compensating bars (*see* Appendix 1) thus:

> It will be in your recollection that in March last with consent of the Chairman and at my suggestion Colonel Forrest sent to the Royal Observatory a set (consisting of three of these bars and their standard). My objects were to try these bars in Greenwich Park and see whether there where [sic] complete. To get all the practical information I could respecting their use. To explain the construction and use of them to the Young Officers from Chatham; and to give Mr Pond an opportunity of measuring a baseline of such length as would be sufficient to determine the distance from the Royal Observatory to Lavendroog Castle.
>
> The first three points have been accomplished the last did not succeed, but it was attempted by myself and the Assistants of the Royal Observatory and only given up because the idle people at Greenwich Fair pulled up the Astronomer Royal's marking stones and so showed the inutility of further proceeding without the usual formality of a guard of soldiers. [90 fol. 325–8] [183]

Figure 6 *Compensating bars.*

In the month of March last (1830) ... the necessity of subjecting the apparatus lately brought out by the *Roxburgh Castle* to a trial previous to its leaving England...

The chairman desired me to wait on Mr Secretary Peel [Sir R. Peel] and obtain from him a party of Metropolitan Police to keep off the curious, to hire a spot called Lord's Cricket Ground in St John's Wood Road, to select from Col Astell's Corps of Royal East India Volunteers a party of 20 men to get a set of tents constructed of the kind I required, and as cheap as I could, to employ Messrs Barrow and Taylor in assisting me and generally to incur such expense as I might find necessary preserving always a strict regard to economy.

I got a set of tents constructed by one Merriman an upholsterer in Leadenhall St for the use of which he charged £30 to having permission to sell them when the work was done. Lord's Cricket Ground was hired at the rate of £8 per week for three weeks. The coach hire, lodging hire, and other expenses of Messrs Barrow and Taylor cost about £17 the extra hire for each of the volunteers was 3d per hour and the contingent charges for wood, stones etc came to about £20. So that this experiment with every attention to economy on my part could not have cost less than £130 or 1389 Sicca Rupees.

The tents were constructed in the simplest possible plan and I must observe that the Gentlemen of the Irish Survey who came to visit and assist me, spoke highly in their praise.

They consisted of six pairs of frames each pair forming an isosceles triangle of which the angles at the base were 51° 31' 42" their slant height was 12 feet 5.8 inches and when they rested on the ground the distance between them at bottom was 15 feet 6.2 inches.

As there were 6 bars of 10 feet 3 inches long I constructed 6 pairs of frames of 12 feet whereby there was a projection of 18 inches at each end; the sides were covered with canvas and were made to fold up so as to admit the light on the shady side ... [263 (265) 92–5]

In April 1830 Everest arranged the set of bars at Lord's Cricket Ground with the assistance of Mr Barrow, Mr Western and the Astronomer at Madras, Mr Taylor, together with Lieutenants Drummond and Murphy who had taken a prominent part in the measurement of the Lough Foyle baseline, and a party of military volunteers. He measured nine sets of bars both ways with a difference of 3/40th inch. In a distance of 567 feet that was equivalent to 0.7 inches per mile or 1 in 90 720. (When set out together, six bars measured 63 feet.) He felt that this difference was most likely due to the thick clay underlying the ground, which retained moisture at the surface, and because for several days of the operation there was considerable rain, there was an alternation of dryness and humidity that was unfavourable. In addition there was the trampling of numerous soldiers' feet which caused small swamps and ankle deep mud and water in the hollows while the higher points were dry and hard. [103 pp. 208–9]

On 8 June 1830 Everest sailed again for India on the *Cornwall*.

Chapter 7

INDIA 1830–1834: A period of intensive field work

Dual role of Superintendent of the GTS and Surveyor General

There are conflicting dates as to when Everest arrived back in India, but the most likely is 6 October 1830 at Calcutta, having left England on 8 June. He resumed his appointment as Superintendent of the GTS on 8 October, a post he then combined with that of Surveyor General of India to which he had been nominated on 12 August the previous year.

Within a matter of days of his return, Everest was setting down criteria for his officers to follow. In particular all officers writing to him must keep letters as short as possible by omitting all frills: an exhortation it would be difficult to say that Everest himself adhered to.

For the maintenance of field books he stipulated that angles 'never are taken by repetition but all the points are observed in each position of the zero in succession beginning from left to right, a method which obviously saves much time when more than two points are in question'. [263 (267) 37–40] The term 'repetition' here was the method where an approximate value was found for an angle. Then, by suitable observing techniques, a multiple of the angle was recorded without noting the intermediate values. The final value was then divided by the number of times the angle was multiplied. Thus the readings would be on station A then station B then A then B repeatedly. For the next angle it would be the same process for B and C and so on. It was elimination of the duplication as at B that saved the time. Thus the method adopted was akin to that used today.

> It is never permitted in the Great Trigonometrical Survey to reject any observation arbitrarily – the observer should at the time of observing fully satisfy himself of the correctness of the intersection, if he cannot do that he should not read off the micrometers but an observation once read off and recorded must not be expunged and must be used unless some very sufficient cause of error should afterwards appear. [263 (267) 59–60]

This admirable policy applies as much today as then and still needs as much ramming home to the operators.

Everest was now well equipped with a wide range of new instruments including a set of compensating bars, standard bars and numerous theodolites. The problem was that he was still without

> ... [an] individual on whom he could rely for rendering him efficient aid in conducting any part of the operations in accordance with the ideas of accuracy and attention to minutiae which for the last twenty years have become the order of the day in Europe.
>
> ... nobody who could aid me at all in getting the most out of them that they were capable of, or even in doing anything with them that could be relied upon. All these he had to form and train, and where to meet with persons suitable to the purpose – that is, persons capable of being taught and willing to learn, who were prepared to devote themselves zealously, to so arduous a task – was a difficulty which seemed at one time appalling.
>
> But it would take a great deal of trouble and verbiage to describe the mental anxiety I had to endure until I had those about me in whose judgement, and ability to take part in the operations I could place implicit reliance...
>
> This difficulty was enhanced by ... my now holding two situations ... That of Surveyor General called for my constant presence in office to conduct all the business and correspondence relating to the Geographical Department ... The other, Superintendent of the GTS, required me to be perpetually in the field, to instruct ... and to take an active part in, observing. [116 p. 10]

As Superintendent Everest had not only to build up and equip new establishments of all grades, but also to execute personally the operations of the Great Arc, devise new methods to meet new and unthought-of problems and take a major part in the reconnaissance and observations. As Surveyor General he was responsible for the field survey and mapping of Madras and Bombay (where he had no deputies) and the now greatly extended presidency of Bengal. His only official deputy was on the full-time task of supervising the revenue surveys of the western provinces. [231 p. 343]

It was in the light of these massive difficulties that Everest spent most of the first two years of his return to India in Calcutta getting to grips with the organisation and paper work of his two posts. Essentially a field man, such restriction did not come naturally to him nor did he welcome it. In particular he must have been very impatient to try out his new equipment.

Back in the field

By November 1832 Everest was able briefly to be back in the field on base measurement, but in the interim he had not been wholly away from the practical aspects of the GTS. We read that at the end of 1830 (i.e. soon after arrival back in India) he was worried about the instability of observation towers. Then in 1831 he introduced a new form of tape. The new tape was called Chesterman's self-acting Measuring Tape which had a spring roller to wind itself up. Eight such tapes of four different lengths were sent out in 1831 for trials. He also overhauled the new instruments and devised others from old stock. In March 1831 he read a paper to the Asiatic Society on the compensating bars.

In November that year he set up a separate computing office in Calcutta under de Penning. [231 pp. 136–47] The need for a separate computing office arose from the backlog of work that had accumulated whilst Everest was on sick leave combined with the increasing number of field observations that required processing. After making the necessary detailed case for such an establishment a computing office was finally sanctioned and established at 35 Park Street, Calcutta. (It was moved to No. 37 a year later when the Surveyor General transferred his office to the foothills of the Himalayas.) [231, pp. 332, 338] De Penning took up the post of chief computer on 20 November 1831 on the understanding that it would mean that he no longer had to go on field work unless he requested it for himself. The office was soon well organised and had eight trainee computing officers. Everest however, ever the stickler for protocol and decorum, was unhappy at the mode of dress worn by these officers. Whilst he appreciated there were religious beliefs to be accepted he felt that a person's nakedness should be sufficiently covered and that shoes should be well blacked. [263 (266) 192]. Those unable to comply should perhaps be kept in a separate work room.

Reconnaissance for the Calcutta base

Everest's reintroduction to field work came with the baseline measurement at Calcutta.

This was to be the first use of the compensating bars in India, and after this no further bases there were measured by chain (for details of the bars see Chapter 6 and Appendix 1). The measurement and associated work took from 5 November 1831 until 28 January 1832 to complete. The longitude series of which this formed part was completed on 2 July 1832.

Why was a baseline measured at Calcutta, which was nowhere near the Great Arc? While Everest was in England on sick leave, no one was considered suitable to step into his shoes on the Great Arc, but instead a longitudinal arc was observed in latitude 24° N. By the time Everest returned, this arc, which had been observed by Joseph Olliver, had been completed from the vicinity of Sironj on the Great Arc to within 100 miles of Calcutta. To close it out and have a satisfactory check on the observations, a baseline was required. Since at that period the headquarters of the Surveyor General was in Calcutta, it was very convenient to measure a baseline nearby and in so doing test out the new equipment.

So enthusiastic was Everest that, within a few weeks of returning from England, he was seeking permission to find a suitable baseline site, but it was over a year before he was in a position to make the actual measurements. [263 (266) 73–5] Initially he had his eye on a site between the towers of Paintal and Dilakas, some eight miles apart and 20 miles north west of the city. His young assistant, James Western was given the task of surveying the possible route. [263 (264)]

In a letter to Lieutenant Western of the Bengal Engineers in January 1831, Everest explained how to carry out the reconnaissance, to take details of all boundaries, of fields, the nature of cultivation, names of owners, value, etc. and 'if the depth of any hollow be such as to occasion a greater difference of level than 30 inches in one set of bars or 63 feet it ought to be filled up, but not otherwise'. Lieutenant Western was reminded to consider paying a fair price for trees that would need to be cut down, and to note all houses, rivers and the like that would be on the line and when the fields were likely to cease being swampy. [263 (267) 49–51]

When the possible cost of this route was analysed, it was felt to be too expensive, and an alternative site along the Barrackpore road seemed to be preferable although it would need two towers to be constructed. In late January 1831 Western produced a map of this second route. [231 p 48n5] The base was to be such that it could be connected to the observatory that terminated this series.

Although Colebrooke had had an instrument for astronomical observations in Calcutta since 1795 [228 p. 191-3], it was not until 1822 that Hodgson agitated for more equipment to allow the establishment of a small observatory there. [263 (40)] The observatory was still in Calcutta when Everest took over. The required towers were to be of sufficient height to allow observation over all obstructions that were permanent features. He was sensitive to the problems caused by the local transport in the vicinity:

…the dust renders it impossible to see, …but I shall not object to the road being kept open provided the following points can be rigidly enforced.

That 150 feet to the north and south … of the part of the line actually occupied by the apparatus, carriages and passengers of all kinds must go at a slow walk… The dust will incommode me extremely, and loose cattle, sheep or pigs, or dogs, or led horses, must on no account be permitted to travel at all by the road between the 5th and 11th milestones. Further, as it is indispensably necessary that the line of vision should be quite uninterrupted from one end of the baseline to the other, carriages, and passengers of all kinds whether mounted or on foot should be strictly restrained beyond the middle of the road. [263 (266) 73–5]

To accommodate the small but growing number of instruments available, alterations were suggested for No. 37 Park Street, part of which both Blacker and Hodgson used as their personal residence. Alterations were agreed and in addition Blacker was able to take on to the staff Saiyid Mir Mohsin Hussain to maintain the instruments. [229 p. 188]

Blacker originally found Mohsin working with an eminent jeweller in Madras and was particularly impressed by his apparent intelligence and acuteness [263 (402) 82–95]. In addition to repairing instruments Mohsin was taught how to take astronomical observations.

Figure 7   *Calcutta HQ of the Survey of India*

Later he was to be employed by Everest who found him an excellent workman and in 1842 managed to get him appointed Mathematical Instrument Maker. [229 p. 485] When Everest moved office from Calcutta to Dehra Dun and the Park Estate, Mohsin went with him and a special workshop was set up at the Park. [231 p. 458] (*see* Appendix 2).

Measurement of the Calcutta base

It was not until August 1831 that Everest was able to set operations in motion and organise the manufacture of tents and towers. In a letter to Casement of 12 November 1830 Everest said:

> Two lofty towers are indispensably required … such that the instrument when standing on the summit, shall be placed mathematically perpendicular along it (middle of base of tower) …said towers must have openings in the direction of the baseline, large enough for measuring apparatus and its attendants to pass freely through, that there must be light enough to enable the observers to bisect with extreme accuracy the dots… [263 (265) 85]

By 18 March 1831 Everest was detailing to Casement the number of staff he felt necessary to measure the baseline.

6 bars at 2 men each	12
6 camels* at 1 man each	6
7 microscopes at 1 man each	7
To trace line of pickets	3
Natives to carry 15 frames of tents	54
To drive pickets	6

(* the name given to the brass trestle upon which the bars were supported. [103 p. 206])

In addition, he requested two sergeants, two corporals, four bombadiers, twenty-two gunners and a company of gun lascars under charge of an officer of the Regiment of Artillery to be placed under his orders 'on account of the complicated nature of the apparatus with which the base of verification will have to be measured in October next, it is my humble opinion advisable that a party of European Artillery men should be appointed to aid me in using it'. [100]

On 1 July 1831 Everest wrote to Rt Hon. Lord Bentinck (Governor General 1827–35) that this

> ... is a base of verification, for the longitudinal series of triangles carried out by Mr Olliver during my absence in England and to shew the necessity of such a base, it will perhaps be sufficient to mention that the series in question, is nearly three times as long as the longest distance between any two bases measured by the late Lt Col

Figure 8 The Calcutta baseline measured in 1831–2.

Lambton or myself consequently so long a series will have no claim to confidence unless it be verified.

After discussing the search for a site he continued

...a straight part of the Barrackpore road might be given up for my operations, the old Chitpore road temporarily repaired, and two towers built of masonry for connecting the line with the series of triangles – I accordingly examined the road and had a survey made of it, the Result of which was that the part between the 5th and 11th milestones could be used with the addition of two towers, although there would even then be an obtuse angle in making the connection, which as it violates symmetry is to a certain extent objectionable. [100/3,4]

When writing to Lieutenant Western on 19 October 1831 Everest detailed his requirements thus:

Your orders will emanate from me and I wish you to report to me daily either in writing or verbally making your reports in as few words as possible. In cases of quarrels or other disorders I do not wish you to inflict regular punishment but to confine until you report to me, except in small matters wherein without such reference you can sentence to half an hour's walk under the inspection of the sentry or half a dozen rattans, bearing in mind that in general much more is to be done by kind treatment than harshness, and that the less frequent the chastisement the more effectual it is likely to be. [263 (267) 78]

Measurement of the line did not commence until 23 November, but before that an intensive comparison was made between the compensating bars and the standard bar A. These comparisons, so essential to sound results, were carried out over a period of 10 hours on each of several nights.

In a letter to Lieutenant Western of 7 November 1831, Everest stipulated that:

It is my intention that during the ensuing week a series of experiments shall be made to determine the relative values of the measuring and standard bars.

They will be commenced at 7 o'clock in the afternoon of Monday the 8th and be continued until the morning at 5 o'clock.

The standard bar is first placed under the two fixed microscopes and its length noted as well as the states of its two thermometers, and the time. It is then moved out of the shed and the bar A [not to be confused with standard bar A; this refers to one of the compensation bars] is brought forward in like manner in succession terminating with the bar designated H which completes one set of comparisons.

The first set being completed, a second, third etc. are gone through consecutively, the state of thermometers belonging to the standard being only noted when that bar is under the microscopes.

The light is thrown by means of two reverberatory lamps placed so far off as not to communicate a sensible degree of heat.

Great care must be taken to prevent anything striking or touching the fixed microscopes except at the graduated heads, which must be moved round most gently. The standard bar must never be touched by any person's hand. [263 (267) 82]

The conclusion celebrated

The measurement took 45 days at an average of 750 feet per day or 12 sets of bars but, towards the end of the measuring speeds of up to 24 sets per day were recorded. The base length reduced to sea level was 33 959.9174 feet of the standard bar A and was at 22° 40' N, 88° 25' E. It had required 539 sets of bars. The discrepancy between the measured value for the base and the value calculated through the 671 miles of the longitudinal chain from Sironj was 7 ft 11 inches. There was a vertical discrepancy of 200 feet compared with a connection to the Hoogly river. [231 pp. 50, 58, 94]

Once all was satisfactorily completed an appropriate ceremony took place. James Prinsep, the Assay Master in Calcutta and Editor of the Journal of the Asiatic Society of Bengal, writing in 1832 described it thus:

> Through the politeness of Captain Everest, the Surveyor General and Superintendent of the Trigonometrical Survey, we enjoyed the advantage of an invitation to witness the remeasurement of the first day's work, with the view of ascertaining what might be the probable amount of error; on which occasion the President of the Physical Class of the Asiatic Society and many distinguished officers of the Engineering department were present. An elegant breakfast was laid out in tents after the ceremonies of the morning were concluded. While contemplating with admiration the order and precision with which the whole process was conducted, we took an opportunity of sketching the apparatus as it stood, that the readers of the Journal might be better able to comprehend the nature of the operation of measurement… [243]

When writing to Casement on 23 January 1832 Everest referred to the completion of the baseline.

> the base of verification was brought up to the North Tower on the night of Wednesday the 18th Instant – It then remained to remeasure the first 11 sets of bars with which the operation had commenced in November last in order that by a comparison of two results made under totally different circumstances, it might be judged whether the bars were truly compensated for temperature. Each set of bars being 63 feet long, the length of 11 sets is 693 feet which was the direct measure of November – by the reverse measure the distance is 1/40 inch more than this, and as the case is an extreme one, it may be concluded that the error in the whole base of 6½ miles cannot exceed 1¼ inch and is in all probability quite a microscopic quantity. [263 (283) 14]

He commented that 1/40 inch was one third of the discrepancy at Lords Cricket Ground and 'a surprisingly small quantity'. [231 p. 50] Prinsep elaborated on this by saying that the error equated to 12 feet over the distance between Calcutta and Delhi or 125 feet in the diameter of the great globe itself. [243 p. 72]

On 5 January 1832 Everest wrote to Robert Shortrede, of the Bengal Infantry, about baselines.

> that its length is computed to a greater degree of accuracy than any eye can detect, not because the small quantities so found are tangible individually but because the computation of every minute value is much more likely to obtain the true value of the whole for the required degree of exactness etc. than where quantities of an inferior order are rejected ... An imperfect base line then, merely tends to create confusion or doubt or had better not have been measured, or it is a truth which cannot be too often repeated or too much kept in mind, that in an extended survey the system pursued should be uniform or invariable. [263 (267) 93–5]

At that time Shortrede was in charge of trigonometrical surveys in Bombay Presidency.

Field operations

In February 1832 Lieutenant Western was carrying a series of principal and secondary triangles up the meridian of Parasnath. In detailing how some of the observations should be taken Everest wrote:

> It not infrequently happens that people who go to choose a station, select a wrong place – in that case, if there is obvious neglect, it is usual to make the defaulter pay a fine, if on the contrary it should seem that the man has done his utmost, there is no remedy save patience, and sending another person; but where a station of difficulty is selected with great cleverness it is always usual to present the party with a rupee and charge that in the contingent account. [263 (267) 105-8]

With regard to theodolite observations he detailed the operation:

> For this purpose you will fix the instrument so that zero shall be under micrometer A and in that position observe each station in succession noting down the readings of each microscope – you will then turn the telescope over in altitude, and the horizontal circle half round in azimuth and observe each point in succession again. Repeat this operation so that by having two readings at each position of A (viz. 0° and 180°) you will see whether any obvious mistake has been committed. ... bring 10 under the micrometer ... and so on for 20–30 up to 50 whereby you will bring every ten degrees successively under one or other of the microscopes... [263 (267) 105-8]

This Everest then expands in some depth and indicates how the mean should be extracted and the distribution within a triangle of any discrepancy after spherical excess has been allowed for. This he did in proportion to the probable error in each angle unless there were overlapping triangles, in which case the procedure had to be modified.

He discussed the observation of vertical angles and then said:

> About 3 or 4 in the afternoon is generally found to be the instant of minimum (refraction) and the same hour after midnight to be that of maximum – in certain conditions the difference is immense.

> It is on no account admissible to leave any angle of a principal triangle unmeasured, so long as there is not an absolute impracticability of access to it. Whenever that is the case, it is better to reject the station and choose another.
>
> The angle book must on no account ever be suffered to fall into arrears – it must be compared with the rough record, without loss of time, and must be attested by two persons as a correct copy.
>
> Of all things be careful of the health of your followers and yourself - it is quite a mistaken notion to run a race against time, for the greatest quantity of work is sure to be accomplished in a given time by those who follow a steady, undeviating system of order and method. [263 (267) 110–13]

Comparison of measuring chains

Soon after the completion of the base, Everest launched into the effort of comparing all the chains of his Department with 10 lengths of standard bar A and of experimenting with the coefficient of expansion of that bar. Before this the standard values used were somewhat uncertain. The chain used by Lambton for his baselines was 'set off from Ramsden's bar at 62°F'. The chain by Worthington and Allan which had come out in 1802 and was used for comparison only, had been set off at 50° from Ramsden's bar [228 pp. 253–4]. Because he found variations in his chains, Lambton twice compared the second one against his 3 foot brass scale that had been laid off by Cary from the scale of Alexander Aubert. [231 p. 46]

When corresponding with Lieutenant Shortrede in Bombay, 15 September 1832, Everest commented at length on the comparison of the chains.

> The bar is furnished with two thermometers whose bulbs are let into the body of the metal, and these are noted at the time of deciding on the position of each microscope, so that the distance between the first and eleventh microscope, is the length of ten bars at the mean of all these temperatures ... You did not mention to me ... what weight you used to give tension to your chain, but it is the longest chain I have ever seen in this country or in England. Of the chains which I have in this office there are six by Gilbert, one by Worthington & Allan, one by Birge and one by Ramsden...

	Width of chain (inches)	Weight (lbs, oz)	Tension used (lbs)
Ramsden	0.30	32	68
Old standard by Berge	0.27	23, 12	68
New standard by Worthington & Allan	0.30	31, 2	68
Gilbert	0.36	46, 4	98
Bombay by Cary	0.57	46, 1½	196

As to the chain used by Mudge, it appears to have weighed 18 lbs and to have been stretched with a weight of 56 lbs. [263 (267) 147–8]

One of these comparisons had been made at Bellary in 1813. This chain was of blistered steel, with 40 links each of 2½ feet length. Each of two brass register heads had 6 inch scales for verifying short measurements made by microscopes. The chain was said to be identical to that used in 1787 on the Romney Marsh baseline. Lambton laid out a distance of 100 feet by means of the 3 ft brass scale and, when the chain was compared with this, there was seen to be a surplus of 0.034 inches or an extension of approximately 1 in 35 300. Some of the cause of this discrepancy was attributed by Everest to accumulated rust and the subsequent cleaning that altered its length. [96 p. 132]

It does not appear that the standard brass scale was ever compared directly with any subsequent standards for the Indian Survey although an indirect comparison might have been possible. [281 p. 1]

Thus by the time Everest was returning from leave in England there was considerable need for a reliable standard, and he had two 10 feet bars and two 6 inch brass scales constructed in England by Troughton and Simms to his own specifications. They were officially compared and certified before their departure from England.

On 15 August 1832 Everest wrote to Casement:

When I left England, one of the Iron Standard bars (of which there were two) was left with Messrs Troughton and Simms in order that it might undergo a series of comparisons with the Tower Standard. Lt Drummond RE of the Irish Survey promised with Mr Simms to undertake these comparisons and to send the bar out immediately.

It is of great importance that the bar with the relative length determined should be in my possession. Want of time prevented me going through the comparison myself before I left England, and if it be not forwarded, it will be necessary for me to take the Standard bar now here to England when I return, until which time there will be an uncertainty hanging over the relative measurements of the Indian and foreign Surveys. [263 (283) 149]

Standard bars

The bar and scale A reached India in 1832 while those labelled B arrived in February 1833. Before their shipment, bar B was compared with the Ordnance Survey 10 ft standard bar designated O_2[*] by Lt Murphy at the Tower in London in 1831. [281 p. 2] Details of this are given in the *Account of the Lough Foyle Base*. [293]

Writing to Lieutenant Alexander Boileau 14 December 1830 Everest said:

I have brought with me from England the means of doing this [comparing the chains against a standard] and without it the operations of the Great Arc both by Col Lambton and myself cannot admit of that rigorous comparison with other similar arcs. [263 (267) 48]

---

[*] For consistency of results all linear measuring equipment has to be standardised against a bar of known length otherwise unquantified errors will be introduced into the measures.

Bar B was twice compared by Everest with iron standard A in the office at Dehra Dun during November 1834 and then in February 1835 under tents at the baseline site. [281 App 2 pp. 4–5] The 6 inch brass scales were compared by Everest in India in June 1835. Both the A bar and scale remained in India and were used in conjunction with the 10 baselines measured by Colby bars between 1832 and 1869.

The B bar and scale were taken back to England by Everest when he retired in 1843 and then passed to the Ordnance Survey. In 1846 the 10 foot bar B was compared with the Ordnance Survey 10 foot bar O_1 and scale B with the Ordnance Survey 6 inch scale.

Everest quoted the results of his comparisons in [116 pp. 433–9]. From them it can be seen how very tiny changes were of importance if the best results were to be obtained. In summary they were:

13–15 November 1834	10 foot iron standards A and B at Dehra Dun
	B – A = 0.000 075 02 inch
11–16 February 1835	10 foot iron standards A and B at Dehra Dun
	B – A = 0.000 022 50 inch
Weighted mean value	first series: 45 observations
	second: 56 observations
	B – A = 0.000 045 90 inch
11–13 June 1835	6 inch brass standard scales A and B at Hathipaon
	B – A = –0.000 100 8 inch
29 September – 15 October 1831	10 foot standard bar B with Ordnance Survey O_2 at the Tower
	B – O_2 = 0.000 011 95 inch
13–14 July 1846	10 foot standard B with Ordnance Survey O_1 at Southampton
	B – O_1 = –0.000 966 44 inch
20 February – 22 August 1846	6 inch standard scale B with Ordnance Survey 6 inch scale at Southampton
	B – O_1 = –0.000 154 16 inch

The 10 foot bar B was later taken to Russia. where Friedrich Struve compared it with several continental standards. Then in 1865 it was back at the Ordnance Survey where Clarke compared it with new 10 foot standards destined for India. [281 p. 2]

Everest demonstrated his meticulous nature when the triangulation in the Bengal Presidency came under his wing in 1831. The work had started in 1828 under Shortrede. Everest questioned him on the corrections applied in the measurement of the base in the Karleh (Karli) plain, about 40 miles east of Bombay. [229 p. 131] The observations for this were taken between 12 December 1828 and 16 January 1829 and gave a length of 4.065 miles. One section of about 1000 feet of this was across the steeply banked river Indrawni (Indrayani) and was not measured directly but by triangulation. [263 (518) 202] Shortrede replied that he had used the chain that Cary had adjusted to be exactly 100 ft at 62°F 'I have therefore assumed it to be correctly of the Parliamentary Standard length.' [263 (323) 48–53]

This did not satisfy Everest who had it compared with the new standard bar A and concluded that:

You have made no allowance for friction and … have taken the length of the chain as being exactly 100 feet. Does that length mean the Parliamentary Standard or the scale of Alexander Auber? There is some difference…yours (error in base) therefore may have amounted to about 1/10th of a foot, which term is omitted entirely. [263 (323) 25–9]

After various other critical comments he said 'at the ends of the 5th chain, you observed a ramrod placed in line, but you do not say what means you took to ensure its perpendicularity, neither … what precaution you took to avoid the effect of light on one side, and shadow on the other…' [263 (323) 25–9] In this particular instance Everest seems to have accepted the subsequent explanations although in general he found Shortrede's work to be 'unworthy of confidence'.

Coefficient of expansion

After the measurement of the Calcutta baseline and intercomparison of the standard and compensating bars, the point to come under close scrutiny was the expansion of the materials. The comparisons had taken place at temperatures considerably in excess of those used in England for comparisons, so a correction was necessary. But what value should be used? The values given by different experimenters at that time varied from 1.001 44 by Troughton to 1.001 18 by Dulong and Petit: an unacceptable difference in this situation.

Thus James Prinsep, under Everest's guidance, used a steam pipe to heat the metal uniformly to boiling point. Henry Barrow constructed the apparatus such that the micrometer was sensible to 1/20 000 inch. It was considered that the apparatus would be good to about the 6th decimal of the coefficient.

Seven sets of readings were taken on the 10 foot standard bar of iron over the period 20–27 November 1832. The factor chosen as the most representative was 1.001 215. This is in the sense of 'Dimension of a bar at 212° whose length at 32° is 1.000 000' which is equivalent to 0.000 006 75 per degree Fahrenheit. [244] See also Appendix 1.

Later the same year we see the first indications that Everest would like to get away from Calcutta and have offices nearer to his first love – the Great Trigonometrical Survey. On 15 September 1832 he wrote to Casement:

> no measure is so likely to forward the operations of the Great Trigonometrical Survey, as my proceeding in person, in the ensuing cold season, to some place on the Meridian of Dodagoontah (either Agra, Meerut or Suharanpore, according to circumstance) and there establishing the Head Quarters, until the operations of the Great Arc are brought to a termination. [263 (283) 153–4]

Movement of headquarters

By 15 November 1832 he is again writing to Casement:

> it is my intention to despatch such part of the establishment of the Surveyor General

and the public instrument as I propose transferring, on the 15 December 1832 – they will go by water to Ghur Mookbesur and there disembarking establish themselves at Masoorie where I shall hire an Office ... until such time as the two Northern Sections of the Great Arc are brought to a satisfactory termination.

I propose separating from the Establishment of the Surveyor General's Office, in such time as to enable me to leave Mirzapore on the 1st of February in order to proceed across the country to examine my various field parties which will then be engaged in their operations.

I will first go to Sugur,... Lt Macdonald's party, thence (examining the operations of Mr Rossenrode's party in my way) shall proceed along the line of the Great Arc to Agra, Lechli and Saharanpoor by which means I shall have an opportunity of seeing how Lt Boileau proceeds...

Having performed this tour of inspection, I shall depute Capt Wilcox to remain at Agra in order to afford such aid in the furtherance of my arrangements as will be required, and I shall proceed myself to rejoin the office, which by that time will I hope be at Musoorie... [263 (283) 201]

By 8 December 1832 Everest was able to report to Casement:

... the subject of the Field Office, I find that I can procure one at Musoorie without putting Government to any expense further than that now incurred for the house at No 35 Park St (Calcutta) which I propose giving up... [263 (283) 227]

The remainder of the story of his office at the Park near Mussoorie is recorded in some detail in Chapter 12.

The arc across the Doab

One of the first tasks that Everest put his mind to after settling in at the Park was to seek a site for a baseline in the vicinity of Dehra Dun. From early November 1833 Everest inspected several possible sites to the north of the Siwalik range and along the south side of the Asan river to the west of Dehra. This area had the advantages of being free from jungle and yet with a plentiful supply of timber and there would be little interference of the local population.

The baseline was not measured until a year after its selection, and in this time Everest embarked on the task of filling in the triangulation between Agra and Dehra. As he left the baseline site in November 1833 to travel south so it is recorded that he had with him 'Boileau, Olliver, and three sub-assistants – 4 elephants – 42 camels – 30 horses – and about 700 natives'.

On 13 December 1833 Everest began working his way northward from Agra towards Delhi. Once he reached Delhi all vestige of elevated positions disappeared, and points in the flat plain of the Doab, between the rivers Ganges and Jumna, would have to be utilised. By 7 January 1834 Everest was marching towards Bahin and the beginning of his problem area. Not only was the area dead flat for 200 miles, but

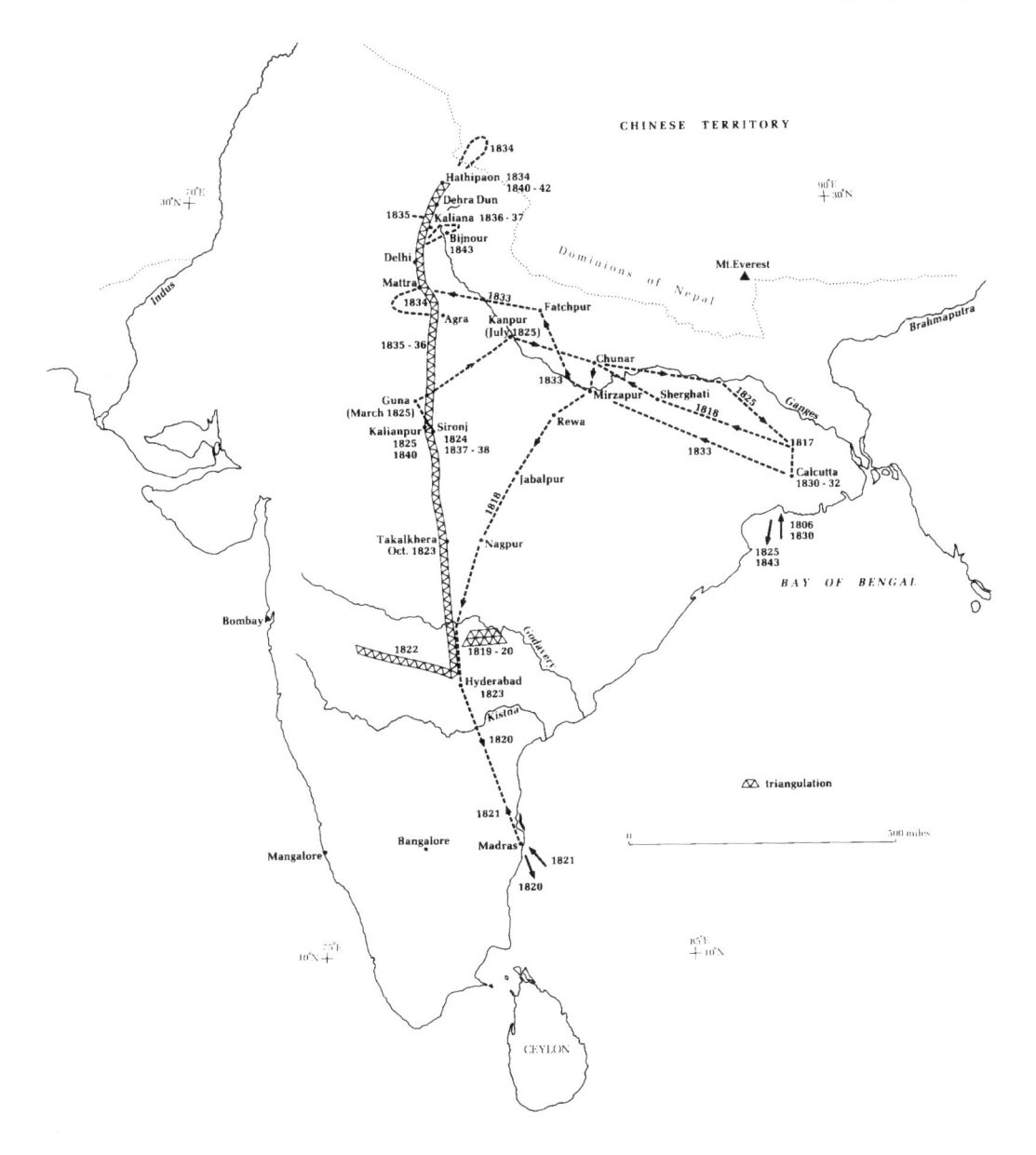

Figure 9 *The main areas of Everest's activities from 1806 to 1843.*

there was considerable smoke and an extensive population. Between the stations at Aring, latitude 27° 30' and Nojhli at almost latitude 30° there is a rise of only 268 feet in some 160 miles. This presented such considerable difficulties in station selection that Everest felt that he had to do the reconnaissance himself. [231 pp. 26–8]

The Doab was described as a flat alluvial formation, totally devoid of all natural elevations, but it had many mounds which appeared to have been raised by the inhabitants as a protection against inundations and attack. This area was very highly populated and cultivated and the villages so thickly scattered that it was difficult to trace a line in any direction. In addition it was quite impossible to expect to see through intervals in the trees which, although few in number, would probably interrupt lines of sight. [108 p. 206]

The slight mounds could possibly be utilised, but Everest had to think more in terms of masonry towers and thus had to be certain of the station selection before starting construction. For this confirmation he made sure that at least two angles in each possible triangle were observed to a minute or so of arc to ensure line of sight. To accomplish this he designed a portable mast capable of supporting the instrument that would stand 30 feet above the ground. Although he wanted a completely isolated observer's platform, he found some cross strutting was necessary, but this was no hindrance so long as observations were limited to calm periods.

Everest found it necessary to supervise every aspect of the construction and dismantling of the mast until much of the chain had been reconnoitred, by which time some of the local staff were becoming proficient. For the target end of each of the test lines, Everest erected bamboo masts some 70 feet high, on top of which he arranged to display a blue light elevated a further 20 feet. For these high masts he was obliged to have special stout bamboo sent up from Calcutta. [116], [265 (II)]

Problems in the field

Not everybody was as co-operative as they might have been: for we find Everest wrote to Nicholas Kallonas on 19 December 1833

> You have detained my party for three days here looking in vain for your heliotrope, and I intend to hand you up to Government ... as your shameful negligence and misconduct deserves.

He was not the only one to receive a stern rebuke. On 14 January 1834 Everest wrote to Henry Keelan:

> You are mismanaging sadly; when instructed to turn your heliotrope to Bahin, you turned it to Pahera, and kept it there ... and I have been straining my eyes to pieces yesterday and today, and all my people have been worried to death in trying to catch the rays from Deri ... Now mark what I tell you. The heliotrope is always to be directed to the station where the principal instrument is ... but you must mend these defects, for I cannot pardon neglect or inattention. [263 (321) 96–7]

Others also had troubles with their lights in these difficult circumstances, although Charles Murphy seems to have had what luck that was going and received congratulations. Besides being readily seen from the mast, his lights were also observed to gradually get brighter and higher in the sky as the night wore on. Just one of the many instances of the assistance given by refraction in the Survey of India.

The problems of having the lights burning at the appropriate times and for long enough periods were highlighted when Everest wrote to Keelan on 30 January 1834.

> I can see your heliotrope well enough when you choose to turn it this way... Then you took it down for more than an hour, and then turned it to me again just for about three minutes of time, as if you were trying how much you could annoy me. I do not know what you mean by all this folly... What object do you propose to yourself by playing these silly puerile tricks? [263 (321) 112]

By early February Everest had reached Delhi from where he commented on the problems of ray clearing. Once the correct direction had been calculated, all intervening obstacles had to be cut down and the blue lights mounted on a high pole for burning. He said 'The labour of clearing these village stations is quite terrific, but it is to be accomplished by perseverance.' [263 (321) 133–5]

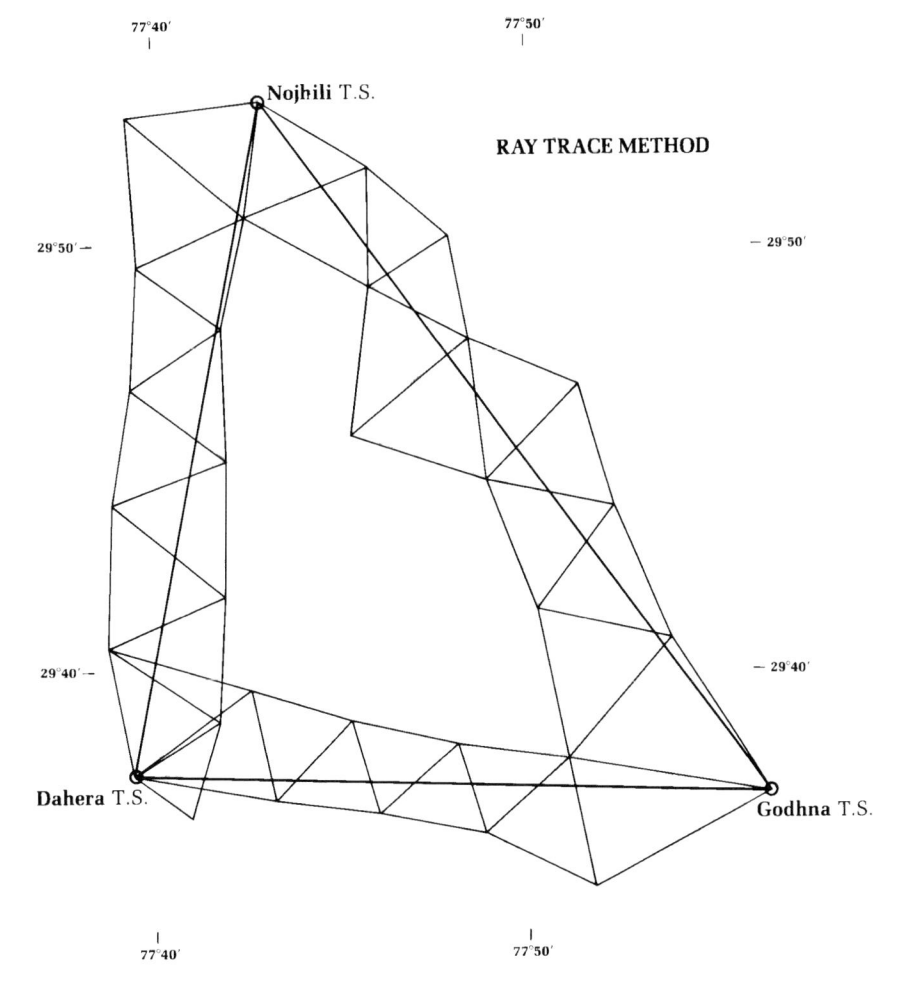

Figure 10 Example of a minor triangulation executed on the Ray Trace method by Babu Radhanath Sickdhar in 1840.

Such were the problems with trying to get a minor series of triangles through the low lying country around Delhi that Everest devised a new system for locating the forward positions. Termed ray tracing, it amounted to a rapid perambulator traverse between the stations that were not immediately intervisible. (See Figure 10 and Chapter 8 for more detail.) This method soon proved its worth but Everest was at pains to do the computations himself. 'Draw up the observations exactly as they are registered in the angle book, and be careful not to substitute A where B should be … I can take out the differences myself, and like to trust my own computations…' [263 (321) 183–5]

Once the traverse between them was complete, plotted and calculated it was readily possible to determine the angle to be set off at one to point towards the unseen one. Then it was a case of laborious cutting of the line until the blue light or heliotrope on the far point could be seen. This signal could, of course, be similarly pointed in the correct direction after the traverse computation.

Delhi

Delhi proved a real headache. Not only was the smoke a problem, but at that time, February, fog also delayed observation.

The proposed use of the Pathan Mosque did not meet with Everest's approval. Whilst it seemed an appropriate point to use he considered that it was such a poor structure that it would not support the weight of a 4 foot high pillar with 4 foot diameter stone on top for the instrument.

The situation made him very irritable again and on 2 February 1834 Everest wrote to one of his staff.

> You are wearing me to fiddlestrings about this Delhi ray. My heliotrope was directed the whole of yesterday at the proper elevation,… Are you not aware that if a heliotrope sends its rays at a particular elevation in any direction whatever, if it is not seen at the precise spot, it must be going either to the right or left, unless there is some high object in front which obstructs it?
>
> Go to the right and the left, and to the front, until you do see my Gulistan heliotrope, and note how far it is out … If you do not take some pains … you will never succeed, and I may be detained here for the next six years… [263 (321) 121–2]

By 6 February he had succeeded in getting the Delhi ray with the assistance of his brother who was at that time practising his ministry in Delhi.

From time to time his remarks were very scathing. Poor Mr Rossenrode was in the firing line next. For a while it appears that Everest's lines went very well when Rossenrode was not around. 'This morning also I arrived at Talkatala, and still thanks to the absence of my Principal Sub Assistant, found all my heliotropes duly directed…' [263 (321) 121–3]

By 9 February he was back in Delhi and suggesting to Olliver that he take on some of the computations as he had others to help him. Everest said 'You must look on me as a worker of miracles'. Then he had another go at the Delhi station where he considered that after all the trials, the gateway to the Ram Dhoara (which they had called the Pathan Mosque) was by far the most obvious and sound point to choose and Rossenrode should have recognised that. Subsequently, even that point was rejected and one was selected at the Tower of Pirghyb (Pir Ghaib), the survey marks for which were described by a visitor in 1951. [231 p. 34n5]

In addition to all the problems with the reconnaissance of such an area Everest had his correspondence to keep up to date as well. In early February no less than 76 letters reached him at one go. This load, together with the constant interruptions he received, affected the reliability of his computations. With such a task it is essential to be able to concentrate for long periods without distractions and this was denied him. Blunders began to occur in his work and this did not improve his short temper 'it is enough to drive one mad'.

Everest kept his staff at full stretch in the pursuit of the reconnaissance up to the foothills. He was insisting on them being 'untiring and alert', and he could not brook the delay of easy going gentlemen. A good example had to be set to his heliotropers who were by now becoming adept at their task. [263 (333) 28–9]

Difficulties continued in March 1834 as the work progressed towards Dehra Dun. Even stores caused a problem. Supplies of digging implements and rope seemed to disappear at an alarming rate.

> I do not know what has become of the digging implements... I fear some has been made away with. Likewise the 1200 fathoms of cotton rope which was supplied about six weeks ago. I cannot imagine where it has all vanished, and everybody is calling out – No Rope! No Rope! Give me Rope.

The dust and rain storms arrived by March, and those at each end of a possible line had to be alert to succeeding periods of calm atmosphere when good sighting might be possible. One difficulty with this was that the conditions might not be the same at each end of the line so the personnel would be in different states of unpreparedness.

This situation was highlighted in a letter to Keelan on 17 March 1843. Everest berated him thus:

> You are certainly most irregular. Who but a half-crazy person would have chosen a time when it was blowing great guns to burn his blue lights in utter defiance of my orders, and you certainly did this on the morning of the 17th... The khalasie tells me you began at four when I was obliged to hold on with both hands to save myself from being blown off the scaffolding and neither lamp nor taper could show its nose. [263 (321)197]

By early April 1834 the parties were nearing the Siwalik Hills to the south of Dehra Dun. Whilst the advent of hills would have been expected to improve the siting of the stations it was not plain sailing. Here there was the added need for the stations to be suitably arranged to link with the baseline and this was troublesome. In addition, however,

> there are no villages whatever from which supplies can be obtained. No water except after heavy rains – no guides to show the paths if any existed – no names by which the divers peaks are known –... no vestiges whatever of anything in the shape of humanity.

> Wild elephants range... leaving their traces in every direction. Where a vagrant male, or stray female who has lost her young, encounters a passing traveller, he must owe his escape to his dexterity and good fortune, for destroy him they certainly will, if they can ... As to tigers, it is the regular den of those which infest the Dun ... alarm ... spread amongst my followers – the terrific stories... told of hairbreadth escapes – ... rendered it incumbent on me to be nowise chary of personal exposure. [263 (286) 276–374]

Perhaps it was these frustrations that fired him into yet another acrimonious missive, this time to Wilcox, who was struggling on the points just north of Agra and, in Everest's opinion, taking far longer than he would have done himself.

> whilst you have been occupied on these two rays – I have advanced from Delhi to Hardwar, having visited 13 different stations, cleared 8 rays, and taken 22 regular sets of angles, all with blue lights, which you pronounce to be impossible at this season of the year ... not a moment has been thrown away ... Though weary for want of sleep, and worn out with watching, I have never indulged in rest until all that depended on my immediate exertions was fulfilled... [263 (321) 224–31]

Early in May 1834 he wrote to Taylor thus:

> The spoon would certainly stand upright in the atmosphere now, – it puts me in mind of those delightful pea soup days when dankly gleaming through the windows of Fleet Street ... Cockney clerks are seen at their desks supplying the absence of day light by that emanating from muttons. [100 p.8]

Reconnaissance for Dehra Dun baseline

Early in May 1834 Everest was able to write to Lord Bentinck that the stations were all selected to join the Sironj base of 1824 to the proposed Dun base. He had by then sketched out the sort of edifice he required at 13 sites to get the lines of sight through and had also planned the form of crane needed to raise the 3 foot theodolite up the towers. The towers would have to be one of 35 feet, three of 40 feet and nine of 50 feet height.

> The Dhun furnishes the most favourable ground near the Great Arc for the measurement of a baseline, because it can be effected without putting the state to any expense or annoying the inhabitants, the tract selected being a waste situated between high natural elevations. [100 p.9]

Figure 11 The west end terminal of the Dehra Dun baseline occupied by modern equipment.

He finally completed the reconnaissance and was back at Hathipaon at 7089 ft and Banog at over 7400 ft on 12 May 1834. After observations there, he was able to 'relax' at his nearby Park Estate during the rainy season and make preparations for the next field season, bringing all his office work and computations up to date. When one looks at the chain of triangles he produced it is surprising how symmetrical he was able to make them, in fact truly remarkable in the light of all the difficulties.

In August 1834 he wrote to Taylor:

it is a great anxiety to me for there are so many little items to be thought of, the neglect of any one of which would leave a vacancy in my data, that I am constantly on the *sin vive*. That I shall get triumphantly through in the end after a little bother I have no reason to doubt. If not, I shall have much more fortune than has usually attended me. [100 p.10]

The same month he endeavoured to put Lord Bentinck's mind at rest.

I examined the tract minutely before I fixed on the site, and am happy to assure your Lordship that your apprehensions are without foundation. There is little or no jungle in the way excepting grass, which is of no importance whatever, and I shall be able without putting the Government to any extra expense excepting perhaps some 10 or

15 Rupees for sickles and some dozens of hatchets, to make the whole line as smooth as the palm of my hand in the course of a fortnight.

In respect of the approximate series generally … it has been carried through two of the most tremendous seasons that have ever been known. The one as remarkable for extreme drought as the other for the early commencement and extreme severity of the rainy season. [100 p.11]

He then referred to the sufferings of Rossenrode in 1832 when travelling from Calcutta to Chumbal, north of Sironj.

The sufferings of that party were severe, several men were struck dead by heat and thirst.

Lieutenant Boileau's reconnaissance was ably conducted and was of great service, but it was all perambulator and compass work and consequently did not serve to determine what I wanted viz, the position in which the meridian of Kalianpur would fall. To settle this question I found it indispensable to take the field myself – I first explored the Dun in November last and having from the windows of my office, then in the hills, well examined the tract below with a telescope, I was fortunate enough at the first trial to pitch on the best site near Sipsoohara.

The series has been carried above houses and trees by means of an instrument raised in mid air, together with its tent, and it is my persuasion, that for so vast an undertaking, the cost incurred has been inconceivably small.

After a few failures such as I was prepared to expect in an undertaking quite new to me as well as to all about me, the very idea of which many people laughed at as preposterous, I found my plan answered admirably – I found to my inexpressible delight that the mast on which my instrument stood, instead of shaking like a leaf as many persons had predicted, was, if people were made to seat themselves quietly, as steady as could be desired – excepting always in stormy weather…

The instrument concerned here was usually a small theodolite.

I established in my camp a regular manufactory of blue lights which I soon taught my own establishment to make of the very first size and quality for very little more than 8 annas each…

It was hard work, very hard work, and such as I should not be able to stand frequent repetition of – but the result is that the agreement of the several parts of the work, though it does not show an accuracy sufficient for a final series of triangles is yet in my opinion, such as to give geographical data of much greater value than any the Hon Court of Directors are in possession of.

When that time arrives, the work of throwing a set of triangles over the whole of India, though it must always be costly, will proceed with an expedition and accuracy unparalleled. The project will cause to startle by its vastness, its feasibility will no longer hang on the life of any individual and if the system of uniformity which it has been my constant effort to enforce be attended to the operations conducted by any one person will be fully intelligible to every other without a glossary. [100 pp. 8–11]

In late 1834 preparations began for the base measurement on the banks of the Asan river near Dehra Dun and with the attendant preliminary and post measurement comparisons of the bars with the standard occupied from 12 November 1834 to 2 April 1835.

Minor triangulation series

Although not directly involved in the observations, Everest was nevertheless in overall control of other chains of triangulation. During this period there were several such series.

1. South Parasnath meridional series

At about 86° longitude, this arc, if produced southwards would link up with earlier mapping by Buxton near the eastern coast. The series started on 6 February 1832 under the control of Lieutenant Western. Unfortunately, before he had observed at more than one station, disturbances called a halt to operations. To add to the problems, Western fell ill on 10 December 1832. As Everest had found also, there was so little proven experience in the parties that Western was unable to trust the large theodolite to any of his colleagues.

Everest was very unhappy with the results of that season's work and found it to be full of serious errors. Although he decided to give Western another chance, the following season of 1833–4 proved no better.

2. The Budhon series

This was a meridional series running up the 78° 30' longitude line from north of Saugor (latitude 24°) just east of Kalianpur, through Jhansi and Gwalior (26° 20' N) then through the Chumbal and Jumna valley, through Mainpuri (27° N), Etah and Aligarh (latitude 28°) to the Ganges and then onwards to the Himalayas and closing on to the Great Arc at its northern end. Thus it was parallel to the Great Arc and about 1° to the east of it.

Started in 1832, the arc was not completed until 1843. Early in its life, Everest visited Macdonald at the scene of operations in March at 1833 at Saugor after leaving Calcutta on 24 December. From there he went on to visit Rossenrode at Kolarus near Gugbara (25° 10' N, 77° 40' E), on 28 March. From here he travelled up the proposed line of the series to reach Muttra (27° 25' N, 77° 35' E) on 13 April. [263 (283) 199–201]

By 1 June 1834 the arc was in the vicinity of Gwalior where an attempt to connect to the Great Arc was unsuccessful and operations ceased for the season.

3. The Ranghir series

This was another meridian arc, following the 79° 30' E line. Waugh started the series near Ranghir (latitude 23° 30') on 6 January 1834 and by the end of July finished for the season at Phara, from where they continued in October.

4. The Amua series

This series was parallel to the previous two and more or less up the 80° 30' E line. Again starting from the Calcutta longitudinal series near 24° N it went through Fatehpur and Cawnpore, to the west of Lucknow and on north to meet the North East longitudinal series around 29° N.

This party was under Renny and began operations on 13 January 1834 near Jubbulpore. When it was approaching the end of the season he was troubled not only by the rains but by sickness among his camp and so began his recess at Cawnpore on 1 July.

5. The Bombay surveys

It will be recalled that Everest was working on the Bombay longitudinal series when Lambton died in 1823, and work then ceased. During 1828 it was decided to provide survey control for mapping in the Bombay area, and this was given to Shortrede to carry out; however, it was not until March 1831 that the work of Shortrede came directly under the control of the Surveyor General. [231 pp. 64–72]

As soon as he had control, Everest took considerable interest in this work. Although he was reasonably satisfied with the base measurement at Karli, he was scathing of all the other work by Shortrede. In particular the triangular misclosures were far greater than Everest could accept if this work was to be assimilated into the Great Trigonometrical Survey.

In October 1835 Everest wrote about Shortrede:

> [his work] was found to abound with errors of such magnitude as to render it unworthy of confidence. A list of the most glaring of these errors was made out, and forwarded to Lieut S., from whom no adequate explanation… has been received yet, and as he has been employed since September 1834 on a duty quite foreign to that of his own proper avocations … it is not easy to say when he will be able to account satisfactorily for the great discrepancies … [263 (286) 228–35]

After Shortrede had been diverted on to local revenue survey work, the triangulation was continued by Lieutenant Jacob.

Chapter 8

INDIA 1834–1839: Better instruments, and remeasurement

Into the Himalayas

Once Everest had completed the reconnaissance of the northern end of the Great Arc and planned the organisation for the actual definitive measurements, he was for pushing the arc ever further. In August 1834 he set out into the mountains to the north of his Estate. His intention was to carry the chain as far as it would go into the Himalayas. The previous year he had commented to Shortrede 'These splendid snowy peaks rise like an impenetrable barrier, but from what I have seen as yet I judge that I can turn their flank on the west side'. [263 (323) 53–4]

He got as far as Nag Tiba, 31 August–5 September, at 9915 feet; Kederkanta in latitude 31° N longitude 78° E, 17–23 September, at around 12 540 feet; and the Chaur at latitude 33° 10' N and longitude 77° 05' E, 4–10 October, at 11 966 feet. This was an area that had been visited by Hodgson in 1816, when he took numerous observations. While Everest was on Chaur he was visited by the traveller Godfrey Vigne and gave him welcome hospitality. Vigne in volume 1 of his *Travels in Kashmir ... and the Himalayas* [277], described the meeting:

> On the way, Mr. Lee Warner, my companion, and myself, received an invitation from Major Everest, Surveyor-General of India, at that time on Chur, conducting the grand trigonometrical survey, and of whose hospitality I shall ever retain a grateful recollection.
>
> The camp of our host was perched as near as possible on the very top, and our chief object was to keep ourselves warm. The tent in which we dined was furnished with a lighted stove, and the entrance carefully closed against air, whilst we drank our wine, and talked to a late hour above the clouds.
>
> On the huge granite rocks that formed the very apex of the mountain, the labourers in attendance had formed a platform of loose stones, purposely carried thither and in the centre of it they planted a mast, as a mark for the survey. Several that they had previously raised on other summits were visible only by the aid of the theodolite; and a powerful heliotrope (in use at Saharanpur in the plains) might, it was supposed,

have reflected the sun's rays towards us from a distance of sixty miles. [277 Vol. 1 pp. 34–5]

Vigne met up with Everest again *en passant,* when the Dehra Dun base was being measured. He commented very favourably on the manner in which Everest was operating, but whether or not Vigne understood the finer points of such work is doubtful. [277 Vol. 1 pp. 39–40]

The Dehra Dun base measure

On 12 November 1834 Everest began, in the grounds of his office at Dehra Dun, 50 comparisons between the standard bar A and the compensating bars. The base was then measured from west to east; a further set of 61 comparisons followed on 5 February and then the base was measured again in the reverse direction, this time by Waugh. The whole was then followed on 2 April 1835 by a final set of 66 comparisons.

At the time of the measurement the site was on Government land, but within a few years it had been allotted to Lieutenant Henry Kirke. Everest had crossed swords with Kirke in late 1834 when there had been trouble between him and Everest's Registrar, Mr Morrison. In his true fiery fashion Everest took up the cudgels and complained to Colonel Young who put Kirke in his place. Somewhat surprisingly,

Figure 12 Dehra Dun headquarters of the Survey of India, c. 1876.

Kirke later named his tea estate in the Doon, Arcadia, in recognition of the Great Arc. Tea growing in this area was a new venture at that time. From waste land in 1835 the site was 'a glowing mass of fields and orchards' by the time Everest left Hathipaon in 1843. [231 pp. 52, 168]

The countryside around the site was plentifully supplied with tigers and very long grass so that transport was essentially by elephant. The back-up facilities on this part of the arc consisted of 4 elephants, 42 camels, 30 horses and 700 followers. It must have made an impressive, almost awesome sight.

Before the first measure of the line was made, the east terminal had to be moved further from the river to avoid the overspill which would have put some bays in the water. The preparation for the modified route delayed the start, but by 1 December everything was ready.

The overall length of the base, reduced to sea level, was first found to be 39 183.973 29 feet and on the second measure, 39 183.773 57 feet. An agreement of 2.4 inches in 7½ miles. From computation through the triangulation from Sironj the value was 39 183.273 feet, or an average of 39 183.673 feet. These were reduced on the assumed height above sea level of the east end of the base which was 185.2 feet higher than the west end. The average elevation was taken as 1980 feet; however, this had been somewhat tenuously arrived at from earlier work by John Hodgson and James Herbert, both of the Bengal Infantry, for mapping Saharanpur. [108 p. 208]

As a means of checking his result, Everest placed two intermediate points along the line which, together with the two terminals, he observed from four stations in the Sewalik hills. This allowed him to intercompare the sections of base to his satisfaction. Since he had considered the error in his linear measure to be 1.6 inches, it is to be wondered what he expected from the comparison of sections. It would not have been easy to improve on 1.6 inches using the techniques suggested, but the definition of what he meant by the 1.6 inches is not given. Since he said it was more than he expected, it could not have been simply a theoretical figure.

While Waugh was measuring the reverse direction, Everest did the base extension net to three surrounding Great Arc stations. He had then intended to press on with the triangulation observations to the south, but once again he was overcome with fever. This, together with the non-completion of the towers, closed down field work for that season.

Laid low again

Everest was confined to bed for the latter half of February 1835 and again for most of the period from May to the end of October. In the brief period in April and early May, when he was able, Everest first visited Amsot and took some of the necessary angles there and then observed the base terminals, together with Amsot and Dhoiwala, from Banog. This mountain, at 7433 feet, is the highest part of its range and has an excellent all around view. Its summit was good for both camping and observing and, so that the large theodolite could be got to the station without damage, Everest had over 2 miles of road built up it, which still exists today.

He then had to succumb to his illness until October when he endeavoured to observe an azimuth at Banog but was too weak to stand at the instrument to do so. So seriously ill was he this time that on one occasion he was bled to fainting with 1000 leeches, suffered 30 or 40 cupping glasses (This was the application of a heated glass vessel to the skin to draw out blood. Whether this or leeches was worse is debatable!) and numerous doses of nauseous medicine. Each successive attack was worse than the previous one and caused considerable alarm.

Back on the Arc

Observations with the large theodolite were completed at Amsot in October and November 1835 by Logan and Peyton. Then on 22 October 1835 Everest left Dehra to restart on the triangles of the section that required 14 huge masonry towers that cost Rs.30 000. The requirement for towers was because the Jumna Plain was so flat and smoke-filled with no hills in sight. There were two at 40 feet, one at 60 feet and eleven at 50 feet in height. [231 p. 82] The towers had foundations some 20 feet deep and had portable cranes to hoist the instruments to the top.

Initially for reconnaissance Everest had timber masts about 35 feet high with bamboo scaffolding around. The observer stood on the scaffolding with the theodolite on the mast and took approximate angles to distant points. The results were not very satisfactory. His men were quite unfamiliar with scaffolding so once again Everest had to take intimate control of all aspects.

It was on this part of the arc that Everest developed the method of ray tracing to improve the rate and accuracy of forward reconnaissance. Ray tracing was the determination of the direction between two points that would be intervisible if it were not for removable obstacles such as trees and huts. It could take three different forms:
- minor triangulation with no linear measures,
- traverse with both angular and linear measures,
- simple alignment of flags.

The first was the most accurate but most laborious. The second was less accurate but simpler and less laborious. The third had sweet simplicity and could be carried out by illiterate local staff. [265 (II)]

By 11 November the party was at Nojhli awaiting the arrival of the large Cary theodolite which had been renovated under Everest's direction by Barrow. There had been much delay in getting the work completed because of problems with engraving the azimuth circle. Although the workshop part of the reconstruction was completed in July, that was in Calcutta. De Penning then had considerable difficulty getting it aboard a steamer, his initial attempt meeting a blank refusal to accept such a large crate.

Thus, although Everest was adamant that he required the theodolite by 20 September so as to be able to test it prior to starting field work on 1 October, it did not reach him until 12 November. It was found to need a few minor adjustments. Everest trusted these to Mohsin Hussain, and then it was ready for use. [116 pp. xxvi–vii]

As a result of the purchases he made in England while on sick leave Everest now had two 3 foot theodolites that could be used on the Great Arc. The one which he had recently acquired weighed 16 cwt when packed for transit. Although it had, by then, been in the country since 1832, Everest had not had the opportunity to test it out until February 1835.

On 11 December Everest arrived at Begaruzpur, and by early January 1836 he was at Saini. For the next month he was restricted in his observations by thick smoke and haze. By 1 May he and Waugh had worked their way down the two sides of the chain to just north of 26° latitude, on the line Juktipura to Pagara, some 60 miles south of Agra. [263 (344) 91–104]

In this area Everest was using an Argand lamp with a parabolic reflector as target and this was very effective even in the worst weather. The lamp was encased in a wooden box with a tin chimney and a circular glass aperture in the door. During a violent storm which lasted three days and nights without interruption, accompanied by much thunder and lightning, the rain falling in torrents and wind blowing in violent gusts, the stations to which he was observing never flickered.

It was while he was still in the same area that Everest noted some very unusual refraction effects. In many instances a distant heliotrope would, in the morning, not appear as a disc but as a tall chimney. On one occasion this vertical extension was equal to 440" or equivalent to nearly 200 feet in height. On another occasion, a target that in the morning was at 4' 32" elevation was, on the following morning at the same time, at 4' 36" depression. [118 p. 209]

The onset of the rains heralded a return to Mussoorie for five months before Everest could set out again.

Salary and expenses

In the early 1830s Everest had protracted correspondence over his salary and allowances. At times this resulted in the need for apology after some of his typically acid paragraphs.

> my predecessors in the office of Surveyor General … drew allowances, and never left Calcutta on duty… I have been put to much extra expense… Moreover I have been obliged to maintain a large marching establishment, for since 1st January 1833 … a period of 28 months, I have been actually under canvass 18 months. During that period I have marched over a distance of full 2000 miles… As a fixture in any place I shall be of much less use than at the Presidency. [263 (286) 195–7]

> If in the course of advocating my pretensions I have given offence, I beg to …

disclaim all intention of doing so… no person is more sensibly alive than myself to the necessity of subordination or would more resolutely discountenance any want of diffidence to superiors in authority. [263 (286) 409]

At the time his monthly salary was of the order of Rs.1800 plus allowances of some Rs.425.

Preparing Kaliana astronomical station

During May 1836 Everest constructed the necessary pillars at Kaliana for it to be used as an astronomical station. Later that year a small building was erected over the pillars in the hope that it would protect the astronomical circle better against the high winds and storms than did a tent at Kalianpur in 1825. As Everest began his trek south in October 1836 to continue the arc he stopped for a while at Kaliana to determine an azimuth. By 10 October he was able to continue his journey south and reached Dholpur by 2 November. Here he was delayed more than two weeks because of his own stubborn insistence on the appropriate protocol being observed when crossing a border.

Problems in Gwalior State

The River Chambal was the boundary between Dholpur and Gwalior State and because of difficulties experienced by Rossenrode some years earlier Everest insisted that his party be met and accompanied through the State. Before setting out on his journey Everest had taken the precaution of writing to the Resident (British Government representative) of Gwalior who had promised to give what assistance he could. Everest had requested 50 horsemen of the Maharaja's contingent to meet him and to accompany his scattered parties. [263 (346) 163–4] Unfortunately, the Resident felt unable to offer all that Everest requested and suggested that he come prepared with labourers or at least be prepared to pay an attractive price for such. This approach did not meet with Everest's approval.

When he duly arrived at the River Chambal, Everest found no welcoming party. He was furious and immediately wrote to the Supreme Government complaining of the discourtesy shown him by the Resident. He expressed in his forthright manner that:

There is no remedy but to wait on the boundary until all obstacles thrown in the way of my progress shall be either removed … voluntarily or … by decided orders from the Supreme Government … I am spoken of in his [The Resident's] communication with the Muktiar as 'one Major Everest engaged in measuring', and my assistants … in the same unceremonious style…

This is the first instance of rudeness and opposition which I have experienced on the part of a British functionary … [263 (346) 202–5]

The retort from the Resident was suitably low key but finished '…there is a dictatorial tone pervading your letter … which I cannot think that the Surveyor General of India is justified in using towards the Resident at Gwalior'. [263 (345) 252–60]

Everest stayed at the river for two weeks until a State official appeared complete with adequate provisions for the survey party. The Maharaja expected Everest to spend four days with him so that he might be received properly. Everest pointed out the delay he had already suffered and stressed that he must go ahead the next day – a very difficult man to please. [263 (286) 406–8]

Aside from such problems as this, the actual clearing of the physical lines to be observed required not only considerable negotiating skill but also courage. In his Report for 1836 Everest referred to the locality of Bhatona, where a 30 foot wide swath had to be cleared.

> The houses which stood in the way were selected by… Mr. W. Rossenrode with extreme care, so that no needless injury might be inflicted, and … duly valued by a panchait… of several tahsildars, peshkars, kanungoes, and others, …so that the proprietors, on the consideration of … ready money,… were … satisfied. The list of dwellings destroyed, however, is disastrous, and I hope it will never again fall to my lot to have so disagreeable a task to discharge…
>
> 5 huts, thatched, in Ramnagar, crushed by the fall of trees, 37 flat roofed houses, 52 huts… of mud, razed to the ground in the town of Bhatona. 12 huts, thatched, with mud walls… Daherpur…
>
> How Mr. Rossenrode contrived to effect this severe operation, and reconcile all parties… surprises me. The town of Bhatona is inhabited by Jaths, who have the character of being a very turbulent… race. Amongst these he ventured unarmed and without a single weapon of defence or any show of force – and, though the weather was unusually cold, and hardly a night passed without a severe hoar frost, yet he had influence enough to persuade the owners to relinquish the houses which furnished them a comfortable shelter…
>
> After the clearance… the blue lights burned at Bulandsheher (28° 25' N 77° 55' E) were at last seen.

(The observation point was Dateri 28° 45' N 77° 45' E, which was almost due east of Delhi. *Panchait* means village court; *tahsildar* means district official; *peshkar* means headquarters official; *kanungoes* means revenue officials, so there was plenty of officialdom involved in obtaining agreement and fair assessments.)

By 20 December 1836 Everest had completed observations at Shergarh and was about to move on to Kansri. After reaching Sironj he first visited the nearby observatory at Kalianpur that had been constructed some years earlier and found it completely ruined. He immediately ordered a new one to be constructed in as durable manner as possible.

Then steps were retraced to Aring where the party arrived on 11 March 1837. His reason for revisiting this point arose from doubts over the steadiness of the three-storeyed masonry tower. Perhaps some of this was due to the alterations Everest made initially. After driving holes through each roof so that a mark could be sunk at ground level yet observed down to from the top of the tower, he then built on the upper roof a cylinder of masonry surmounted by a 4 feet diameter slab for the instrument. [263 (287) 254–5] To overcome his fears Everest got Waugh to construct a pillar that went from the ground to the top so that the instrument would be completely isolated from the area on which the observer stood.

By 16 March the observations had been satisfactorily achieved and Everest was on his way to Kaliana where he arrived on 24 March. He took with him on this journey – of some 150 miles by bullock cart! – two stone slabs from quarries near Aring so that he could have very stable pillars.

The new astronomical circles

At Kaliana the two new astronomical circles were waiting for his inspection before their use for observing zenith distances of stars of known declination when close to the meridian. To give as stable a support as possible the slabs were mounted to the pillars to take them. When the circles were set on wooden supports in Calcutta, Everest had noticed a certain unsteadiness about them which he had hoped the use of pillars would correct. Unfortunately, there was no improvement. He and his sub-assistant Said Mohsin made all possible adjustments and observed with the utmost care but to no avail. The circles were useless as they stood. [263 (344) 108–84]

Everest dismantled the instruments and measured every component. Such was his knowledge of the engineering of instruments that he was able to assert that the Y supports for the telescope were too slender for the weight they had to support. In addition, he found that the horizontal circle and tribrach were both too weak. These faults led to severe vibration when there was rotation to the instrument. [231 pp. 131–2] The two circles were not equally at fault so Everest distinguished between them by calling one Troughton and the other Simms. Simms was the worse of the two.

He had waited all these years for the best equipment and when he got around to trying to use it there were defects. His attitude was predictable. He arranged for George Barrow, his Mathematical Instrument Maker, to come up from Calcutta to effect the necessary repairs. This, however, was not a five minute journey. It was October 1837 before Barrow reached Kaliana, and it was to be another two years before the circles were fit for use. It was only during this contact with Barrow that Everest learned that such errors were considered to be inherent in instruments of this type despite all the skill that Troughton had used to endeavour to eliminate the problem.

Remeasurement at Sironj

The season of 1837–8 started with the observations for azimuth at Banog then the party moved south to Kaliana on 1st October. Everest waited there for the arrival of Barrow in order to detail his required modifications to the astronomical circles, and then he moved on southwards to reach Sironj on 17 November 1837, ready to remeasure the baseline there.

The remeasure was required because when, after determination of the Dehra Dun baseline, Everest computed through the chain of triangles to Sironj (a base that had been measured in 1824 by chain) he found a discrepancy of almost 3.5 feet. [263 (344) 76–80] He was certain that this large difference would mostly disappear when the new compensating bars were applied at Sironj. However, fever had returned to wrack Everest again and the measurements had to be entrusted to Waugh who managed to complete the line by 18 January 1838 as 38 413.367 526 0 feet. [116 p. 277] The new value exceeded the previous one of 1824, of 38 411.899 12 feet; a difference that Everest put down to poor knowledge of the true length of the chain and of temperature. This latter value was modified after recalibration of the chains in 1832 to become 38 410.543 feet. [116 p. xxxiii] Agreement with the value computed from Dehra Dun then became only 7.2 inches. [210 p. 90] The measured value of Dehra Dun was 39 183.873 and its value calculated from Sironj was 39 183.273 feet. [116 p. xlii]

Once a few doubtful angles had been reobserved on the return journey northwards the section of arc from Sironj to the Himalayas was complete. But Everest was not satisfied. Now that he had better instruments and had improved techniques, he felt it necessary to reobserve all the triangulation from Sironj as far south as Bidar. This would cover work that both he (1823–5) and Lambton (1817–22) had done previously. It was certainly the only way that one could be certain of consistency between all the observations over the section from 18° to 24° and onwards to 30+°. [231 p. 41]

After measuring the Sironj baseline Everest left the equipment there in anticipation of remeasuring the Bidar base. He left meticulous instructions for its care.

> The best means to prevent it [rust on the bars] is to cover them with mercurial ointment, hog's lard, or mutton fat. If either of the latter two are used, it must be applied in the state in which it is taken from the animal... If mercurial ointment is used, care must be taken... that none comes upon the brass bar, as it will create verdigris,...
>
> To change the hog's lard,... the old should be taken off and... the bars near the silver dots must be cleaned with a ... soft brush and barley water, ...One of the greatest evils to be guarded against are the white ants, and you will order the tindal ... to examine every box daily. [263 (371) 148]

The bars managed to be damaged by the application, and later removal, of lacquer but this fortunately did not materially affect them. When transported to Bidar in 1840 the equipment required 34 camels and 3 elephants. [231 p. 55]

Repairing the astronomical circles

Leaving Sironj on 19 January 1838, Everest and Waugh spent some days at Shadaora near the station of Deadheri. Waugh received detailed instructions on 26 January 1838 for the remeasurement and then proceeded the 650 miles south to Hyderabad. All the triangulation between Bidar and Sironj was completed by Waugh and Renny by June 1839. In the retriangulation by Waugh there were numerous examples of changes in the angles of several seconds, even up to 12 seconds. The discrepancy between the computed and observed value of the old 1815 Bidar base was 6 feet 7 inches. [231 p. 16]

On 9 March 1838 Everest was back in Dehra supervising the instrument repairs. At the onset of the rains the work was transferred to the Park [231 p. 132] where there was a workshop available. For much of the next 18 months Everest would have preferred to have been out in the field observing but felt obliged to stay near the astronomical circles. Details of the changes made to these astronomical circles are in Appendix 1. After two and a half years of exhausting work the circles were completely rebuilt and ready for use. Phillimore describes the achievement as 'one of the outstanding triumphs of Everest's professional career'. [231 p. 135]

Minor triangulation series

While all the work on the Great Arc was progressing Everest continued to have overall control of operations on other minor series. Several of these were started around 1830 but took many years to complete. Among these were:

1. South Parasnath meridional series

This series continued to be dogged with problems. First Western was dispensed with; then his successor Bridgman died soon after taking over, and then Thornton took charge just for a matter of months before the reins passed to Boileau at the end of 1835. Before long Boileau was sick and went to China for six months to recuperate. When he returned in December 1837 it was again a short-lived appearance before fever struck down 62 of his 107 followers many of whom died. By December 1838 Boileau had had enough and withdrew from the Survey.

Although the chain was completed by June 1839, it was not to Everest's satisfaction. Poorly conditioned triangles and high triangular misclosures made the results unreliable. Much was blamed on the theodolite they were using which was an 18 inch by Gray. Everest had found it somewhat top heavy and, although Barrow made a fine job of rebuilding it, there were still problems with a poorly divided circle and poor quality optics. [231 pp. 59–61, 143]

2. The Budhon series

During the 1834–5 season the Budhon series progressed across the Jumna and the

Ganges. The method of forward selection of points by the ray tracing method, so successfully adopted on the Great Arc, was also used here. Unfortunately, the leader of this party, Macdonald, fell ill and never recovered. His place was taken in November 1835 by Lieutenant Edward Ommanney of the Bengal Engineers, who was unable to make much material progress until the following season. There was a need for towers but permission was not readily forthcoming. During the 1836–7 season progress was made as far as Gwalior but on 31 May 1837 Ommanney resigned and Olliver, his only assistant at that time, took over.

In his first season in control, 1837-8, Olliver was told by Everest to change tactics. Instead of a chain of single triangles, he was instructed to use centre point polygons with sides between 8 and 15 miles long. These were preferable to quadrilaterals because in the low lying plains the long diagonal of quadrilaterals could cause severe problems. On the other hand all sides of a polygon can be of the same order of length. [263 (371) 127–8]

In March 1838 the party was withdrawn to work on the Great Arc and the series was dormant until November 1839 when Renny and Murphy carried it on. [231 p. 64]

3. Ranghir series

By the time the party took the field for the October 1835 season they were still awaiting permission to use towers and so had to content themselves with ray clearing. Waugh was called away and Armstrong left in charge for 1835–6.

Permission did finally come for the use of towers and work was started on 10 masonry columns that had been designed by Waugh. A further 14 were to follow. Misfortune struck this group on 10 April 1837 when the scaffolding on one tower caught fire and the theodolite, which had been left at the top, was damaged so severely as to cause work to be abandoned for the season.

Season 1837–8 started with a new theodolite, but progress was slow principally because the lines set out in the plains were too long for easy observation, and much was left to the vagaries of the weather and atmospheric conditions. 18 to 22 miles was later considered excessive for the plains and 11 miles was a far better target figure.

1839 was interrupted because of the severe weather conditions, so progress was again slow. [263 (379) 325–8]

4. The Amua series

By the season of 1836–7 Renny was observing in the vicinity of Cawnpore but poor visibility postponed work during the dry season. After the first rains the atmosphere cleared.

During October 1837 Renny was detached to assist with the Sironj baseline and the arc was left under the guidance of Charles Murphy and Charles Lane who took it as far as the Ganges. Renny returned by March 1838 and pressed on to Sitapur before again being called to assist on the Great Arc. By the end of 1839 all the series was completed.

5. Bombay longitudinal series

When William Jacob returned to Bombay in March 1837 after spending some while with Everest at Sironj, he put forward the idea for surveys in the Bombay area. This was supported by Everest and Jacob was successful in obtaining staff. Excellent progress was made until the rains intervened, but Shortrede was able to complete the series in the 1838–9 season. All of the 1839–40 season was spent on the computations, but these were only provisional until the remeasure of the Bidar base by Waugh in 1841.

When Everest was in a position to assess the relative worths of various series he considered that this one by Jacob was the best by far.

6. Karara series

In 1838 a series was started in the vicinity of the meridian of 81° 18'. In the south it took off from the Calcutta longitudinal series and then passed through Rewah and north to the Ganges to the west of Allahabad. Initially it was as a double chain of triangles (akin to centre point polygons) crossing the Kaimur Range and then to the north of the Ganges it was mostly single triangles through the plains of Oudh to the Nepalese border near 28° N.

Operations began after completion of the Sironj baseline and were under the charge of William Jones. They started from Karara on 1 March 1838 but had to be abandoned by mid-June because of both the rains and the onset of sickness. By 1 October they were ready to recommence but had hardly begun again when jungle fever struck the whole party. Jones lost his assistant who died on 18 November, and all others were prostrated to the extent that Everest had to abandon the work for that season.

Nothing further was done until 1841. [231 pp. 67–75]

May I have a CB please?

On 18 November 1838 Everest wrote a memorandum to the Honourable The Court of Directors of The East India Company. It was an exercise he was to repeat several years later. He had studied the Gazette for the award of honours and noted that fourteen Lieutenant Colonels and eight Majors, many of whom were his contemporaries, were amongst those awarded the Companion of the Order of the Bath (CB).

In bringing this to the notice of the Court he begged to suggest that not one of them had exposed himself to greater hardship, risked his life more freely, or devoted himself more steadily and faithfully to the service of the Court than himself.

He stressed that the qualities required in the operation of the Great Trigonometrical Survey of India were precisely the same as those needed by military men: habits of subordination, of fortitude, of arrangement, of combination; 'the undertaking is one of unparalleled magnitude, calling for the exertion of all the talent that your Honourable Court's service can command'.

In the light of all this, he respectfully urged the Court to consider his position and take such measures as it judged appropriate. [105]

He was unsuccessful.

Chapter 9

THE JERVIS AFFAIR

*F*or extended periods during the years 1837–39 Everest was very ill. As a result of this, and his previous serious illnesses of 1835, there arose an unpleasant situation concerning Major Thomas Best Jervis (1796–1857) of the Bombay engineers, and the East India Company.

A successor designate

At the time of illness Everest had written to the Company about his poor health but in such terms that the Directors were obviously worried that he might well die in post with no successor lined up. Their reaction, without taking any apparent guidance from either the Government in India or Everest himself, was to appoint a successor designate, dependent on either the death or enforced early retirement of Everest. The appointee was Thomas Jervis and this, together with the terms of such appointment, was made known in a letter to Jervis of 2 September 1837.

The background to what then became a controversy stemmed from work that Jervis had done over the period 1819–30 in South Konkan, the coastal area to the south of Bombay. This had consisted of a statistical survey for revenue purposes, a geographical survey including baselines and triangulation in the coastal area around Bombay, a map covering the area from Goa to Daman and an incomplete survey of the Deccan.

Early in 1836 [231 p. 449] the Directors were informed of this work by Jervis and the accompanying letter referred to the professional skills and talents of Jervis. Such praise was, however, contrary to the sentiments of others. In 1832 John Jopp, the Deputy Surveyor General, Bombay, described the work of Jervis as 'not of a high quality, so many [villages] appear wanting, and there are discrepancies between the original and compiled maps…' [229 p. 126]

Everest's estimate of the unreliable character of Jervis' work shocked the Directors.

> We cannot but express our surprise at the opinion given by the Surveyor General on the Konkan Survey… which occupied Major Jervis for upwards of ten years … We

expect our surveyors to perform their work on an approved and well understood system so that at any stage... the survey may be intelligible to a qualified surveyor,...

Neither Everest nor Waugh would accept the work for geographical purposes. In fact Waugh reported many years later – in 1856 – that there was an error in the triangulation that rendered it almost useless for distances, and that many places had incorrect latitudes.

However, the Directors were very impressed by the sheets that Jervis had submitted. It is probable that they judged them more by their appearance than from any check on their accuracy. They considered them to be suitable for incorporation in the proposed Atlas of India [231 p. 450] and sent the material to the Surveyor General for that purpose. A selection of the material reached Everest personally, and he found it impossible to make anything of the disjointed items.

Everest, of course, was not so much interested in the decoration but in the accuracy of the presentation. He decided that there was no evidence of the sheets being based on accurate control, nor did they bear the signs of careful survey of the detail. His conclusion was to reject the material for use in the Atlas although he was willing for Jervis to take the material and turn it into a map of his own.

Everest hears the news

It was apparently not until Everest met the Governor General in early 1838 that he learnt of the provisional appointment of Jervis as his successor. [231 p. 317] At that time Jervis had yet to give the lecture that was to lead to the troubles, but Everest was already familiar with the quality, or lack of it, in the survey work that Jervis had been praised for.

On the strength of the initial praise of his work by the Directors and their appointment of him as provisional Surveyor General, Jervis went to England and made the acquaintance of many influential people, particularly members of the Royal Society. In fact, so much did Jervis impress these eminent gentlemen that he was invited to address the British Association in August 1838.

In this address he outlined the progress of survey in India and put forward plans for it in the future. In effect he was saying what he would do when he became Surveyor General and how the existing state of affairs could be improved. In addition he put forward various ideas relating to pendulum observations, experiments on magnetism and tides and the comparison and verification of standard bars. Such was his confident, persuasive and almost arrogant manner that many urged the Company to pursue the ideas put forward by Jervis for improving the survey of India and completing a topographic survey of the country within seven years. He went further, having the lecture printed for private circulation and sending a signed copy to Everest.

Everest had not taken immediate open objection to Jervis' appointment since the Directors had made it clear to him that, so long as he was in office, they looked to him alone as their responsible adviser on all matters relating to the Survey. However the act of sending him a copy of the lecture sparked off a furious display from Everest who tore the lecture to shreds and emphasised his contempt for the mapping that had set all this in train. [231 p. 11]

The document, Everest's response and supporting items

The document was in eleven parts and compiled effectively as if Jervis were already Surveyor General. The whole can be found in a few remote locations but is also in printed form, bound in with Everest's detailed response, in *A series of letters to the Duke of Sussex*. [107] In this version the responses by Everest occupy 147 pages and the contribution by Jervis 46 pages. Then follow a variety of supporting items:

- A letter from Captain W.A. Tate replying to one from Jervis,
- An extract from a Memoir by the late David Scott,
- A recommendation adopted by the British Association,
- A letter from Captain F. Beaufort, RN Hydrographer to the Admiralty,
- A letter from Colonel Colby RE,
- An extract from the last report of Colonel William Lambton,
- A document dated 14 July 1838 from the Royal Society signed by the Duke of Sussex (=Augustus) and 38 scientists most of whom were Fellows of that Society. The signatories were

  W. Whewell Pres. Geol. Soc.
  Wm. Buckland VPGS
  A. Sedgwick VPGS
  Charles Daubeny FRS FGS
  W.R. Hamilton FRAS. Pres. RIA
  T.R. Robinson FRS
  David Brewster V Pres. RS Edin.
  John Phillips FRS Prof Geol London
  F.R. Chesney FRS
  C. Wheatstone FRS
  L.F. Boscawen Ibbetson FGS
  R. Sheepshanks FRS, MRAS
  G.B. Airy Astronomer Royal
  Francis Baily Treas. & VPRS
  S. Hunter Christie Sec. RS
  J. Walker Pres. Inst. Civ. Eng.
  H.F. Talbot FRS
  J.F.W Herschel
  Edward Sabine
  Jas. Clark Ross FRS

  Augustus FP Pres. Roy. Soc
  T.M. Brisbane Pres. RS Edin
  W.J. Hamilton Pres. RGS
  James Ivory FRS
  John Steetley
  Chas. Lyell VPGS FRS
  G.B. Greenough VPGS VPRGS
  R.I. Murchison VPGS FRS
  Geo. Peacock FRS
  F. Beaufort Hydrographer
  J.W. Lubbock FRS
  W.H. Smyth Forn. Sec RS
  G.H.S. Johnson FRS
  Woodbine Parish VPRGS
  John G. Children VPRS
  M. Faraday
  H.T. De la Beche
  W.H. Sykes FRS V Pres Hist. Soc. London
  H. Lloyd
  [186 p. 97], [12]

- A letter from Major Jervis to the Secretary to the Court of Directors.
- A memo on the preliminary recommendations made by Jervis.

The whole, when combined with Everest's response occupies 225 pages. Such is the varied content of this publication and the manner in which it is presented that it is pertinent to treat it in some detail. Much of the content referring to Everest's life and work is not recorded elsewhere and so is of considerable historic importance.

Everest commented on many aspects of his professional work and in so doing justified some of his actions and mocked many of the comments in the paper by Jervis. Unfortunately Everest neither dated nor indicated from where he was writing each missive but for publication to have taken place in London during October 1839 they must all have left India by early 1839 and so must have been written during the 1838–9 field season.

The address by Jervis was on 26 August 1838. The item signed by the 38 learned gentlemen was dated 14 July 1838. The address was printed privately which could not have been before September 1838, and it took at least four months for material to travel to India. The letters give few clues as to their date although, as Phillimore indicates [231 p. 438], there is a suggestion that letter VII was written about April or May 1839. It must have been very soon after that when all were despatched to the printer, since the British Museum received their copyright copy by 19 October 1839. To summarise then it would seem that the letters were all written after February 1839 and before June 1839.

All this, however, is complicated by the fact that the Duke of Sussex ceased to be President of the Royal Society on 30 November 1838, although he did continue as a Vice President until 30 November 1839. One would have expected this information to have reached Everest especially since he was a Fellow. Where did he address the correspondence? Where are the originals now? Were there any replies sent by the Duke since there is nothing pertaining to the affair held at the Royal Society, nor, until very recently even a copy of the published letters. Or were they really open letters that were never actually sent to the Duke before circulation of the published document?

Everest was particularly incensed that the Duke of Sussex, as President of the Royal Society, and many other eminent scientists should have been duped so easily by Jervis. In addition he took these eminent persons to task for exhibiting their lack of knowledge of any aspect of life in India. Despite his normal hasty nature, which he undoubtedly exhibited in the letters of response, one can but sympathise with him in many ways for the manner in which he was treated and the lack of recognition given to his efforts in the address by Jervis.

Before summarising Everest's responses it is appropriate to discuss in chronological order, the other 10 items mentioned above (Jervis' address and the nine supporting items).

In June 1838 Jervis wrote to both Colonel Colby RE and Captain F. Beaufort RN asking for details of how officers were appointed to the Ordnance Survey then working in Ireland, and Nautical Surveys respectively. Colby indicated that the Duke of Wellington had devolved upon him the task of both appointing, and firing if necessary, all officers and others employed by him. Beaufort similarly said that it was essential for the selection of every individual to depend upon the person in charge of the work. He indicated that their Lordships had never interfered with any appointments and he advocated that the same approach should be used in India.

On 14 July 1838 38 eminent gentlemen signed an Address to the Honourable Court of Directors of the East India Company. It began

> We the undersigned ... view with great interest the objects which Major Jervis FRS (who we are given to understand has recently been appointed by the Honourable Court to the charge and superintendence of the Great Survey of India) has submitted to us for the extension of science; the improvement of the geography of India; and of the countries stretching between its frontiers and the Caspian Sea...
>
> Major Jervis's proposition for the organisation of an establishment of men and officers ... seems worthy of the encouragement and favour of the East India Directors...

The document then continued to itemise the specific objects thought particularly desirable. These included:

- an improved topographic map of India engraved in the style of the British and other European Surveys, accompanied by other information similar to that of the Survey of Ireland and adopting a uniform system of orthography,
- improved geography of the north-western frontiers of India,
- experiments on tides,
- experiments connected with magnetic dip, intensity and variation,
- verification of the standard unit of measurement of the Indian Survey compared with that of the British Survey.

Sir John Herschel recommended that no pains or expense should be spared in procuring as perfect a unit of measure as possible, comparable with the British measures. It would be desirable also to have a facsimile of each of the continental standards as a means of comparison. Herschel also recommended that an arc of longitude comparable in extent to Lambton's meridional arc should be measured trigonometrically. Baily recommended pendulum observations at the principal stations, at very great elevations, on the sea coast and on the intermediate tableland.

The document concluded with the statement 'Lastly; it is desirable that in all these pursuits Major Jervis should avail himself of the advice and aid of Sir John Herschel, Mr Baily, Mr Whewell and Mr Airy...' [12]

Is it any wonder that Everest took such extreme exception to a string of suggestions that were quite impracticable in the context of staff and resources available? These ideas may have seemed superb to eminent gentlemen sitting in an ivory tower in London but it would be quite hopeless to achieve them when considered in relation to the country and conditions under which they would be required to be performed.

On 6 August 1838 Jervis penned another statement to the Honourable Court detailing what he felt was immediately necessary for the organisation of the men and officers for the Survey of India. After listing those authorities he had consulted he stated that:

> They all agree that the Irish Survey far excels any other... but the credit of discovering and pointing out its peculiar applicability to the wants and circumstances of India, is indisputably and exclusively my own ... In appointing me to this arduous and responsible office, the Honourable Court may rest assured they have made no injudicious choice.

He went on to intimate that the document signed by the 38 had already been written but had yet to be presented – it was dated three weeks before. He then appended a list of ten measures which he considered immediately desirable. These particularly referred to the staffing of the organisation and the training requirements.

On 26 August 1838 Jervis addressed the British Association for the Advancement of Science on the subject of *The Survey of India and the Geography of those vast and interesting regions*. Right at the start he referred to these regions as '[those] in which I am destined to direct the important geodetical investigations for the determination of the Earth's figure'.

He discussed many topics including geology and geography, early mapping, the qualities of various personalities and aspects of the survey. He particularly criticised the topographic mapping by saying it bore no comparison with that in Europe and hinted that this was something he could rectify. 'I am justly proud of succeeding to an office which has been so ably filled by Rennell and Lambton, and Everest.'

Yet again, before the end of his 46 pages, Jervis referred to

> When I was appointed to this responsible office ... I sought the best information I could get... and in Ireland, I came to the conclusion that the method practised in the latter is wonderfully calculated to fulfil all those desiderata which I have enumerated...

The following day, 27 August 1838, The British Association made various recommendations in a resolution. These included the need for a longitude arc to complement the meridional one; that Indian and English standards should be compared; that observations should be made of the pendulum, of refraction and of magnetic phenomena and that there should be a large-scale topographic map of India.

On 12 September 1838 Captain W.A. Tate replied to a request from Jervis regarding the ability, quality and fitness of the local Indian people for work on survey.

The other two items were extracts from earlier material. Everest quoted several pages of detail from the 1822, and last, Report prepared by Colonel Lambton. This referred to various aspects of the Great Arc. The other was an extract from a Memoir by David Scott, printed in 1832 and related to the sub-division of labour as advocated by Colonel Colby in Ireland and its promotion in India.

Had Jervis misunderstood the terms of his appointment? Obviously not, since in a letter of 26 December 1837 to the Chairman of the East India Company (EIC) he said '…in selecting me provisionally as the successor of a Rennell, Lambton and Everest' then again in another memo to the EIC he said 'some points of great moment connected with the honourable post to which I have been provisionally appointed'. [186 fol.12, 44] He was, however, obviously making detailed plans around his own ideas and in a letter of 6 August 1838 to J.C. Melvill of the EIC [186 fol. 76–83] he talks of what is 'immediately necessary to the organisation of the men and officers for the Survey of India'.

Further, in a letter of 12 June 1838 to the EIC, he drew unhappy comparisons between the excellence of the Irish work of Colby and that in India. Such comparison was quite explicable because of the differences in terrain, conditions, extent and just about every aspect of the work in the two countries. Half inch or one inch to the mile would suffice in much of India whereas Ireland was covered with minor triangulation and chain survey for mapping at six inches to the mile.

The Duke of Sussex

Duke Augustus Frederick (1773–1843), the sixth son and ninth child of King George III and Queen Charlotte was born at Buckingham Palace on 20 January 1773. Because of his delicate health he spent much of his life up to 1804 on the continent, where he went to the University of Göttingen. While in Rome in 1793 he met Lady Augusta Murray, second daughter of Earl of Dunmore. They married secretly on 4 April 1793, but it was declared void by the King in August 1794 when, on the birth of a child, the marriage became common knowledge. Prince Augustus became Baron Arklow, Earl of Inverness and Duke of Sussex in 1801.

He was President of the Royal Society from 30 November 1830 until 30 November 1838. He gave splendid receptions in Kensington Palace to many men of science but the expense they incurred induced him to resign the presidentship and use his money to improve his library. He was somewhat eccentric and wayward although he had well meant intentions. He died on 21 April 1843 of erysipelas and is buried in Kensal Green cemetery. [21]

The series of letters

The document by Everest is entitled *A Series of letters addressed to His Royal Highness The Duke of Sussex as President of the Royal Society, remonstrating against the conduct of that learned body*. By Lieutenant-Colonel Everest. London. William Pickering 1839 [107]

On the title page Everest had a very apt quotation in the light of the privations and hardships he endured for many years. It read:

Well saith Solomon, "much reading is weariness unto the flesh". How many hundred studious days and weeks, and how many hard and tearing thoughts, has my little, very little knowledge cost me, and how much infirmity and painfulness to my flesh, increase of painful diseases, and loss of bodily ease and health.

How much pleasure to myself of other kinds, and how much acceptance with men have I lost by it, which I might easily have had in a more conversant and plausible way of life.
Baxter's Dying Thoughts

His short preface also is worthy of verbatim quotation:

To expect that the public in general will interest themselves in a polemical discussion on subjects of abstract science, is, I am aware, unreasonable; yet perhaps there may not be wanting those who, on the bare principle of justice, will consent to peruse the present series of letters; and who make it the rule of their lives to frown down the strong when combining to oppress the weak and the absent.

To such persons I address myself – I ask no advocate – I court no favour; I complain of wrong inflicted by a body of men, powerful from their influence, their learning, their rank; and all that I ask is a fair and impartial hearing.

He began by indicating that he had just seen a pamphlet written by Major Jervis with which was bound several other documents including one which was an address to the chairman and Court of Directors of the East India Company,

bearing the signatures of many of my countrymen most distinguished for their attainments in science, and at the head of them that of your Royal Highness as President of the Royal Society.

The pamphlet and documents bear on subjects in which I am most intimately concerned, and therefore I have made bold to address your Royal Highness...

After introducing himself he spent nearly three pages summarising his work on the Great Arc up to that time and indicated that full details were contained in his publication of 1830. [96]   He then felt able to comment that if in the Address to the East India Company

I find myself treated as a thing gone by, and unworthy of further note, your Royal Highness will assuredly admit that I have just cause to complain; and though there are certainly no direct symptoms of positive disrespect, where my former labours are alluded to, yet to my present labours, in which I have been unremittingly engaged since my return to India in 1830, and am still hourly occupied, not only is no allusion made, but the gentleman selected by my employers, to succeed only *eventually, and in case of my being compelled by ill health to leave India*, is spoken of as already installed, and I am out of office.

In so appealing to His Royal Highness, Everest described himself as 'a brother Mason, one of your Royal Brother's Lodge, the Prince of Wales Chapter…'.

Everest then proceeded to take apart the Address signed by the 38 eminent gentlemen. At the first paragraph of that Address he commented:

it is so obviously incompatible with my continuance in my present situation, that, considering I am one of their body, who have had no bed of down to recline on in following up my pursuits … it would have been commonly civil and decorous, to dispose of me decently before thus needlessly assuming that I am set aside to make room for a successor.

He then detailed his activities since his appointment as Surveyor General and said 'At the time … there was not a single person who had the slightest experience of Geodetical operations, except three of my sub-assistants, who were not scientific men'. He stressed the vast difference between working over flat country and hilly country. Particularly since he would not entertain any old series of triangles as adequate. All angles had to be more than 30° and less than 90°. Thus in this very flat area he had to devise new methods, become proficient in them, and instruct others. He highlighted Mr J. Peyton, Lieutenants Waugh, Renny and Jones – the last three had been his pupils since 1832 – as the only ones proficient with new equipment.

He described the factors that led to a near fatal illness in 1835. By the end of which year medical gentlemen said recovery was past all hope unless he left India immediately. Eventually he recovered, but news of the situation had already reached the Court of Directors who searched for a successor and selected Major Jervis. Everest reminded them that 'so long as I remained in office, they must look to me alone as their responsible adviser, in all matters connected with the survey of India'.

He used these facts to substantiate that the address to the Society by Jervis was untrue.

> It may perhaps be urged, that the learned Fellows… were informed that Major Jervis had been appointed to the charge which I have held for so many years, and that I was actually removed to make room for him by expulsion, – by death, – by incapacity, – by sickness, – by disinclination, – or some other cause; but please your Royal Highness, this is not the mode in which reasonable beings usually proceed…

He states that such indiscretions are unbecoming in learned men.

> How would Professor Airy and Captain Beaufort feel if put in my position and were thrust so unceremoniously off his seat because he had incurred a serious illness while in office?

> There are those, who might say that I publish nothing, but is there one among them who can say that he ever asked for information from me that was not forthcoming? "and information that is not worth asking for is not worth having". Of the 38 only 2 made passing reference to the GTS at all during my stay in Europe 1826–1830.

> … my settled aversion to jumping to conclusions, over the head of facts; the constant impulse which is within me to perform my work first and talk about it afterwards…
> [107 pp. 4–13]

Thoughts on the Royal Society

Everest laid accusations against the Royal Society as one reason for this attack. 'The habitual spirit of selfishness and monopoly of the Royal Society, which prompts that body perpetually to form a little clique, knot, or coterie of a particular set at their head, within the compass of which all is gold – pure refined gold – without it all dross – mere dross'.

In a second accusation he was saying that there was 'the evident tendency of persuading those who are subject to its influence that they know all about a matter of which they in fact know nothing'.

In Letter III, he returned to his promise to illustrate the points he raised about the Royal Society and used Lambton as an example. Lambton commenced the GTS in 1799 and until allied armies occupied Paris (1815) he never received one word of encouragement, sympathy, assistance or advice from the Society. Until then, none of the results had appeared in the Transactions of the Royal Society. Yet when one of Lambton's assistants went to Paris he was known and well received by the celebrated scientist Delambre who knew all about the Arc. The outcome was that Lambton became a corresponding member of the Institut (French Académie des Sciences) in 1817. Upon the story reaching the ears of the Royal Society they too then made him a Fellow 1817–18 – after 18 or 19 years of toil. Everest then instanced how it was the French who sent expeditions to Peru and Lapland to decide the figure of the earth while the Royal Society did nothing.

At this stage Everest took Jervis to task over his assertion that the large theodolite was sent originally as a present to the Emperor of China; he said the fact was otherwise. There does seem to be some question as to the correct situation in this regard. In his volume of 1830 Everest referred to a paper by Lambton in the Asiatic Researches in which it is stated that various instruments were originally sent to the Emperor of China as a present. The Emperor refused to accept them from Lord McCartney's Embassy so they were taken to Calcutta by the astronomer Dr Dinwiddie from where they were purchased by Lord Clive who was at that time Governor of Madras. [96 p. 50] The confusion probably arose from the term 'various instruments' which was assumed, wrongly by some, to include the great theodolite. The instruments listed as the intended gift included a zenith sector, steel chain, level and chronometer but no theodolite.

In Jervis' address there was a very similar version to that of Lambton indicating that the instruments had been purchased in Calcutta for the Government on Lambton's suggestion. [187 p. 14] Everest refuted that version and instead said that because of some enhancement in the price of its construction, the great theodolite was purchased from Ramsden by Colonel Twiss of the Board of Ordnance. The Court then rapidly had a facsimile made by Cary which was taken in its passage to India by the Pièmontaise French frigate, landed at Mauritius but gallantly forwarded by the French Governor to Madras. [107 pp. 20–1]

> Hence then, considering the supineness and long lethargy in which the Royal Society had indulged with regard to this question [of arc measurement] … it might have been supposed, that when at last they aroused themselves from their trance, they would have considered it incumbent on them to make some amends … Those who have

formed such an expectation, however, may now undeceive themselves. [107 p. 22]

Because of this he resolved to become a Fellow of the Royal Society and on p. 22 he stated why he joined the Society and then said '…this august body seemed to be perpetually involved in angry squabbles and discussions about jobs and matters of patronage, hard words, jealousies and fears … I sat a silent observer'.

He returned on page 24 to the question as to why the Society had not been involved in arc measures. What about the Cape arc that he commented on in 1820 and with which Professor Airy disagreed? Why had none of the eminent Fellows who had experience of the Great Arc never been asked by other learned Fellows to bring their influence to bear?

He commented on an officer (Jervis) charging his master of misdirection of the funds of their constituents, and published that letter not only to his masters but to the world at large, 'and that too in a pamphlet with which is bound an address with 38 eminent signatures… is a decided breach of subordination and decorum, striking at the very root of discipline'.

He commented that the 38 learned Fellows knew nothing of the practical fitness of Jervis.

> In the meantime the Court, who are well known for caution in placing their confidence, will hardly have allowed it to escape them, with what ominous silence my name and experience are passed over and set aside.
>
> Is there any mode of escaping the conclusion that my absence is the cause? that I am well known to be a person having an opinion of my own, which I will never yield but to conviction?

He asked what would have happened to all his data if he had had to go to England in 1834–5 on sick leave – and yet might still happen? Would some other have purloined it as their own as better known people in the past have; and here he instanced Halley and Newton with respect to the work of Flamsteed.

At this stage Everest became even more vociferous about the injury that the Royal Society had been party to and set out the conditions he felt should be applied to Major Jervis. Nothing that had been executed by or under the direction of Everest should be received from the pen of Jervis or anyone else except his trusted assistants, Waugh, Renny or Jones. Jervis should confine his descriptions solely to matters that had been executed by himself or under his guidance. Since Jervis would almost certainly have to avail himself of arrangements, inventions and persons trained by Everest, he should make full avowal of such circumstances. [107 pp. 31–5]

He suggested that Jervis, if in post under the terms suggested, would be a puppet of the like of Herschel, Baily, Whewell and Airy. This he considered to be 'rather a strong order' since they were only plain English gentlemen of solid common sense, implying that they knew nothing of the circumstances pertaining so far away in India.

Whilst various recommendations were made regarding geology and botany for example, there was no reference in the Address to past achievements in these areas. Various employees of the Government of India had done considerable work in geology and botany over many years including members of the survey staff, such as Waugh and Renny. These men were both highly distinguished men of science and both brother masons, yet their names have never been mentioned within the walls of the Royal Society. [107 p. 46]

Related scientific experiments

Everest mentioned that among the activities that Jervis said should be done was the observation of tides, magnetic dip and intensity. His response was 'utter impossibility of depending upon amateurs for doing aught but what suits their own pleasure'. Thus without large expense nothing could be done.

In his humble judgement nothing short of the extreme uncompromising rigour of registry and observation as pertained in the Great Trigonometrical Survey could be tolerated for an instant. 'I have always avowed that experiments either incautiously made or negligently recorded, are far worse than no experiments at all.'

By page 51 Everest was launching into comments on the pendulum and pointing out that the reference made to Lambton and to an uncompensated invariable pendulum was incorrect. In the middle of this discourse on the pendulum Everest indicated that he was particularly hurt that his former tutor, Sir James Ivory, should have been among the 38. [107 p. 53] After discussing pendulum experiments at length he questioned whether the 38 learned Fellows had examined the details of the experiments with due caution, and went on to talk of those 'gentlemen who sit quietly in their easy chairs, before their comfortable firesides in the winter, or may gaze at will on the verdant lovely meadows, golden corn fields, and majestic ocean which surrounds the shores of adorable England ... what can it signify to them, or why are they to take the trouble to ascertain what goes on in India?'

Again he sniped at the 38 gentlemen by indicating that during 1830 he had invited anyone that was interested to inspect the new instruments that he was about to take to India. For 20 days at great cost they were on display yet how many of the 38 took up the offer? He came back to pendulums and referred to his paper in the Memoirs of the Astronomical Society [99] that now seemingly was cast aside as trash and rubbish, yet originally met with a cordial reception at the Astronomical Society – and so cannot be so undeserving. His theories there had prompted Edward Troughton to produce a new form of pendulum.

> It is not the Court but I who am to blame, for the omission of pendulum experiments, for the implements are actually at my disposal as far as material is concerned ... and all that... has been wanted is the personnel to put them into action.

> Better it is to do one thing effectually than many things imperfectly. Pendulum experiments cannot be superintended by me because I have too many calls on my time. [107 pp. 49–62]

Everest next launched into the areas of meteorology, botany and orthography. Here he asked whether the 38 gentlemen imagined that all the natives of India either spoke or wrote English [107 p. 65], and proceeded to use several pages to explain the situation as it was. He also mentioned that he had invented and constructed a barometer pump, whereby barometer tubes may be filled without the slightest apprehension of breakage. It had been used successfully on several occasions.

He admitted to not understanding the reference in the Address to 'Not that it is contemplated that it would be necessary to begin the work *de novo,* but merely to examine it in order to render it more complete, and to combine it effectively with future labours'. He asked what would be done with any errors found and had any consideration been given to the cost of such an exercise let alone its tedious and harassing nature.

In the context of the suggestion for detailed survey from India to the Caspian Sea he referred to the idea as like trying to survey the Moon or Jupiter. He said that 'all things which are desirable are not attainable and that there never was a Government which less needed to be stimulated than that of the E I Company, to undertake rude initial surveys of the kind herein adverted to'. He then detailed why it would be ridiculous to try to mount an expedition to such areas for the purposes of geography or surveying.

He came to the defence of the Court of Directors thus:

> even if they had not to boast of having been the principal champions of St George in the field of Geodesy, in connection with the figure of the earth, for the last forty years; during which long period they have never been greeted with one sympathising cheer of approbation from those bystanders who were most bound to encourage them in their arduous contest. …even so late as 1819, in my first essays, I was occasionally necessitated to carry on my operations in the territories of the Nizam (of Hyderabad) at the point of a lance… [107 pp. 74–82]

He then goes on to consider his achievements since the 1824–5 season.

Jervis' field work

Towards the end of this catalogue of events he said:

> Major Jervis's Survey of South Konkan – the coastal area to the south of Bombay –, for example, is in my office, and at this instant lying on my desk, and it does not contain much to boast of… Did the thirty-eight learned Fellows themselves examine into Major Jervis's proficiency as a Surveyor before they publicly proclaimed that he was entitled to their confidence and to that of the Court of Directors? [107 p. 84]

In 1841 the Directors responded thus to Everest's condemnation of Jervis' work 'We regret … that Lieutenant Colonel Everest has … found the records … in unsatisfactory state…' [231 p. 309]

In Letter VIII Everest turns to the assertion by Jervis that it was he who had discovered the benefits of the Irish Survey and that credit for discovering the advantages of the Irish system should be his alone. In fact, Everest pointed out, it was he (Everest) who had submitted a document to the Court in 1829. Why had not Jervis made an elementary check on prior investigations before laying such unfounded claims? [107 p. 87]

Jervis had made claims as to the manpower and time required to complete the survey of Ireland. Working on a ratio of the areas of Ireland to India of 1 to 35.5162 Everest calculated that the round period of 72 days was that within which it was clearly Major Jervis' opinion that the survey of Ireland ought to have been completed. [107 p. 89]

In drawing comparisons he indicated that whereas the Irish survey was within but a few hours of the most civilised nation, the Indian Survey was carried out in a land little short of barbarous. Then in Ireland all instruments were carried in spring carts while in India they were conveyed on shoulders and heads. [107 pp. 102–3]

In relation to the collection of statistics Everest emphasised the necessity of learning the languages of India, such as they are spoken by the cultivators. 'The dog language which is acquired at Addiscombe Seminary... is of as small use ... as the best French would be in Berlin or Naples'. [107 p. 95]

He indicated that the Great Trigonometrical Survey was one of the hardest modes of life in any part of the world, and the worst relatively paid. Why did Major Jervis not step forward when Lambton died in 1823 and offer himself as willing to enter the arena, and grapple with a task which had well nigh put an end to Everest? [107 p. 114] In 1828 and 1833 there were other opportunities for Jervis to show his ability yet he seems not to have availed himself of them.

> In fact ... before this pamphlet was ushered into birth, Major Jervis's name was unknown on this side of India except in my office, and then only in connection with one of the many minor surveys in progress, under the general superintendence of the Surveyor General of India, amongst which it ranked but as a second or third rate performance... [107 pp. 117–18]

His last letter, No. X is mostly devoted to Sir John Herschel. The learned Baronet had made various recommendations regarding standards of measure and the comparison with those in India and that the current one should be sent to London for verification. He then referred to thermometers. In all of these statements it is implied that these things had not been done. In rebuttal Everest referred Herschel to pp. 50–52 of his 1830 work. [107 pp. 131–2] He further referred to pp. 121–4 of his book regarding the chain and calibration against a brass scale. Everest's remedy had been to remeasure the Sironj base with compensating bars.

> If Sir John Herschel had been seriously intent on advancing science in India ... it is sadly to be lamented that he did not, as was at one time expected, pay a visit to the country of which he seems desirous to be scientific dictator... [107 p. 145]

Were the letters an embarrassment?

Knowledge of the letters to the Duke had certainly circulated rapidly because, by September 1839, George Airy, the Astronomer Royal, was beginning to feel that the Jervis affair was an embarrassment. On 20 September 1839 G.B. Airy wrote to J. Melvill of the East India Company [3], that he had recently received a copy of the letters written by Everest to the Duke of Sussex.

> commenting severely on the conduct of myself (by name) and of other persons, in the presentation of a memorial to the Court of Directors of the Honourable East India Company respecting the course which it might be desirable for Major Jervis to pursue, in the management of the Great Survey of India.

> As to the nature of Lieut Col Everest's remarks induce me to suppose that I may have acted on an entire misapprehension of the rank and extent of authority with which Major Jervis went to India, and that apology may be due from me not only to Lieut Col Everest but also to the Court of Directors with the entire management of the Indian Survey, or whether that officer went out merely as locum tenens for Lieut Col Everest during his limited absence from the Survey.

On 26 Sept 1839 Melvill replied to Airy [217] 'In reply I have the commands of the Court to acquaint you that the appointment confirmed by them upon Major Jervis was that of successor to the office of Surveyor General of India *upon the death or resignation of Lieut Col Everest.*'

The next day Airy replied again to Melvill. [3]

> I beg further to state that – as appears from the information communicated by the Court of Directors, the impression under which I (with others) addressed to the Court of Directors the memorial to which Lieut Col Everest's pamphlet alludes, were correct, with this exception only, that I considered the appointment to be positive and not contingent. And if any use has been made of the memorial as containing positive and unqualified recommendations by persons who are aware of the contingent nature of the appointment, I beg to state that such use is without my consent and entirely opposed to my wish. As it would seem from Lieut Col Everest's pamphlet that the memorial in question has been published, I beg to state my opinion that no persons except the memorialists and the Court of Directors is justified in publishing it, and that I entirely object to its publication except with the consent of the Court of Directors and that it was not supposed by me that it was intended for publication.

> Should no objection be entertained by the Court of Directors, I propose within three or four days to communicate the information conveyed in your letter of the 26th inst as to the appointment of Major Jervis, to the other persons named in Lieut Col Everest's pamphlet: viz. His Royal Highness the Duke of Sussex, Sir John Herschel, Professor Whewell and Mr F. Baily. If any objections be entertained by the Court I shall consider myself bound to consult their wishes, or receiving your notification thereof.

As the Court did not raise any objections, by 3 October 1839 Airy wrote to those he named and among his comments were: [4]

Here is a charming appointment for you! Given a …fellow like Everest on the one side and … one like Jervis on the other, I do not think that a better appointment could have been obtained for the purpose of setting them together … and dragging some innocent persons into the quarrel.

Airy, in a letter to Herschel on 3 Oct 1839, commented:

You have probably seen a pamphlet printed by Lt Col Everest (sold by Pickering) in the form of a series of letters addressed to the Duke of Sussex, in which Whewell, Baily and I are slightly abused and you come in for a double … (heaven knows why) I really began to be a little uneasy – not for fear of the said Everest's pamphlet, but because I thought that Major Jervis must have misrepresented his powers and must have thereby led us to do Everest an injustice. So to ascertain the nature of Jervis's appointment, I wrote a formal letter to the Secretary of the East India Company, and received the following answer which I am authorised to communicate to you.

The matter of Jervis's appointment is rather odd, and with a little indiscretion on his part may lead other persons into any degree of confusion; I suspect that he had been so imprudent as to set upon a commission (which is only contingent though in reality it may be absolute) just as if it was ostensibly absolute. I suppose that there is some wonderful secret history connected with this appointment. [2]

Confusion there certainly was, however, because Herschel, in a letter of unknown date and to an unknown person [165], said:

Jervis's appointment was as I understand, a thing *done* and that he (J) was going out to succeed Everest. It now appears that Jervis's appointment is strictly *reversionary* i.e. on Everest's death or resignation. Well! be it so, then let us hope he will live long and enjoy such health as will enable him to retain his office – I see no harm done.

Everest asked Jervis to visit him

On his return to India the enthusiastic Jervis wrote to Everest (13 December 1839) and Everest replied (30 December 1839) from Kaliana. He started by saying:

My dear Sir, I was duly favoured with your letter of the 13th inst. and its enclosures … From this you will be able to judge of my plans, and from your own accordingly. I am well mindful of what Solomon says "Hope deferred maketh the heart sick"…I fear it will disappoint you, but nothing whatever will move me from my post until the whole arc to the north of Beder is completed so as to be invulnerable to all imputations of inaccuracy… and can … sympathise with the annoyance it must cause you to find me so pertinaceously continuing to occupy the station which you were led to suppose was vacant … but my mind is made up to complete before I go all the work of the great arc of which I am anyhow implicated…

He then continued by asking Jervis to seek permission from Major Sir James Rivett-Carnac, Governor of Bombay, to travel to visit him on business, promised him a hospitable reception and a wish to talk over scientific acquaintances. 'It is in truth high time that you and I were better acquainted and knew each other thoroughly. I feel assured that the more that takes place the better friends we shall be..." [231 p. 438]

Later we find Everest applying the finishing touch writing to Jervis from Dehra Dun on 25 August 1841.

> I had made up my mind to retire from the service at the end of this year, and to let my career close with the measurement of the baseline in the valley of the Manjera, under the full persuasion that the court of Directors would allow my pension as full Colonel. Last year I drew up a memorial to them... I went on furlough sick... having been employed on duty in England eighteen months of that period so that by the end of 1841 I should have served in India thirty and a half years. I prayed them therefore, in consideration of my arduous career, to allow the said eighteen months to reckon as service to the east of the Cape, whereby I should be able to retire on the pension of a full Colonel. This memorial has not been granted ... and therefore I must wait till my claim is realized by running the whole period. [189 p. 217]

Jervis requested retirement for himself from 30 December 1841.

Everest's son, Lancelot, in his memorandum relating to his father, concluded comment on the Jervis affair thus:

Though my father's appeal to the Duke was successful it nevertheless left a painful impression which is not easily effaced. He had been placed in an awkward situation owing to no fault of his own, and it was only his stern determination and plucky spirit which carried him through, and enabled him, notwithstanding it and his illness, to remain in India, and continue his work there until its completion and his retirement a few years later. These combined with what must have been an abnormally strong constitution, gained him the victory against heavy odds. [130 pp. 28–9]

Chapter 10

INDIA 1839–1843: Final measurements, and the handover of the Survey to Waugh

Reciprocal astronomical observations

The two seasons following 1838 were tidying up operations where Everest was particularly involved with astronomical observations at the stations of Kaliana and Kalianpur. Although the stations were separated by some 5° 23' in latitude they were almost on the same longitude as each other. In fact Kalianpur is at 77° 39' 18" whilst Kaliana is at 77° 39' 06". Thus, the times of meridian transit of a given star at each station were almost identical. Everest's intention was to endeavour to achieve simultaneous observations at these two stations with Waugh at Kalianpur and himself at Kaliana.

This was not the first time such an operation had been tried. A hundred years previously Bouguer and De la Condamine had used it successfully on the Peruvian arc. There the conditions were, if anything, much worse than in India because the route between the stations lay along the top of the Andes.

Contact between the two would not be easy since they were 38 days' march apart, so careful planning was necessary. Operations were to commence with Waugh leaving on 9 October 1839 for Kalianpur. He would arrive about 24 November to be followed a little later by Said Mohsin who would give Waugh assistance with mounting the astronomical circle and then return to Everest at Kaliana. In fact, it took Waugh, his assistant and sub-assistants nearer six weeks to complete the journey. Everest and Renny were at Kaliana by 12 November and made preliminary observations to eight stars, as did Waugh. So important was the role of Said Mohsin that he was allowed to travel between them by dak (relays of riding ponies) which was one of the most expensive modes of transport. [231 p. 170]

The requirement for the instrument mechanic to be at both stations was because of the effect that the long rough journeys could have on such delicate instruments. Particularly on the 400 mile journey to Kalianpur it would be possible for almost any part of the instrument to work loose. Everest's worry was that something mechanical could go wrong with the instrument, delay follow at that station through any inability to remedy the problem, and the whole operation get out of synchronisation because of the difficulty in communication between observers.

The main observations were planned for the period 4 December 1839 to 4 February 1840. [263 (344) 267–72] For this they chose 36 stars from the catalogues of Piazzi and of the Royal Astronomical Society, and each of these was observed on 48 consecutive nights at each station. For almost 85% of these observations the weather was good and the sightings simultaneous. Then the conditions deteriorated at Kaliana. This resulted in Waugh completing his work on 22 January whilst Everest had to continue until 3 February. [263 (402) 247]

After completion of the observations at Kaliana and Kalianpur the two teams were each to move south. Everest was to go from Kaliana to Kalianpur while Waugh was to go from Kalianpur to Bidar. Waugh did this straightaway so that he could prepare the new observatory at Damargidda near the Bidar base to the north west of Hyderabad. Everest waited until after the rains to set out from the Park and, again with Renny, arrived at Kalianpur on 20 November 1840. Plans were for the simultaneous observations to begin on 24 November. Having yet another bout of fever on this appointed day, Everest had to let Renny do the initial observations. As luck would have it, Captain Shortrede turned up at Kalianpur and was able to share the observing with Renny until about 8 December when Everest was again able to take over.

This time 32 stars were selected and operations were completed on 11 January 1841 having been simultaneous throughout. Everest considered that his observations at Kalianpur were better than those he took the previous year at Kaliana but those by Lieutenant Waugh at Damargidda presented some initial difficulties. This turned out to be due to the effect of humid air on the cross-hairs. Once this was corrected by ensuring the access of dry air and not damp air at noon, the problem disappeared.

The method of observing and star selection was such that there was an interval of some 4 minutes between successive stars. For checking the collimation, Everest observed on to a transit instrument set up in front of the observatory with its object glass towards the astronomical circle. He modified the transit by inserting a piece of mother-of-pearl in which was drilled a fine hole. Over this he stretched two spider webs at 60° to one another. He arranged for this to be illuminated either by lamplight or daylight. This he observed three times on each face both before and after each night's operations.

The circle had oil lamps to aid the reading. Each instrument had two, each of three inches diameter. They had five nozzles that threw light into the telescope via its axis. Two nozzles illuminated the reading microscope and two the ends of the bubble. [263 (402) 298–368] In the intervals between stars the micrometers and ends of the bubble were read and the instrument reversed in azimuth. The stars were alternately north and south with face changed between each. The resulting amplitude for Kalianpur to Damargidda agreed well with that of 1825 but Everest considered this to be purely fortuitous.

compensation of errors so frequently occur that some men … look on precision of principle as less worthy of regard than it merits. That is a fatal mistake and, where errors combine instead of compensating, we learn… the true value of prudence, and a rigorous attention to accuracy in principle as well as practice. [231 p. 101]

Everest was able to leave Kalianpur on 15 January 1841. His route back to Dehra took him to Bharatpur, Alwar, Gurgaon, Delhi and Kaliana so that he could map positions of the hills on which the stations of Chapra, Par and Meoli were located (these lie between Agra and Delhi). [231 p. 43]

Remeasurement at Bidar

1842 and 1843 were fully occupied with computations, reports, charts and correspondence. The final piece in the fieldwork jigsaw was the remeasurement of the Bidar base which Waugh completed in late 1841 after being able to do some of the preliminary construction in April 1840. Waugh was unable to find any trace of Lambton's Bidar base of 1815 so was forced to select a completely new site some 8 miles south west of the old one. As with the astronomical observations, Said Mohsin was again in attendance in case of problems with the instruments and was also employed at the microscopes when Jacob and Olliver were ill. [231, p. 55]. It was 11 October 1841 when Waugh began 57 comparisons of the compensating bars and standard bar. Then on 19 October the measurement proper began. This was completed on 4 December.

The measured length of this base was 41 578.536 ft reduced for an elevation of 2030 feet above the sea. This was only 4.296 inches different from that derived through the triangles from Sironj – a distance of 425 miles through 85 triangles.

Further comparisons were made on 9 December, and by 11 January 1842 the party was on its way back to Dehra where it arrived on 1 April. The main items of equipment were deposited in a safe, dry room in Agra under the supervision of the Ordnance. [231, p. 56].

At long last we now find Everest admitting that he had staff who were capable and willing to perform as well as himself.

It would have been supererogatory for me to proceed … to the spot to superintend a work which either of these gentlemen [Waugh and Renny] was qualified and willing to perform fully as well as I could, and my absence at a distance of upwards of 1000 miles from my office might have caused a great delay to public business… [263 (402) 298–368]

Minor triangulation series

The minor series were also proceeding nicely.

1. The Budhon series

This was resumed in November 1839 under Renny. During the next year the chain was redesigned and taken as far as Moradabad. After filling the gap from the Chambal, Renny and his party recessed at Dehra from 4 July 1841. All the following field season was spent laying out triangles northwards to 30° N and a junction with the Great Arc north of Meerut. Then on to the North Connecting series.

Many of these stations required towers, but all was ready for observation during 1842–3, when it was brought to completion. Everest visited them in the field before their departure for Dehra in mid May 1843.

2. The Ranghir series

During 1841 the series progressed well to the north of Almora and joined the North connecting series of which it became part. Severe snow many feet deep stopped all observations at the mountain stations. During the 1842–3 season Waugh connected the Ranghir and Amua series.

3. The Karara series

This restarted on 1 September 1841 with Shortrede in control. His results were uninspiring and progress very slow. He had left Dehra on 11 October. After delays, he reached Allahabad on 17 December and Rewah at the end of January. Many weeks were spent at his base station because of difficulties with visibility and fire.

By April Shortrede gave up his reconnaissance and returned to Allahabad by the 27th. During 1842–3 better progress was achieved with additional staff and some observations made.

On 27 October 1843 Everest visited the party on his final journey. He was surprised to see them still in cantonments, but Shortrede was unabashed. The series took until May 1845 to complete.

4. North connecting series

During August 1839 Everest expressed the need for a series running along the foot of the mountains. By 1840 he had obtained permission for a series to connect the Great Arc, Budhon, Ranghir and Amua series. One major problem was trying to avoid infringing Nepalese territory. The work was put under James Du Vernet, and he started in September 1841. In total some 300 secondary triangles resulted and at the same time he located many Himalayan peaks.

Season 1842–3 saw Waugh connect the northern ends of the Ranghir and Amua series, requiring eleven towers to do so. During January and February 1843 Everest visited the parties on this series and that by Du Vernet, called the Sub Himalaya Range.

The next season Du Vernet ran a series to connect the northern ends of the Amua and Karara series and continued on until he met Shortrede. These connecting sections were to be later up-dated. [231 pp. 18–20, 71].

5. Bombay longitudinal series

After a season, 1839–40, when no field work was done, Jacob rejoined this series on 13 May 1840 and took over from Shortrede. On 19 October 1840 the party left Poona working eastwards and also laying out a baseline site to the west of Poona.

When the weather permitted in October 1841, work commenced on the Bidar base. The series was completed by 1843–4. Various other small pieces were also observed in this area.

Calculations of the probable errors of observed angles give values for the Great Arc ranging from 0.25 to 0.44 seconds and for the minor series values from 0.79 to 2.23 seconds.

The human computer

Other aspects of the trials and tribulations that beset Everest when he was Surveyor General are worth mention.

Obviously it is one thing to take numerous field observations but another to compute those values into the required results. Lambton had done most of his own computations during the off season and initially Everest did likewise.

In 1818 Lambton had described his situation thus:

[he] always put the computers (having formerly four) two and two together, dividing the work so that two might go over the same ground as a check to each other, and when they had finished, they changed their parts and went through the whole again, so that each two might be a check upon the other two. All this being done, the whole was revised by myself. [263 (144) 195]

In addition to the computation of all the triangles, co-ordinates and heights, Lambton was also engrossed in aspects of the figure of the earth and abstruse aspects of phenomena that affected observations. In 1818 he had produced his own set of parameters for the figure of the earth.

When Everest took over as Surveyor General, he had such a backlog of computations to be done that he set out to form a special computing section. By the end of 1831 he had appointed eight Bengali students as computers at an initial wage of Rs 40 per month. (This might be compared with that of an assistant surveyor of around Rs 450 per month or the Surveyor General, with various allowances, of Rs 3250 per month.) Within a very short while Radhanath Sickdhar (Sukdhara), who had joined on 19 December 1831, had demonstrated that his mathematical ability was of a superior quality and he was appointed as a sub-assistant to the GTS.

With the formation of this computing group under de Penning an able section soon materialised. Not all met with the approval of Everest, however, and he was particularly upset at the mode of attire worn by the staff during the rainy season.

> It is not considered amongst the English consistent with decorum to enter an apartment destined either to business or domestic affairs without such ... dress as is simply sufficient to cover the nakedness of the person. In points essential to their religion or the ... habits of their country, I shall carefully abstain from interfering, ... but in such as are not at variance with these, I have a right to expect that they will in their turn comply with my wishes.

> It is not at variance with these notions to wear a pair of clean stockings, and clean well-blacked shoes, neither does it at all war with any of their habits, ...that I am aware of, to wear trousers or paijammahs (instead of the dhoti)... [263 (266) 192–3]

So strong were his feelings that he even suggested that those who would not comply should be provided with a special room to work in.

Even with the number of computers now in post, the occasional error slipped through and, when noticed, was pounced upon. Nothing but the most rigorous check of every item was good enough. By 1835, there was some agitation amongst them for a rise in pay, and this was supported by de Penning who was well aware of their worth now they were trained and competent. In particular, he singled out Nil Comul Ghose as worthy of special treatment.

Whilst Everest took all this on board and was able to get an increase for Ghose up to Rs. 100, the others were less fortunate and had to be satisfied with their existing Rs. 40. The outcome was as might be expected. By 1838 all except one had left for positions in the revenue department where the remuneration was some eight times better. The process of training had to commence all over again.

Even the star man, Radhanath Sickdhar, was now agitating for better terms. This galvanised Everest into extolling all the attributes of Radhanath to the Government and in the end squeezing an increase of Rs.100 out of the system. [231 pp. 341–2]

De Penning had problems of his own from 1833 when he was not only overseeing the computing office but had other multifarious tasks put his way when the office of the Deputy Surveyor General was moved from Calcutta to Allahabad. So noticeable was the effect on the efficiency of the computing output that Everest had to request that his Chief Computer should confine his activities to those for which he was originally appointed. [231 p. 108]

New formulae and forms

New formulae were derived by Everest, and fresh computing forms and procedures were required as he refined his approach to all aspects of the computations. In particular he was troubled by the procedures for dispersing the errors around the triangles. He devised new equations

> for the purpose of accomplishing the same harmony amongst the sides, which has

hitherto only been for amongst the angles… they [the formulae he devised] furnish a determinate mathematical rule for disposing of small errors which must be got rid of, … and thus take away all excuse for that most objectionable presence of charlatanry called… ' judicious selection'… [263 (324) 176–84]

It would appear that his technique was akin to the method of equal shifts.

In line with his determination not to condone errors in computing, Everest was just as tough on his observers. After commenting on the general untrustworthy nature of some particular field books he said:

I may be told whilst thus criticising the performance of others to look at home, and that my operations … are just as liable to be faulty… but as far as human care and caution could prevail, I have guarded against error…none of the objections…apply to me, …for I have in no case arbitrarily rejected an observation, but have always taken the general arithmetical mean without selection or exception…[263 (171) 359–96]

When any set of observations is computed, the arithmetical mean … must be taken forthwith, and the probability of error computed. When the three sets of observations of any triangle are all computed, the spherical excess should be computed, and the error partitioned agreeably to the probabilities… [263 (267) 59–61]

Thus it would appear that he took on board some of the developments of Gauss and others but there is no indication that he generally applied the rigorous theory of least squares to the observations. By the time he retired, Everest again had eight computers at full stretch trying to keep the work up-to-date.

Wish to retire

Everest gave early notice that he wished to retire as soon as the Great Arc was complete. Field work ended with the remeasurement of the Bidar baseline in December 1841 – a date coinciding with the resignation of Jervis. He had by this time accepted that Everest was, by hook or by crook, going to survive until the Arc was complete, so he decided it was preferable for him to leave India rather than wait to succeed to the hierarchy.

It was Everest's opinion that the two offices of Superintendent of the GTS and Surveyor General should remain combined in the manner in which he had held them. With this, the Court of Directors agreed. [231 p. 318]

In November 1842 Everest formally submitted his resignation and recommended that Andrew Waugh should be his successor. He wrote directly to both the Chairman and Deputy Chairman of the Court of Directors with regard to the attributes of Lieutenant Waugh.

He stressed that while, in 1825, he had difficulty in finding anyone to whom he could entrust the post of deputy to him, he now had several very able persons who had trained under him. But of all his military officers he could only class Waugh and Renny as completely successful and had fulsome praise for both. [263 (344) 108–84]

When he was unable, through illness or pressure of other work, to supervise particular projects it was to Waugh that he had turned most often.

> His talents, acquirements, and habits, as a mathematician, a scholar, a gentleman, and a soldier, are of a high order, and as such I feel that in recommending him as a fit person to succeed me, I do but perform the last essential service … to … my masters from whom I have received so many acts of kindness. [263 (438) 122]

> … there are but few officers in India who have ever had any experience in Trigonometrical operations, and … those who have only been accustomed to the ruder modes of surveying would have much to unlearn before they could begin to learn aught … to select a person whose sum total of practical skill and theoretical attainment, powers of endurance, and all other essential qualities were a maximum, Lieut. Waugh would be the very person of your choice. [263 (402) 399–406]

His glowing reference had the desired effect and in a letter of 3 May 1843, whilst recording their appreciation to Everest for all his work, the Court appointed Waugh to succeed him.

Henry Lawrence commented:

> Measuring an arc of the meridian is an achievement … which people in general cannot be expected to appreciate, aware as they are only of the vast expense, and seeing no tangible results in the shape of maps. In this stupendous work the Surveyor General has surpassed the European astronomers, and the result is of vast moment to abstract science; but unless his arc is used as the backbone of a web of triangles to be thrown across the continent of India, it is of little practical value.

> Independent as he seems of all local authority, and unshackled as to his expenses, had he been as anxious to supply a general and accurate map of the country as to astonish the savans of Europe with a measurement exceeding all others, as much in accuracy as length, he might have combined … the Revenue and Trigonometrical operations, and furnished a map of India as correct as there is of any part of the world.

> The Superintendent of the Survey is undoubtedly an able man, as well as a first rate mathematician; but, forgetting that real talent shows most in simplification, in applying the depths of science to life's ordinary purposes, he undervalues everything that is not abstruse. [83 pp. 119–20]

Lawrence was with the Survey of India, from 1833 to 1838, and was particularly involved in the revenue survey before leaving for a political career. [231 p. 454] He was to be killed in action at the defence of the Lucknow Residency during the Indian Mutiny of 1857.

Appreciation

The appreciation of Everest's service by the East India Company came in a letter of 3 May 1843 from the Honourable the Court of Directors to the Government of Bengal.

Part 1. The announcement of the intended return to Europe of Lieut. Colonel Everest, and his consequent vacation of the office of Surveyor General and Superintendent of the Great Trigonometrical Survey of India, affords us an opportunity of which we readily avail ourselves, of expressing the high sense we entertain of the scientific acquirements of that officer, and of the ability and zeal which he has displayed in the discharge of the arduous duties entrusted to him.

2. With the measurement of an arc of the meridian of unprecedented magnitude, now finally completed, and with the Great Trigonometrical Survey, the most important portion of which is now in rapid progress towards completion, Lieut Colonel Everest has been prominently connected almost from their commencement, and for the last twenty years they have been under his exclusive management and superintendence.

3. The able manner in which he has conducted these important and scientific works has frequently elicited our approbation, and cannot fail to cause his name to be conspicuously and intimately associated with the progress of scientific enquiry. [261 pp. 253–4]

The wording of paragraph two of the appreciation gives the impression that the Directors and perhaps others saw the Great Arc and the GTS as two separate operations. Whilst it was only the Great Arc that was of importance to the determination of earth parameters it was also an integral part of the national framework upon which all mapping depended, so it ought to be treated as an entity.

Radhanath Sickdhar

Elsewhere mention is made of the considerable assistance given to Everest by Mohsin Hussain as an instrument maker. The ability of a second member of the locally recruited staff also attracted Everest's attention. This was Radhanath Sickdhar. His forte was in computing.

He was appointed as a computer in 1831 and the following year became a sub-assistant on the GTS. During late 1832 he accompanied Waugh and Renny when they left Calcutta to join the GTS. He joined Everest in Mussoorie and went into the field with him in October 1833. He proved a very successful operator of the blue lights that were used across the Jumna valley.

Everest spoke highly of him and of his mathematical ability. He became indispensable with regard to all aspects of computations and, when he was at one point offered a post elsewhere, Everest was alarmed and said it would be like losing his right arm. [263 (348) 207–8] A rise was sought in order to keep him and Everest was successful in this.

His relatives had a copy of Everest's 1847 work [116], the title page of which was inscribed to Babu Radhanath by the Court of Directors of the East India Company.

When Everest retired in 1843 Radhanath moved from Dehra Dun back to Calcutta where he became just as trusted by Waugh as he had been by Everest. He prepared the first edition of the Auxiliary Tables of the Great Trigonometrical Survey published in 1851 and was entrusted with the comprehensive formulae and method of geodetic computation devised by Everest. [231 pp. 110, 462] However, whilst there are some who also credited him with the computation that determined the height of Peak XV (Mount Everest) as 29 002 feet, there is only contrary evidence available. It would appear that at the time of that computation Radhanath was in Calcutta while the computation work on the Himalayan observations was taking place in Dehra Dun. [231 p. 462], [232]

Radhanath had left Dehra Dun in January 1849 to supervise the major computations and report writing in Calcutta. In the Report he submitted in 1856 there was no reference whatever to the North East longitudinal series from which the observations were made to the snow peaks. This was dealt with in Dehra Dun by John Hennessey. Waugh acknowledged this fact when referring to the snow peak computations.

In 1849–50 Sickdhar assisted Colonel H.L. Thuillier and Lieutenant Colonel R. Smyth in the preparation of their *Manual of Surveying* for India although this is not mentioned in the volume.

In relation to Mount Everest Radhanath was said to have expressed his 'gratification at the highest snowy peak known in the world being named after Colonel Everest, our late master. At the same time… it would have been more natural if the local name was adopted'. [272] In 1853 Radhanath became a member of the Royal Astronomical Society. He retired in 1862. In 1865, when writing to Colonel Walker, Everest commented 'and chief of all my pupil and friend Radhanath Sickdhar… I look back on them with sincere affection and gratitude for the facility with which they served under me'. [123]

In September 1843, just two months before leaving the sub-continent, Everest wrote to George Airy from Dehra Dun indicating that he was sending him some documents relating to the GTS that might interest him. He concluded:

> My task being now ended I propose being in England next spring and shall be most happy to afford you any additional information which you may desire for I entertain a sincere hope that the interest which you were formerly so renowned for taking in Geodetic operations is still as great as it used to be. [111]

It is quite likely that it was this information that Airy incorporated almost as a stop press item in his celebrated work of 1845 on the figure of the earth [5 p. 213]

It could be that Everest was so involved in the Great Arc that topographic work took very much second place in the Survey of India for a period of some 20 years. On his retirement however, with large parts of the network complete, it was opportune for Thuillier to bring the mapping element back into prominence. This would have been far more difficult to achieve but for the excellent foundation that Everest had insisted on laying during his term of office.

Everest retired on the pension of a full Colonel and his resignation was gazetted on l December 1843. [263 (453) 99]

Chapter 11

ENGLAND 1843–1866: Retirement, a late marriage, and recognition

Seeking the award of a CB

On 14 March 1845 Everest took up again his quest for the award of a Companion of the Order of the Bath (CB). It was in 1838 that he had first petitioned The Honourable The Court of Directors of the East India Company that it might consider his position in the light of the many others who had been so rewarded. He now stressed that he had brought to a successful conclusion, before he retired, the monumental task of the Great Arc of India. At the same time he had trained an efficient establishment and designed permanent principles for a system that hence only required to be acted upon. His object was to shew that by retiring when he did he acted from higher motives than mere personal or selfish considerations; as a step likely to promote the interests of the Company. Nevertheless, so that the world at large might be aware of his achievements he would like to be gratified by such an honorary distinction. [115] He was unsuccessful again.

Health concerns

During his retirement Everest was as active as he could be in the various societies to which he belonged and the Athenaeum, Oriental and U.S. Clubs of which he was a member. As a Freemason he belonged to the Prince of Wales Chapter (a fact he referred to in his letters to the Duke of Sussex and used it as a reason why the Duke should support his arguments as a brother Mason). But all the while he was dogged with ill health. From July to November 1845 he is said to have been in the United States but there is no trace of his movements there nor of his contacts. His comment was that it was 'for the benefit of my health'. At regular intervals we find him spending two weeks at Bath, Tunbridge Wells or Brighton as a means of combatting his rheumatism and gout. Such was the severity of these that he commented later (in 1860) that 'I am laid on my beam ends with a very severe attack of gout and no hope of attending the meeting next Monday.' Then he referred to being 'crippled with rheumatism'.

He found the waters at Bath to be particularly helpful in shaking off 'a most tormenting internal pain'. To have just a few days without discomfort was apparently a major accomplishment. At one stage (June 1860) he took a house in Tunbridge Wells for two months in an effort to get relief from his pains.

Marriage and children

In November 1846 Everest decided to get married. By then he was 56 years old but his wife was 33 years his junior at 23. In fact George Everest was six years older than his father-in-law.

At the time of his marriage or shortly afterwards he took a long lease on 10 Westbourne Street near Hyde Park. This remained his London home until his death, and Lady Everest is recorded as residing there until at least 1875. Their six children (Emma, Winifrid, Lancelot, Ethel, Alfred and Benigna) were born and brought up in that area. With his young family around him, holidays were generally taken in Tunbridge Wells or Brighton. At Brighton the children were given riding lessons and Sir George often accompanied them and the riding master on to the Downs. Of his social activities little is known but there was an occasion on 5 February 1858 when he and Lady Everest entertained Dr and Mrs Livingstone to dinner, and amongst others invited was Dr Shaw, the secretary to the Royal Geographical Society.

Everest was a firm believer in God and attended regularly the Sunday morning services at St James' Church, Paddington. This was in the time of the Revd Dr Boyd who later became Dean of Exeter. The children were brought up to family prayers before breakfast, at which the servants attended, in good Victorian fashion.

Sir George was also a great believer in the principle of the Golden Mean – the quality of being equally removed from two opposite extremes – and made the children learn by heart an Ode from the second book of Horace, that began *Rectius vives, Licini, neque altum semper urgendo…* ('The proper course in life, Licinius, is neither always to dare the deep, nor, timidly chary of storms, to hug the dangerous shore…'). The sentiments expressed through the whole Ode very much reflect George Everest's lifestyle. His son Lancelot commented that whilst at the time he did not understand a word of it, the whole had remained with him ever since. [130 p. 27]

Living in Leicestershire

At this time Everest and his wife were residing at Claybrook Hall near Lutterworth in Leicestershire. Little Claybrook (Claybrooke Parva) was a small village of only some ten houses and the mansion of Thomas Edward Dicey. The sister village of Great Claybrook (Claybrooke Magna) was somewhat larger. The Hall was built in 1714 although a previous building on the same site was burnt down in 1641. George Everest is listed as a resident of the Hall in 1846 but the connection between him and the Dicey family is not known. It would appear that the Everests did not own the property, so they were presumably tenants of Mr Dicey.

Leicestershire and the neighbouring area features in the Everest story in various ways since, in addition to his residence at Claybrook, the family roots of his close friend Major Feilding were in the Lutterworth area. His wife's family came from the region around Oakham in the then county of Rutland, and Lincolnshire.

Whilst in Leicestershire, Everest was an enthusiastic rider with the hounds:there were several Hunts to choose from. Everest also compiled most of his second historic book, which was published in 1847 in two volumes: *An Account of the Measurement of two Sections of the Meridional Arc of India, bounded by the Parallels of 18° 3' 15"; 24° 7' 11"; and 29° 30' 48"*. [116] With his previous work of 1830 [96] a superb insight is given into the operation of the survey work as well as providing considerable detail of the observations, computations and results.

At the time of his death this volume was described as

> such particulars are set forth as will enable a scientific observer to test the manner of working and the results obtained, and full explanations are given of the ingenious methods devised by the author for the elimination of error, together with tables, plans, and engravings of the instruments employed.

From time to time his correspondence with Dr Shaw of the Royal Geographical Society made comment on the staff at the Survey of India under Captain Waugh. Dr Shaw apparently had the habit of referring to them as Everest's 'children and grandchildren in India' which obviously went down well with the great man as he used it also when writing to Dr Shaw. [119] He continued to correspond with his successors in India and appears to have obtained cloth for his suits from there. He preferred Kashmir woollen cloth or Puttoo to anything of English manufacture. [231 p. 441]

Awards and recognitions

Award by the Royal Astronomical Society

It was for this work, and the long series of operations that went into it, that in 1848 Everest was awarded a Testimonial by the Royal Astronomical Society. A glowing citation was read by Sir J.F.W. Herschel at the time of the award although the actual text was compiled by Mr Galloway. To select but a few of the comments:

> I have next to call the attention ... to a very important and remarkable work...

> In the execution of these operations (the GTS), a series of hardships, privation, and suffering (arising from the unhealthy nature of the country through which the triangulation was carried, and exposure during the rainy season), were encountered to which no similar operation presents any parallel.

> Of the manner in which Col Everest contrived to remedy the defects of the Astronomical Circles ... His ingenuity and perseverance met with deserved success, for the whole of the observations appear to have been made in an extremely satisfactory manner. The merit of the actual performance of a work attended by so

much difficulty is greatly enhanced by the manner in which the results are placed before us.

Viewed by itself, the measurement of an arc of 11½° of the meridian is an operation of great interest to astronomy; but Col Everest's work derives great additional interest from its connection with that of Col Lambton. From the conjoined labours of the two illustrious geodesists we have now a continuous meridional arc, extending from Punnae to Kaliana – upwards of 21° in length – by far the longest line which has yet been measured on the earth.

The Great Meridional Arc of India is a trophy of which any nation, or any government, of the world have reason to be proud, and will be one of the most enduring monuments of their power and enlightened regard for the progress of human knowledge. [167]

Now it is interesting to note that the presentation was by J.F.W. Herschel, who years before had been one of the signatories to the Royal Society 'Jervis document'. Perhaps it was for that reason that, when turning to the next recipient of an award, Herschel said 'It remains for me to notice another more valuable work…'. Even if he privately thought that Everest's contribution was the inferior of the two, surely one would not say so in public unless it were an attempt to apply a retaliatory barb.

Member of the Royal Institution

Everest was not slow to get involved in various scientific societies once he was back in England. Among the first was the Royal Institution. He was balloted for, and duly elected, on 3 February 1845. The citation read simply:

Lt Col George Everest FRS. Engineer in the Service of Hon East India Coy. Late Surveyor General of India and Superintendent of the Great Trigonometrical Survey, of 16 Bury Street, St James. Nominated by

    J. South
    M. Faraday
    Ashley
    J.A. Paris

Others of his family also became members. His brother, the Rev Robert Everest FGS was nominated, by, among others, Sir George, and elected on 2 March 1857. Lady Emma Everest was put up for membership and elected on 7 December 1863. In 1878 two of Sir George's children, Alfred Wing Everest and Lancelot Feilding Everest were nominated and elected on 1 April. In 1881 a further family member, Henry Tryan Wing, Capt 97th Regiment was elected with one of his nominees as Lady Emma Everest.

Subsequently he was elected a Manager of the Royal Institution each year from

1859 to 1866. [248] As a Manager Sir George fell under the following regulation of the Institution:

> The Managers of such an Institution will be above all suspicion of interested motives, their situation in life places them out of reach of mean jealousy of interested competition, and if contrary to all expectation, the effects of prejudice should, in some respect or other, be directed against their laudable exertions, a firm perseverance in their duties must at length remove that ignorance which alone can give them birth. [Royal Institution Prospectus 1803]

By Act of Parliament there were fifteen or more such Managers.

Visitor to the Royal Greenwich Observatory

In May 1859 Everest added to his commitments an appointment as visitor to the Royal Greenwich Observatory. He had been nominated through the Royal Astronomical Society and held the post until his death. [3]

Honorary member of the Asiatic Society of Bengal

In 1860 Everest received the following letter from the Asiatic Society of Bengal.

Asiatic Society's Rooms
Calcutta Nov 19 1860

Dear Sir,

I have the pleasure of informing you that at the General Meeting in October you were elected an Honorary member of the Asiatic Society of Bengal.

The following is an extract from the report of the Council proposing you for election. 'Col George Everest, Fellow of the Royl. Society, formerly of the Bengal Artillery, Surveyor General of India and Superintendent of the Great Trigonometrical Survey of India from 1823 to 1843 and Surveyor General 1830 to 1843. Of the many important works executed under Col Everest's direction the most important and that by which he will be best known to posterity is the northern portion of the Great Meridional Arc of India comprised between the Damargida and Dehra Dun Baselines, $11\frac{1}{2}°$ degrees in length, the account of the measurement of which was published by himself in 1847. The whole Indian Arc is equal to 21° 21' 16", or about 1469 miles. No Geodetic measure in any part of the world surpasses or perhaps equals in accuracy this splendid achievement.

By the light it throws on researches with the figure and dimensions of the earth, it forms one of the most valuable contributions to that branch of Science which we possess, whilst at the same time constitutes a foundation for the geography of Northern India, the integrity of which must for ever stand unquestioned.

Col Everest reduced the whole system of the great national Survey of India to order, and established the fixed bases on which the geography of India now rests. His determination of the amplitudes of the two Northern sections of the great Meridional Arc by means of simultaneous observations taken to the same stars with counterpart circular instruments and his method of determining the celestial Azimuth still practised, may be considered the most perfect modes of obtaining an astronomical element known to Science.'

I have the honour to be
Sir, Your Obedt. Servant
Edwd B. Cowell
Secy Asiatic Society of Bengl. [71]

Col George Everest

A CB and Knighthood

After reputedly declining a knighthood on his retirement – supposedly because he had not been given a CB – the rewards finally came to him in 1861. The CB had at last been granted on 26 February. *The Times* of 20 March recorded an item from St James Palace of 13 March:

> The Queen was this day pleased to confer the honour of Knighthood upon Colonel George Everest CB and FRS, on the retired list of the Bengal Artillery, formerly Superintendent of the Great Trigonometrical Survey and Surveyor-General of India.

Donations

On 16 March 1865 Everest wrote to Edward Sabine, President of the Royal Society thus:

> Some five and twenty years ago more or less I computed the accompanying book of Tables for the use of the Great Trigonometrical Survey of India and found them of extensive use in computing azimuths.

> The Tables have never been printed and this and another copy in manuscript which I left as an heirloom when I quitted my post in 1843, are the only records of my labours in that line.

> If the Royal Society will do me the honour to accept this manuscript and put it on some spare shelf of their library I shall be highly gratified. The day may come sooner or later when perhaps it may be thought desirable to perpetuate the work by putting it in print but that is of no concern of mine, with this sole proviso and exception that as it has cost me much time and toil I hope the original may not be left to the merciless dirty fingers of printers devils and spoiled by their daubings.

Believe me,
Truly Yours,
Geo. Everest. [124]

A Coat of Arms

In 1844 Everest obtained a coat of arms.

Arms: Per Fesse Azure and Sable on a Fesse indented between three Cinquefoils Argent as many Storks heads erased on a second.

Crest: On a wreath of the Colours Upon a broken Battlement proper, a Stork Sable resting the dexter foot upon a Cinquefoil Or.

Motto: *Semper Otium Rogo Divos*. (I always ask leisure from the gods.)

The above arms and crest were granted by Sir George Charles Young, Garter King of Arms and Joseph Hawker, Clarenceux King of Arms, by Letters Patent dated 9 September 1844, to George Everest Esquire, Lieutenant Colonel of the East India Company on the Bengal Establishment, eldest son of Tristram Everest formerly of Greenwich in the County of Kent and late of Gwernvale in the County of Brecon, Esquire, in the Commission of the Peace for the said County of Brecon, deceased, to be borne by him and his descendants and by the other descendants of his said late father.

In the Patent it is stated that these arms and crest were a variation of the unauthorised arms (Sable a Fesse between three Cinquefoils Argent) and crest (a Stork) which had 'for a long period' been borne by his family.

Figure 13 *A copy of Sir George Everest's coat of arms as presented to the Survey of India by the Royal Institution of Chartered Surveyors in October 1990.*

Portraits

Several portraits exist of Sir George. [231] contains at page 442 a copy of the profile crayon sketch made on 30 July 1843. After passing through the Bontein family it is now with the National Portrait Gallery in London to whom it was presented in July 1932. The standard head and shoulders photograph was taken about 1860. [231 p. 434] It was later used by Lady Burrard when she painted a colour version in the early 1900s. In addition there is a further head and shoulders portrait – similar in many respects to the original photograph – and two full length photographs of Sir George taken at the same time but in different poses, one standing and the other sitting at a desk. In 1988 with the winding up of the Survey of India Reunion a portrait was presented by the Reunion to the Royal Geographical Society in London. This was painted by L. Ramos from photographs supplied by the Society.

Figures 14 (opposite) and *15*   Portraits of Sir George Everest.

Figures 16 and **17** *Everest's indomitable character seems to emanate from these portraits.*

His death

On his death in 1866, George Everest was buried in Hove. This is surprising because, although the family had occasionally taken holidays in Brighton, there would seem to be little other substantial connection with Hove. In the Hove churchyard most of the memorials appear to be of a variety of materials, designs, shapes and sizes, but three neighbouring headstones, apparently of the same stone and of the same size and shape, stand out conspicuously. Closer inspection shows that they record the deaths of the following people.

> *Lucetta Everest*   *George Everest*   *Thomas Wing*
>   *Emma Colebrooke*   *Mary Ann Wing*
>   *Benigna Edith*

Lucetta Everest was the sister to Sir George and she died in 1857. Emma and Benigna were children of Sir George and died respectively in 1852 and 1860. Thomas and Mary Ann Wing were his father-in-law and mother-in-law and died respectively in 1850 and 1880.

Thus the first of these to die was Thomas Wing and it is known that at the time he was resident at 21 Marine Parade, Brighton, and that he died of phthisis. His burial in Hove was then quite logical. Although Emma was the next to die, the names of her and her sister Benigna are below those of Sir George on the headstone. This does not, however, imply that they were buried elsewhere and only remembered on the headstone. Records indicate that they were buried in Hove on 17 February 1852 and 28 January 1860 respectively. The headstone must, then, have been made only after Sir George's death. Lucetta died at Marylebone – perhaps even at Westbourne Street – so the reason for burial at Hove is unknown. Then with Sir George in 1866 the idea

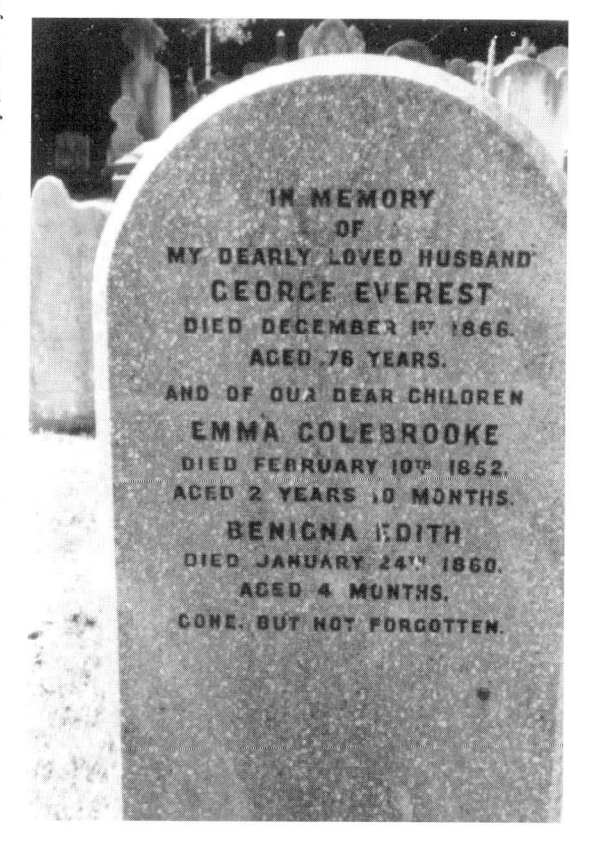

Figure 18 *One of the three family tombstones in Hove, Sussex.*

may have been to be interred in the same churchyard as his wife's family since his own father was buried in Greenwich and his mother in Bromley.

Lady Everest outlived her husband by 23 years and, although she died in London, not far from where the family had resided for so many years, she is not interred with her husband. One slender piece of information suggests that she remarried, but it can be refuted by the reference to her as Lady Everest in the death announcement. She was among the first to be cremated at Brookwood crematorium near Woking but it is not known where her ashes were scattered or interred.

The gold chronometer

At some time before 1930 Everest's son Lancelot gave to Sir Sidney Burrard the gold chronometer which had been presented to George Everest on his retirement from the East India Company. Around 1930 this chronometer was presented by Burrard to the Royal Artillery Institution at Woolwich. It was subsequently stolen from their museum on 2nd February 1974 and never recovered. [130 pp.24–5]

Part II

The Park Estate and Family Matters

Chapter 12

THE PARK ESTATE

*I*n December 1827 a military convalescent depot was started at Mussoorie, and this encouraged Europeans to settle in the area. In late 1832 Everest sought permission to move his office from Calcutta to near the northern end of the Great Arc and to stay there until its satisfactory conclusion.

Move to Mussoorie

He had left Calcutta on 24 December 1832 and collected his brother Robert from Ghazipur before reaching Mirzapore on 1 February 1833. They were at Saugor on 3 March and Agra about 10 April 1833. This was Everest's first visit to this area, which was to be his operational area for the next eight years. [2 p. 434] About 10 May 1833 Everest reached Mussoorie and established himself at the Park, near Hathipaon (or Elephant's Foot) about three miles west of the town and two miles from the Great Arc station of Banog. Hathipaon stands at some 7089 feet above the sea while the bulk of the Park Estate is a few hundred feet lower.

It had taken five months to move the office from Calcutta to Mussoorie. This was accomplished with a fleet of boats on the River Ganges, starting at the Hooghly, the branch of the river delta which passes Calcutta. The personnel included six Europeans who were allocated three 14-oared and one 10-oared budgerows (heavy barges), three 14 ton baggage boats and four cook boats. The Indian draughtsmen were allocated one country boat each with followers and escort using country boats at the rate of around a ton per man. Two 18 ton 'Dhacca pulwars, boarded floors' were used for instruments and records, and there was also a 7 ton pansway. Murphy, one of the senior surveyors, was allocated a 32 ton putillah for himself and the equipment he was taking to Agra. It must have been a most impressive convoy. [263 p. 170]

Charles Morrison, the Registrar, reached Garhmukhtesar on the Ganges – some 60 miles east of Delhi and 120 miles from Dehra Dun – on 8 May 1833. Then after going north west to Meerut took a more or less due north route to Dehra Dun. [231 p. 163] Morrison detailed the journey and all its tribulations in some detail [231 pp.

Figure 19 *Painting of Park Estate showing the buildings before they became derelict.*

170–2]. After disembarking at Garhmukhtesar they required '30 hackries [bullock carts] and 200 coolies at the least ... besides 5 sets of palankeen bearers for ourselves and families'. [263 (284)] On reaching Mussoorie, Morrison temporarily took the Old Brewery – some distance north east of the Park – but Everest said that a better site might be that near Budraj, on which Cloud End house was later built.

This was an estate that Everest had purchased in December 1832 from General William Sampson Whish (1787–1853) of the Bengal Artillery – who had taken possession of the area in May 1829 and built the Park House in 1829–30. He was then commanding the Artillery at Sirhind. At the time the site was said to be in the wastes of Rikholi. [231 p. 439] There is no record of Everest having visited the area before purchase of the Park, so one can but assume that it was bought solely on hearsay, presumably through one or more of his assistants. When the House had been built, Everest had been in England, and from the time of his return to India until moving to Mussoorie he was in and around Calcutta all the time.

At the time of purchase the boundaries were not well fixed. The drawing given to Everest by General Whish showed that the high peak of Hathipaon was included in the Estate and also a considerable part of the eastern slope. Everest got the matter of boundaries sorted out in November 1833. The house was at about 6700 ft above the sea – a few hundred feet lower than Hathipaon at 7089 feet. Colonel Young's men had been there that year and measured the Estate. [231 p. 439] They found it to be

>...850 puckah or 2550 kutcha biggahs, or about 283 English acres. The boundary on the Badraj side [a hill of 7320 feet, 6 miles from the Park] was defined by a private path leading from the Park to the Pioneer Road. On the Masuri [Mussoorie] side the boundary commences at a tree ... towards Lt Tweedale's [1804–83] House. From this tree the Park boundary is supposed to run as a direct line across the dell or khud by the base of Hathipaon towards Dehra. On the Dehra side the precipitous dell below the Budraj range and Banog is the limit to your estate. [230]

Everest wrote to Morrison on 14 May 1833 that:

>The nature of the Park grounds will not admit of my disposing of any part without spoiling the value of the rest. I cannot allow any permanent building to be erected, neither can I give any person any claim on the land of my estate. [231 p. 164]

It took 63 coolies and three tindals to bring all the stores up from Rajpur (at the foot of the ridge 9 miles from Mussoorie). Morrison reported that had he not moved out of the brewery site they could well have been injured when severe snow brought all three brewery chimneys down together with parts of the wall and roof. [231 p. 165]

On 10 June 1833 Everest wrote to Casement:

>The House where I have fixed my office though the only one at the place at all adapted in my opinion for the purpose, labours under the disadvantage common to all other houses at Landour and Masuri of being at a vast distance from water – the consequence of that is, that the Native Establishment can with difficulty be prevailed on to locate themselves in the vicinity... [263 (286) 1, 22–5]

Water initially had to be brought by eight mules from a considerable distance. There is in the grounds near the Park House a deep storage tank dug by Everest and there would appear to be water channels leading to and from it that could still be recognised in 1990.

Other properties

There were besides Park House, two bungalows: one called Bachelors' Hall, and the other called Logarithm Lodge. Many of the sub-assistants had erected temporary dwellings in the grounds, but with the understanding that they had no claims to the land on which they stood. They were liable to two months' notice to quit.

Logarithm Lodge was probably the building in a delightful position overlooking the Dun, about 500 yards to the west, having four rooms with corner fireplaces grouped into a central chimney, and a detached cookhouse. Bachelors' Hall could not be identified in 1955.

When in the process of adding the additional buildings Everest wrote:

I intend the buildings to be made of stone and mortar only using bricks in certain cases where they cannot be dispensed with, because stone is to be had fit for the purpose of building within a few yards of the spot, the whole ridge of Masuri is a formation of limestone which only requires to be burned, and firewood is to be had for the price of the labour of cutting and fetching it. There is plenty of good clay too at hand for making bricks, but the want of water is a serious impediment …

The whole cost of the work proposed will therefore resolve itself into the hire of a sufficient number of Pahari coolies, carpenters and masons … [263 (286) pp. 8–11]

Access to the Park was extremely difficult and over many years Everest was continually improving the route.

When The Park was first built … access to it was by exceedingly difficult and dangerous paths, since which period not only has a safe and broad high road been cut by the Pioneers, which is maintained in annual repair at the public expenses up to the entrance to the Estate, but from the entrance beautiful roads and equable slopes, traced and cut on the most scientific principles, have been made by Everest at his private cost. [231 p. 440]

He used any temporarily inactive labourers available to him for the whole while he was there.

Government edict

The Government refused to let Everest use the Park as a permanent office nor would they initially sanction his idea of constructing a workshop and observatory there. An edict of 1 February 1834 forbade 'all civil and military officers to have their offices in the hills'. [231 p. 165] To overcome the Government edict that his permanent office could not be at the Park, Everest arranged for it to be in Dehra Dun. This took some persuasion but he advanced the need for a recovery area after the heavy toil of the observing seasons and also the nearness to the north terminal of the great arc. On 21 February 1834 he moved the office to Dehra Dun. Luckily there was then no objection to Everest and his assistants spending the rains at Mussoorie.

Morrison greatly regretted the move and expressed the hope that at least Everest would be allowed to reside there during the hot season. [231 p. 165] This he was able to do and also managed to retain the field workshop and computations there under his direct supervision.

In the early days the staff at the Park were: Morrison, the Registrar; Peyton, the computer; Dias and Ross, writers; Scott and Cornelius, draughtsmen; Mohsin Hussain and Shaikh Miajan in the workshop. In 1838 Barrow was brought up with the astronomical circles.

The buildings

In 1838 Fanny Parks paid several visits to the Park. [231 p. 439]. In her writings

[225] she referred to the fine estate and the fact that Everest was at that time building a most scientific road round the great obstacles.

He had requested permission in 1833 to construct an observatory and a workshop but only in 1839 received permission to build. This allowed him to train his assistants out of the field season specifically in the techniques of astronomical observations for zenith distances. In particular also, the workshop, of six rooms, was for the graduation of the new circles. [231 p. 99] Before commencing graduation of the second circle double doors were fitted to keep the outside air from entering the room.

> That portion comprised between the said high road and the panidhar (watershed) to its north was made over to me by Mr Peyton, on condition that he made over to me the portion lying between the high road and the panidhar to the south, so that ... the Pioneer, or high, road is the boundary which separates us.

In 1843 the Registers show Park House as built 1829; assessed annual rent Rs. 1600. It had two large rooms, one 36 × 18 feet, with smaller rooms behind, built on the edge of a crest overlooking the Dun 4000 feet below to the south. Bachelors' Hall was registered as built 1833, rent Rs. 300; Logarithm Lodge built 1835, rent Rs. 600. It was said that the Park House was still habitable in 1951. The observatory was built in 1839 about 100 yards west of Park House. The register was at 20 inches to 1 mile and is referred to as [DDn Mgte files 15, 37,117–18 and map MRIO 185 (11–12)].

In 1866 a room was built at the Park for pendulum experiments. It was subsequently used in 1870 for comparison of bars for determination of the expansion coefficient. [265 (1) App. 6]

In 1840 Everest made an agreement with the inhabitants of Rikholi village that on payment of Rs. 50 they granted to him the sole right of quarrying slate, and cutting grass and wood within the boundaries appertaining to their village for the space of four years from 26 September 1840 to the exclusion of every other person, those of the village excepted.

In 1841, as the time for his retirement and sale of the property drew close, Everest sought confirmation of his ownership. This was after a Revenue survey by Brown where the area of the estate was given as 570 acres. Thus Everest was anxious to get his ownership registered so that there could be no problems when he left.

In April 1843 Everest was complaining bitterly to the Superintendent in Dehra Dun because five of his boundary pillars, erected in 1842, had been maliciously injured and he narrowed the date of injury down to 23–24 March 1843. He wanted this remedied urgently because of the pending disposal of the estate. [263 (453) 28–9] [230]

In July 1843 Everest was concerned because of the lack of a medical officer in the area, the previous one having gone on leave and not been replaced. He indicated that his estate was one of the most valuable in the area and suitable for the accommodation of families but there would be a serious objection if no medical help was available. [263 (453) 67–70]

Figure 20   *Photograph of the Park Estate near Musssoorie, taken in 1990.*

Disposal of the Park Estate

Everest left Mussoorie 16 September 1843 on his way to retirement. The Park had been his headquarters for 10 years and he was loathe to leave. When he came to dispose of it, together with his house in Dehra, the property was very valuable but there was no ready purchaser. He was thus inclined to consider renting it rather than dispose of it too cheaply. At the same time he disposed of all temporary buildings.

It was, in fact, purchased in 1863 (or possibly 1861 [231 p. 166]) through Everest's agent, by Colonel Robert Thatcher (1812–99) of the Bengal Infantry. Subsequent purchasers were Colonel Skinner and then John MacKinnon, who on his death in 1870 left it to his sons the brewers. These sons, Philip and Vincent, had been sent to England to train as brewers and ran the business and the Park Estate up to 1918. Thatcher had a grandson and granddaughter still living in Mussoorie in 1957. [231 p. 440] The grandson, in a letter to Colonel Phillimore on 19 January 1956, from Rosemary Cottage, Mussoorie, said the purchase date was 1863. In 1942 the Development Corporation sold the Park to some Kumaon paharis for a song. It was they who cleared much of the forest and replaced it with potato cultivation.

In March 1842 William Brown started a large-scale (16 inches to 1 mile) survey of Mussoorie and Landour. 'The proprietors are very particular about their boundaries, and … they would prefer as much attention being paid to define them as was consistent with a moderate accuracy … in preference to a triangulation giving a few points … very correctly'. [231 pp. 220–1] Brown's maps still existed as current estate plans well into the twentieth century and made an almost continuous series from Landour and Rajpur to Cloud End and Banog. They showed traverse circuits and boundaries and most of the buildings of that period. [263 (287) 21–2] and [DDn Mgte file 42]

The remains

In March 1951 Phillimore visited the Park with Margot Heaney. They found the house in good order with an Indian owner. Near the main buildings was a rectangular tank about 40 × 16 feet with an acre or so of land sloping gently towards the four sides of the tank. To the west of the house stood a row of servants' quarters and stables of more recent construction although they may be partly contemporary with the house. An isolated contemporary building was in good order with fireplace but only about 14 ft square. It might have had other rooms built around it and have been Bachelors' Hall. Five hundred yards to the west along the escarpment was an isolated ruined building with glorious views and a good connecting road. It had four good rooms with corner fireplaces grouped into a central chimney. A cooking house was close alongside – this is probably Logarithm Lodge. [230]

Figure 21 *The Park Estate near Mussoorie below Hathipaon, 1990.*

In 1955 the Park was a pleasant estate, wooded except for the west end, where there was a golf course. It was said that the nine-hole course was put there by Mackinnon and that Bachelors' Hall was hole number 5; hence that building is now referred to as No. 5. In 1988 the Government of Upper Pradesh acquired 172.91 acres of the Park Estate for some £230,000 and were negotiating for a further 172 acres. The sums of money involved are large and the development will take yet more large sums, not all of which have yet been found.

The future

The views from the site of the House take one's breath away. To the north there are the peaks of the distant Himalayas; to the east, the close-by peak of Hathipaon; to the south, the 1000 m drop to Dehra Dun and the plains of North India; and to the west the Observatory is silhouetted on a small eminence.

In 1990 the Director General of Tourism (Hill Region) put forward various proposals for the restoration and development of the Estate [223]. It was hoped to restore the Park House and the Observatory to their original design and reclaim the areas damaged by quarrying. A range of tourist facilities were envisaged, not just for the Park but the surrounding area from Mussoorie. These included hotels, shopping centres, an amusement park, treks and riding lanes. In addition there were thoughts of rock climbing, hang gliding, and an art and craft gallery and museum. If much of this was in or close to the Park Estate, then the area as Everest knew it would be spoilt for ever. Rather it should be left as isolated as possible and when renovated put to some useful purpose such as a small survey school.

Figure 22 *Hathipaon seen from the ruin of the Park Estate.*

Chapter 13

FAMILY MEMORIES from Sir George's niece, Mary Boole

*M*ary Boole, a niece of Sir George Everest, was a prolific writer of articles, mostly of a religious character, in a range of journals, often from the Indian sub-continent. Long after her death these were brought together in a collection of her works by E.M. Cobham. [67] Very few of the writings bore at all on her uncle George or her other relatives but those that did present the only known published details of aspects of the Everest family. Interestingly they show a quite different side to Sir George from the one that comes out in the works of Phillimore and others and so are a unique commentary.

Figure 23   *Mary Everest Boole*

Her relevant writings were spread over the period 1878 to 1914, by which time she was 82 years old and only a couple of years from her death. It must be remembered that, even when she started these contributions in 1878, Sir George had been dead for twelve years, so memories could be misty and biased towards those aspects of his life that she wished to remember, rather than a balanced view. With no corroborative published material there is no way of deciding how the memories a niece had of her uncle were biased. While she dwelt at length on religious facets that may well have been solely because of her own primary interests.

The following paragraphs are taken from Mary Boole's various writings to show her recollections, true or coloured, of her uncle. The first is the only item written specifically about Sir George; the others are passing references to him within writings on other topics. The principal writing that relates in detail to Uncle George was penned in March 1905 as *The Naming of Mount Everest*. [44]

Very little is known of the English Godfather of that Himalayan mountain. Circumstances, into which I cannot now enter, led to the destruction of nearly all written memorials of his life; and as he married in late middle age, his own family were too young at the time of his death to remember much that can throw light on his Indian career. Nor do I know much of it; but I have a vivid recollection of his character and personality; and perhaps the little that I can tell about him may throw light on the nature of the feeling that prompted those who remembered him to connect his name with an inaccessible snow-peak.

No portrait taken of him reminds me of him as much as does a statue of Moses, the legendary Deliverer of Israel, of whom it was said that 'no man knows his sepulchre'. As a matter of fact, George Everest was buried at Hove near Brighton. But if I were writing impressions of him in metaphor or parable, I could not express them better than by saying of him, as is said of Moses, that after spending his life leading other people out of bondage to within sight of a Promised Land, he went up a mountain to be alone with God; and some Unseen Power disposed of him, and left his friends suddenly aware that they had in reality never known much about him. He was always a landmark, a beacon to others; always utterly lonely, and always misunderstood.

In the year 1806, he being then 16 years old, was sent to India as a cadet. He had received an ordinary unscientific education of a young gentleman of that period, which comprised a certain amount of accurately taught Latin, a very little illogical theology, and a large amount of flogging, fagging and fisticuffing. He was considered an excessively troublesome boy. Not that he was ever accused, I believe, of any action that could be considered really wrong, deceitful or unkind. But he was always in danger. And his escapades kept the ladies of the household in a perpetual state of terror. At last he was found in a quarry, firing against the rock, reckless where the bullets might recoil to. His father was spoken of by people who did not like him as a man who loved philosophic calm more than anything else on earth; he decided to send George away for the peace of the household. Before sailing, George went to take leave of his sister, a pupil in a select boarding school. She took him into the school room; whereupon he blew out the candles, and took advantage of the confusion that ensued to kiss the pupils and teachers indiscriminately in the dark. That is all that I can remember being told of his boyhood; except that he was his mother's favourite child, and her death, which took place soon after his departure, was supposed to have been hastened by grief at losing him. He seems to have been the idol of the whole family, and perhaps his father's decision was not as selfish as it was represented to be. He [the father] was in many respects a wise and farseeing man. He knew his children better than most fathers do; and I am inclined to think that he sent the boy off because he felt that such a boy would be safer facing dangers and difficulties for himself than being coddled and spoilt at home.

I have heard from others that George became a brave soldier. He certainly was a brave man, possessed of a remarkable amount of both physical and moral courage. But I have not the least reason to suppose that he ever cared for his profession, or thought with anything but pain of any use of military weapons against natives of India. I have no doubt that he did his duties as a soldier as long as he had any such duties to perform; but many circumstances conspired to make those duties distasteful to him, and he was, I know, very thankful when a happy accident diverted his energies from the work of conquering India to that of surveying it.

He had an active and eager mind, a thirst for knowledge of all kinds. His first device for procuring mental food and occupation was to get himself taught as many languages as he could, by native teachers. This must have struck, from the first, the keynote of his future career, marking him as different from ordinary English officers, among whom it was not, I believe, at that time, much the fashion to enter into sympathetic relations with conquered countries.

He made the acquaintance of a learned Brahman, who initiated him into the ancient philosophy of the relation of Man to The Unseen Inspirer, to the As-Yet-Unknown Truth, which forms the common underlying basis of several Asiatic religions (including the Brahman, Buddhist, Parsi, Jewish and Christian). His devout mind was intensely impressed with the ancient conception of a relation between Man and The Unknown - a relation which is direct, primary and independent of, and prior to, any religious belief, and of which each religious formula is a more or less inadequate expression. The whole idea of any particular religion being the essential right one, or claiming to be necessarily truer or better than any other, became utterly abhorrent to him. That anyone should attempt to convert anyone else to his special form of worship, or should speak of his own doctrines as essential to salvation, was enough to rouse his contempt and anger to the last years of his life. And I fear he never was on very good terms with Christian Missionaries. Having learned all the languages that he could get anyone to teach him, he next turned his attention to the study of mathematics. And he caught a vision of the Ancient Secret, that mathematical notation is not a mere mechanism for calculating numbers, but the supporting framework of that same organic relation of Man's mind to the As-Yet-Unknown of which religions are various outward expressions.

Now the main difficulty of mathematics consists in the fact that students treat it as a 'secular' subject. As soon as the student sees mathematical notation in its true light, as the supporting skeleton of religious thought, what was before obscure becomes easy to understand. This is the explanation of many a phenomenal career in mathematics. George Everest's progress in the study was rapid and sound.

While he was still quite young, the Surveyor-General of India needed an assistant. My uncle's account of what then happened was this. If the authorities could have found another man in India, who knew mathematics enough, they would not have appointed him, for already they hated him; but there was no one else competent to do the work; therefore they were reluctantly obliged to appoint him.

And so he escaped from the duty of coercing and slaughtering natives to the more congenial task of organising and supervising the work of the native surveying staff, and teaching trigonometry to young Parsis and Hindus.

By what particular mode of offence he had made himself thus early hated by the authorities, he did not inform me. One can infer its probable nature by what followed during the survey.

That he loved difficulty and danger for their own sake became evident when, after his return to England, he told us tales of his hairbreadth escapes from peril. Jumping into a live cactus hedge to escape wild cattle was one of his adventures; leaping from a precipice into the branches of a tree growing from its side in order to avoid being crushed by some heavy vehicle that had broken the traces going up a hill, was another; all these episodes were told as pure jokes and with great zest; but never a word can I recollect of fighting anyone – except the English priests of red tape!

The quarrel that I heard most about was over a signalling instrument. When he took charge of surveying parties, he found that the native surveyors were made to take outdoor observations when it was unhealthily hot and to do calculations indoors when it was cooler. He wished to reverse the arrangement; but the signal could not be seen, from hill to hill, except in the glare of the sun. He invented a more suitable instrument, by means of which signals could be seen in cloudy weather. Having made it a real working possibility, he applied to the authorities for leave to bring it into general use in survey. There ensued a long and angry correspondence; red tape probably could not understand why any implements were needed except those in use; and George Everest was not the man ever to understand why there need be any delay in carrying out any project which tended towards the emancipation of man, woman or child, from bondage or unsuitable conditions. His advocacy of his proteges may have been more zealous than judicious. What is certain is that people who were content because they were taking for granted the legitimacy of their own position dreaded his influence, and whoever felt wronged and oppressed loved him – then, and always, and everywhere, to the end of his life.

In one of his letters to the authorities he made the remark that he knew he was making himself obnoxious, but that would not trouble him; he had taken for his motto: 'Serve God and fear no man'. Remarks of that kind do not tend to make a subordinate popular at headquarters! But they act as rice seeds cast at random into swampy ground; after many days some of them are found growing. A young Parsi employee in some Government office had to copy the correspondence. The story was told to my aunt, either by Mr Dadabhai Naoroji or Mr Fardunji, when they were in England together, and I think the copying clerk was one or other of them. Whoever he was, he was struck by the phrase, and adopted it as his motto. Later on, this clerk realised that Parsi women were being kept in a degraded condition of ignorance. He and one or two friends organised a movement for procuring English teachers for Parsi girls. They had many difficulties to encounter and some persecution and obloquy from their coreligionists; but they had taken for their motto 'Serve God and fear no man'.

I do not know whether the quarrel about the signalling preceded or followed my uncle's long visit to England in the eighteen twenties. During the course of that holiday he went to Germany, taking with him his youngest brother (my father), then latterly ordained a clergyman of the English Church. They went to a great gathering of scientific men somewhere in Germany, travelling in company with John Herschel and Charles Babbage. Knowing as I do know something of what my uncle's influence was on my father then, and after his final return to England on my own mental development, I can read between the lines of Babbage's and Herschel's writings and gather something of what my uncle's Brahminical view of the meaning of mathematical notation must have done to sow in their minds the seeds of an ideal more spiritual than those current in English universities.

He came home finally in 1844. I was then a child of twelve. From the first, he was always the centre of stir, passion, emancipation. No repressive prohibitions that anyone made could be enforced where he was; he simply laughed to scorn the notion of anyone coercing anyone else. Children and animals worshipped him. Every one loved him to a degree that was distracting; yet the grownup members of the family seemed to spend a large part of their existence in wishing that he would not say the things he did say. Perhaps they feared he would spoil us children! If they thought so it was a great mistake. Loyalty to relationships and to accomplished facts emanated from him like a contagion. I can remember nothing which made me so willing to submit to my parents as uncle's naughty-sounding exhortations to me not to let anyone coerce me. He had a power greater than any one else I ever knew, of calling out the spirit of *Liberte oblige,* of making one feel that the claim for Freedom is the outcome, not of dislike to doing what the authorities wish done, but respect for one's self as a child of God. He made one long for discipline for its own sake; he called out in one a passionate revolt against all that could come between one's soul and God; including one's own weakness and selfishness and cowardice. I feel certain that he would sooner have cut his tongue out than have uttered a word calculated to induce anyone in India to real rebellion. But probably the home authorities suspected the contrary, for red tape has the fatal property of strangling in its devotees the power of understanding loyalty. If the home authorities had had the courage to give to such men as George Everest a free hand in India, it might have been, by this time, in a condition of more real self-government than it now is; better able to reform itself religiously, to educate itself intellectually and to protect itself from internal evils, such as famine and plague; a more organic unity within itself, and therefore more actively and helpfully loyal to England than is yet the case.

An episode in my uncle's life after his return to England may serve to throw light on the reason of the strange sense of unrest which surrounded him. He took a country house at Ascot, and soon afterwards was visited by a deputation of neighbouring gentlemen who had come to express their desire that he should represent their party in Parliament. He agreed to stand, and asked what steps he was to take. He would have to pay a certain sum into a local Bank. 'Very well. But what is to be done with it?' 'It is for election expenses.' 'Yes; but how will it be spent?' 'Make it payable to the order of your agents, and do not ask for details.' My uncle rose and rang the bell; when the servant entered, he said, 'Show these gentlemen out.' Such was his account to me of the one attempt of his countrymen to enlist him in the service of his country in the capacity of Member of Parliament. The whole question as to which party won at an election, and who was and who was not in power, would always have seemed to him trivial in comparison with the harm that would certainly be done to the country by the slightest dishonourable action. And it is not difficult to guess why a man who took that kind of tone was obnoxious to men accustomed to treat as 'necessary election expenses' a certain amount of bribery done by agents with the sanction of the principals!

He helped and befriended Bishop Colenso, and everyone whom he heard of as trying to reform or broaden the English Church or any other Church; but he never allowed his name to be publicly connected with any movement of reform, because he would not allow himself to be supposed to belong to any party in particular. 'I don't believe in Colenso's view more than any one else's,' he said; 'I back up whoever is attacked, because I can't bear to see a man ill-used.'

This religious impartiality gave pious people the impression that he was indifferent to religion. Never was there a greater mistake. On one occasion during the last year of his life, I was in his company, along with two pious ladies. Beside his couch was a small kerchief. When the door had closed behind the other ladies, he took up the handkerchief, beneath it lay a Bible. The next few minutes were among the most awful of my life – awe-ful in the literal sense of the word. I realised then what a struggle his life had been, torn between his childlike and humble devoutness, his passionate love of God the revealer; his longing to be in religious communion with his friends, and his intolerance of any religion which claimed to be superior to all others.

When I undertook to write this sketch, I thought it would be easy to say a great deal about the man whose name, while he lived, was synonymous, wherever he went, with the sense of help for the weak, of hope for the hopeless, of cheer for the sorrowful, of courage for all who were struggling in any good cause; and of bewildered terror to whoever was content to take for granted the rightness of his own claims, rights, religion, opinions, beliefs, or moral codes. But I find there is little I can say. And perhaps, after all, this incapacity of mine to find much to say about him is the true interpretation of the motives of those who named the mountain after him. Perhaps he filled the background of their minds as of ours with the sense of inexplicable inspiration and power, without giving them many intelligible facts to record. They gave his name to the highest thing they knew of, and left it to the Eternal Silence.

The following are short extracts from a variety of Mary Boole's writings that comment on her uncle or on her father Thomas, brother of George Everest.

Home-side of a scientific mind [40]

The late Lieutenant-Colonel Sir George Everest, Surveyor-General of India, who made his way in the world entirely by his mathematical knowledge, has also told me that his own choice of geometry as the occupation of his leisure was determined by lack of power to procure in India books on other subjects.

Scientific research by natives of India [45]

This means of course, the need for a readjustment of all the relations between England and India in matters of Physical Science. That such a readjustment must come, sooner or later, somehow, has been known to a few Mathematicians here, ever since my uncle, George Everest, came home (before 1830); and more specifically, since De Morgan wrote his preface to Ram Chundra's book. But Dr Bose's performance struck the conviction home to the very heart of our Physical Science.

The Preparation of a Child for Science [42]. To Ethel Gertrude Everest (Daughter of Sir George Everest):

My dear cousin,

Three-quarters of a century ago your father, during a visit to his native land, infused into the minds of a few young mathematicians, among who were Charles Babbage and J. Herschel, certain ideas about the nature of man's relation to Unknown Truth which underlay all science in ancient Asia, and which he had learned from Brahman teachers. The seed which he then sowed has borne abundant fruit in English Mathematics. Of his subsequent work in India some have sought to express their appreciation by giving his name to a great inaccessible snow peak. You and I think that we shall more truly fulfil his ambitions by making as accessible as we can to little children in all parts of the Empire that open gateway to the Unseen at which he stood in perpetual adoration to the last hours of his life.

M.E. Boole.

Figure 24 Ethel Gertrude Everest, third daughter of George and Emma Everest.

Indian thought and Western Science [46].

> As to my own family, whatever one's opinion may be of the taste displayed by the English in altering the ancient name of the great mountain, there can be no doubt that the choice of my uncle's name in connection with this queer kind of vandalism was meant as a full recognition of the services rendered by him to engineering science. If, therefore, at your request I tell the people of India some facts which I happen to know in the history of modern European science, I do so not as one appealing to men of another race for recognition denied by her own; but because I venture to hope that for the sake of my uncle and my husband what I say may at least gain a thoughtful hearing. When you have read what I have to say, I ask no more; I do not wish to convince anyone against his instincts; judge for yourselves.
>
> My uncle, George Everest was sent to India in 1806 at the age of sixteen. Things were different in those days from what they are now; there were neither competition Wallahs nor Officers trained in England by 'Army Coaches': the boy went out ignorant, unspoiled and fresh. He made the acquaintance of a learned Brahman who taught him – not the details of his own ritual, as European missionaries do, but – the essential factor in all true religion, the secret of how man may hold communion with the Infinite Unknown. This my uncle told me long afterwards. Some time about 1825, he came to England for two or three years [actually five years], and made a fast and lifelong friendship with Herschel and with Babbage, who were then quite young. [But both were only two years younger than Everest.] I would ask any fair minded mathematician to read Babbage's *Ninth Bridgewater Treatise* and compare it with the works of his contemporaries in England; and then ask himself whence came the peculiar conception of the nature of miracle which underlies Babbage's ideas of Singular Points on Curves – from European Theology or Hindu Metaphysic? Oh! how the English clergy of that day hated Babbage's book!
>
> My uncle returned from India finally in 1844. He never interfered with anyone's religious beliefs or customs. But no one under his influence could continue to believe in anything in the Bible being specially sacred, except the two elements which it has in common with other Sacred Books: the knowledge of our relation to others, and of man's power to hold direct converse with the Unseen Truth. In 1846 my father, Uncle's younger brother and pupil, published a paper on the names of God in various languages, in which he attempted to show that Odin, God, Theos etc. were names for spiritual vitality imparted to man…
>
> I end as I began. Tell Hindus to read De Morgan's Preface to Ram Chundra. Tell them that it is the voice of Mount Everest calling to India to awake and arise, and recover the treasures of its past. With heartfelt gratitude to the memory of the Brahman who taught my dear uncle (not his special religion, but) the underlying principles of true progress.

This was a letter from Mary Boole to Dr Bose which began

> Dear Dr Bose, Nivedita (Margaret Noble) conveyed to me your request that I would explain what I meant by speaking of the unfitness of the English people to undertake the education of such people as the Hindus …

It was written in 1901 from 16 Ladbroke Rd, London. The reference to Ram Chundra was elaborated thus: [67 p. 948]

> You know that Professor De Morgan caused a *Treatise on Maxima and Minima* by Ram Chundra to be published in England, in order to prove to English men of science that the Hindu mind masters, without the aid of the differential calculus, problems which among us had hitherto been solved only with the help of the calculus…

(*A Treatise on Problems of Maxima and Minima, solved by Algebra* by Ram Chundra was published by W.H. Allan & Co. London 1859 from the Calcutta edition of 1850 published by P.S. d'Rosario.)

The forging of passion into power [47]

> George Everest – went to India at sixteen years old, in the service of the East India Company. Put himself under the tuition of natives in India. Learned from them Oriental languages, religion and philosophy, and taught himself European mathematics from books. Became Surveyor-General of India.

A child's idyll [48] and *At the foot of the Cotswolds.* [49]

The subject of the following paragraphs is George Everest's brother, Thomas Roupell Everest – see also chapter 2.

> My father and mother were profoundly in love with each other through all their married life. For that reason each turned to the other that side which was most pleasing to the other; and each hardly suspected what the other was really like when they were apart.
>
> Father was one of the victims of the terrible epidemic of influenza which swept England in the year 1837, when I was not quite five years old. It left him a wreck. For ten years he was unable to do any parochial duty. He put a curate in charge of the parish and went to France to be near Samuel Hahnemann. He was unable to bear children near him except for short periods, and we were left a great deal to servants. We lived mostly in lodgings in various parts of Paris and its neighbourhood, and two or three times spent a few months of summer in Cornwall where Mother's family lived. On those occasions we lodged at St Mawes, then a small fishing hamlet. [Mother was Mary Ryall] My father was a passionate lover of the sea which my mother hated…
>
> We sometimes attended day school, sometimes had a governess and sometimes mother gave us lessons.
>
> My father as I knew him after I was fifteen, was one of the last men in the world to try to cramp a child's originality, genius or will. To me he was more like a lover than a father. He spoilt all the girls in the parish and especially his daughter. But during his long period of ill-health he was absorbed in the personality and ideas of that Draconian philosopher Hahnemann…
>
> One of the winters that we were at Poissy, the Seine was frozen over. I saw wagons crossing it. That winter, complicated by Hahnemann's drastic and rigid rules about

'cold baths' and 'walks before breakfast' was hard on us children, and I see now, harder still on Mother and Nurse, who must have suffered severely from having to do the cruel things they were ordered to do. The little boy used to scream at the bath (from which Nurse had just cleared the ice, sometimes by the help of a stout stick) I was too proud to scream...

In the year 1840 my father took the Chateau de l'Abbaye at Poissy, the house long afterwards occupied by Meissonier the painter...

Hahnemann practised homeopathy and one of his ideas was that the patient can gain strength only on the condition of bearing cold. In particular this stretched to cold baths, going on walks and playing in a cold room. [67 Vol. 1 p. 180]

The religious philosopher as a social harmonizer [43]

At Queen's College the writer was considered rather successful at teaching philosophy to young girls, in a manner which made each more, not less, interested in and loyal to the religion of her own parents, while promoting religious harmony and mutual helpfulness between persons of different sects. As I am sure that any power which I may have possessed in this direction is largely due to what my uncle, George Everest, learned from a Brahmin nearly a century ago, it will only be a natural expression of gratitude – if I can hand on the help to any native of India, who may now be puzzled owing to the apparent conflict between Eastern and Western ideas.

The Building of the Idol [41]

My father was a country clergyman. When I was almost an infant, he became ill, was supposed to be consumptive, and was ordered to give up work and live abroad. He had at that time private means, and took the opportunity to study, under Samuel Hahnemann, the mutual interaction of body and mind – a subject to which, at that time, neither physicians nor clergymen had as a rule paid much attention. He also collected books on ancient occult science. After some years, the true cause of his illness was discovered and removed; and shortly afterwards a financial crisis in the country deprived him of his fortune. Had he not been a clergyman, he would have had to face poverty and begin life all over again. As it was, he came back to his parish, to a rectory cottage, and a small but assured stipend. ... He took me from school, taught me a little in an erratic fashion of his own, and set me to work to assist him in teaching my young brother, and in the parish. As soon as I was old enough to understand at all the attitude of the neighbouring clergy, I perceived that my father was considered an anomaly amongst them. He was censured for not being loyal to the Church, for neglecting to maintain the dignity of his order, and for spoiling his poor parishioners and putting them out of their proper place. Hints of this kind puzzled me at the time.

Chapter 14

RELATIVES AND FRIENDS

Numbers within curly brackets refer to the numbers on the family tree.

{1} Tristram Everest (Everis)

Married 26 November 1682 to Ann Fisher at Charlton, Kent

Died Will dated 5 November 1717, proved 17 April 1721

Of Church St, Greenwich. He was a butcher. Witnesses to the marriage were Martha Everest and George Everest. There were 14 children: eight daughters and six sons. Was this the same Tristram Everest as is recorded in Tonbridge as baptised there on 14 January 1656?

{2} Tristram Everest

Baptised 20 January 1688 at Greenwich

Married to Mary Hack?

Died Will dated 2 December 1751, proved 9 April 1752

Of Greenwich. Like his father he was a butcher. They had four daughters and four sons.

{3} John Everest

Born 20 July 1718

Baptised 31 July 1718 at Greenwich

Married 5 April 1741 at St Mary Magdalen, Old Fish Street, London, Susannah Cole (b. 1718) of St Olave Southwark. (By licence)

Died Will dated 24 January 1769, proved 20 February 1769

Of Greenwich. He was an attorney, initially articled in 1736 to William Radley of Greenwich. There were ten children: five daughters and five sons.

{4} William Tristram Everest

Baptised 22 November 1747 at Greenwich

Married 2 August 1786 to Lucetta Mary Smith at Greenwich

Died 18 May 1825 Will dated 22 November 1820. Buried at Greenwich

Of Greenwich and Gwernvale House, Crickhowel, Co. Brecon. JP and Deputy Lieutenant Co. Brecon. He was a solicitor. He was one time solicitor to Greenwich Hospital and to Chelsea Hospital. Lucetta died 17 May 1809 and was buried at Bromley, Kent.

166 Everest: The Man and the Mountain

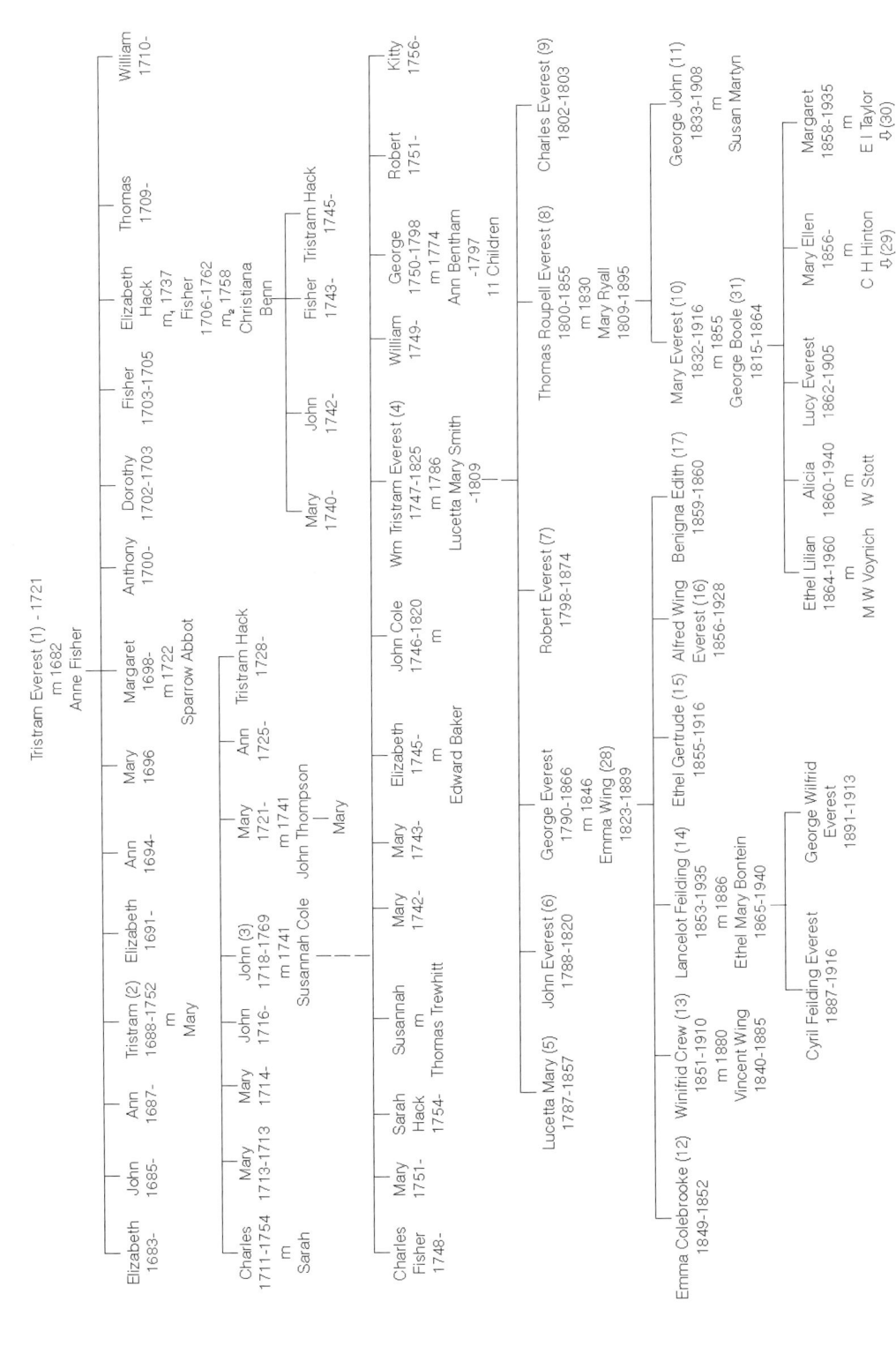

Figure 25 *The Everest family tree.*

{5} Lucetta Mary Everest

Born 14 July 1787

Baptised 11 August 1787 at Charlton, Kent

Died 10 January 1857 at Marylebone. Buried at St Andrew's (Old) Parish Church, Hove, Sussex.

She was the first daughter of Tristram and Mary Everest. She never married.

{6} John Everest

Born 7 December 1788 (or 15th?)

Baptised 12 January 1789 at Greenwich

Died 8 May 1820 Oxford. Buried in the cloisters of Corpus Christi College, Oxford

He was the first son of Tristram and Mary Everest. Fellow of Corpus Christi College Oxford and a Barrister at Law. He never married and died in his rooms at the College. The record indicates that '… the usual procession [for a College funeral] round the quadrangle was dispensed with, and such only of the Society (as thought proper) attended in the chapel and followed the corpse to the grave'. Why this should have been probably stemmed from his rejection of a Fellowship. The records [70] read:

1818 May 11. Mr Everest, Barrister at Law, Probationer elect of the College (according to our Statutes) as Senior Master – desirous to wave [sic] his admission thereto, and to continue Discipulus (in order to prosecute the profession of Law) having appealed to our Visitor to know whether he might not apply to himself on this occasion a Determination of Bishop Men in the case of Mr Francis Dickins, not dissimilar from his own, of which the present Society entertained some doubts, in that such determination had never been acted upon at any subsequent period, to the year 1697 and was then a new case.

The President communicated at a full meeting this day the visitor's Decision on the question proposed to him by Mr Everest, viz. 'That he does not see sufficient ground for departing from the said Determination, and is of the opinion it should continue to be observed, and that Mr Everest should be allowed to have the benefit of it'. Accordingly the President, upon receiving notice from Mr Everest that 'he accepted the Bishop's Decision, and *pro hac-vice* (for this occasion) waved his admission to his Fellowship …'

The small flagstone covering his last resting place is simply inscribed J.E. 1820.

{7} Rev Robert Everest

Born 15 September 1798 ([231 p. 442] gives 19 January 1798 and [130] says 19 December 1798)

Baptised 25 March 1799 at Greenwich

Died 7 February 1874 ([231] p. 442 says 1870). Buried at Ascot

He was the third son of Tristram and Mary Everest (George was the second son). MA Oxon. Ordained and appointed chaplain in the Service of the East India Company (Bengal Establishment) 1 April 1829.

{8} Rev Thomas Roupell Everest

Born 7 December 1800 at Greenwich

Baptised 25 March 1801 at St Alphege Church, Greenwich

Married 27 September 1830 to Mary Ryall

Died 15 June 1855 at Wickwar. Buried at East end of Wickwar Churchyard

He was the fourth son of Tristram and Mary Everest. He obtained an MA Cantab. and became Rector at Wickwar, about 5 miles north of Chipping Sodbury 1830–55. His wife Mary Ryall (1809–8 May 1895) was a sister to Dr John Ryall, Professor of Greek and Vice Principal of Queen's College, Cork where George Boole was Professor of Mathematics. In 1837 Thomas was a victim of the 'flu epidemic and as a result went to France for almost 10 years, leaving his parish in the charge of a curate. In her writings, Mary Boole variously described her father thus:

> My father always taught me to think of him, not as my father, but as the servant of the poor, whom it was my privilege and duty to assist for the time being. ... My father T. R. Everest, was a learned occultist in days when occultists were few. He did everything in his power to call attention to the dangerous reaction which must come if the clerical and medical professions persisted in ignoring the phenomena of Mesmerism, Trance and Clairvoyance. ... My father lived in an atmosphere of medical reform and had a strong feeling against certain kinds of medical practice in vogue in his day. I had heard once a doctor recommending for me a certain mode of treatment, insisting that it was harmless and answered well. To cut short the discussion father said, speaking metaphorically: 'It may answer for other people but: it could not for my child; Hahnemann would rise from his grave to prevent it'. [67 pp. 65, 82, 339–40]

{9} Charles Everest

Born 13 December 1802 at Greenwich

Baptised 31 January 1803 at Greenwich

Died young

Fifth son of Tristram and Mary Everest.

{10} Mary Everest Boole

Born 11 March 1832

Married 11 September 1855 to George Boole

Died 1916

She was the daughter of Thomas and Mary Everest. Her husband was George Boole who died in 1864. She was a prolific writer and earned considerable praise from G.K. Chesterton. Among her *Collected Works* [67] were many items with a religious theme that appeared in a range of Asian journals. Of the others was one called *At the Foot of the Cotswolds* [49] and another *The Naming of Mount Everest*. [44] The latter was a short biography of Sir George Everest. See Chapter 13.

{11} Rev George John Everest

Born 29 July 1835

Married Susan Martyn

Died 1908

He was the son of Thomas and Mary Everest. He became MRCS in 1857 and later emigrated to Canada where he was ordained in Toronto in 1873.

{12} Emma Colebrooke Everest

Born 10 March 1849 at Hove

Died 10 February 1852 at Dover. Buried at St Andrew's (Old) Parish Church, Hove

She was the first daughter of George and Emma Everest. It is interesting that the Everests should have used the christian name Colebrooke. This almost certainly referred to William Colebrooke {32} who effectively introduced George Everest to survey work.

{13} Winifrid Crew Everest

Born 13 September 1851 (or 17)

Married 18 November 1880 to Lieutenant Colonel Vincent Wing

Died April 1910

She was the second daughter of George and Emma Everest. Her husband, Lt Col. Vincent Wing was of the Royal Horse Artillery. He was a distant relative to her since his father and her grandfather were brothers. They lived for some time in Dorchester, until he died, when she moved to Lyndhurst.

{14} Lancelot Feilding Everest

Born 28 May 1853

Married 3 June 1886 to Ethel Mary Bontein

Died 1 April 1935 at Hampstead

He was the first son and last survivor of the direct line from Sir George and Lady Everest. Educated at Harrow, he was an MA, LLD (Cantab) and Barrister at Law. He was also a Fellow of the American Geographical Society of New York. Ethel Mary Bontein (1865–1940) was the daughter of John Sims Bontein {35} – formerly an officer in the Royal Marines – and granddaughter of Colonel Bontein and of General Sir William Chalmers. It was this Colonel Bontein who at one time owned the sketched portrait of Sir George Everest that had been made on 30 July 1843 by a British Army officer. They had two sons, Cyril Feilding Everest (1887–1916) who was in the Machine Gun section of the Canadian Infantry and died about 9 October 1916 in France; and George Wilfrid Everest born 13 October 1890 and died 11 October 1913.

Lancelot's second christian name, with its unusual spelling, was derived from that of one of his godfathers, Colonel Feilding {34}, a cousin of a former Earl of Denbigh, who was a close friend of Sir George in India.

{15} Ethel Gertrude Everest

Born 1855

Died 21 March 1916 at Chippens Bank, Hever. Buried at Rustall near Tunbridge Wells

She was the third daughter of George and Emma Everest. She never married. She wished to present her sixteenth-century, timber-framed house of Chippens Bank, near Hever, to the National Trust but it was not of sufficient interest nor well enough endowed to be accepted. Hence, when she died, it became a 'Home of rest for any person or persons, men or women, of such position and social standing who desire temporary peace and rest'.

{16} Alfred Wing Everest

Born 3 August 1856

Died 30 October 1928 at Letchworth

He was the second son of George and Emma Everest and lived for some while at the Orchard, Norton Way, Letchworth. He was educated at Eton and Christ Church Oxford where he obtained a BA. He never married and went to Canada and lived a life of hardship.

***Figure 26** Chippens Bank, Hever. The home for many years of Ethel Gertrude Everest, a daughter of Sir George Everest.*

{17} Benigna Edith Everest

Born 3 September 1859 at Hove

Died 24 January 1860 at Paddington

Buried 28 January 1860 at St Andrew's (Old) Parish Church, Hove

The youngest daughter of George and Emma Everest. Such was the Everests' use of the names of friends and relatives in the christian names of their children – Colebrooke, Wing, Feilding – that one cannot but feel that the unusual names Crew and Benigna must also have some significance that is yet to be deciphered.

{18} Vincent Wing

Born 9 April 1619

Baptised 11 April 1619

Died 20 September 1668. Buried at North Luffenham

Of North Luffenham, Rutland. He was a famous mathematician and astrologer who forecast his own death. He was the author of seven treatises including *Astronomia Britannica* of 1665. He was said to have died of consumption through doing survey work in all weathers. He had little formal education and was self-educated in astronomy. He began his career as a surveyor, compiler of almanacs and astrologer. In 1649 he published *Urania Practica* and in 1664 wrote *The Geodaetes Practicus: or the Art of Surveying* which was probably the first work written as a handbook. He used logarithms and trigonometry so did not rely wholly on the protractor and scale.

Figure 27 Alfred Wing Everest, youngest son of Sir George Everest, with Mrs Charlotte Sheffield, daughter of Thomas Twining Wing.

{19} John Wing

Born 1643

Buried 26 November 1726 at Pickworth

Of North Luffenham and Pickworth. He was coroner of the County of Rutland. As Editor of *The Art of Surveying* by his uncle Vincent Wing, he augmented the work in 1700. In 1710 he became Editor of the *Almanack*.

{20} Tycho Wing

Baptised 9 May 1696 at Pickworth

Died 16 April 1750. Buried at Pickworth

Of Stamford and Pickworth (and Thorney Abbey). As a philosopher and astronomer he was named after the Danish astronomer Tycho Brahe. The Rev. Dr Stukeley, antiquary, of Holbeach, Lincolnshire, born 1687, in his *Diary* said he spent many agreeable hours at Stamford with Mr Tycho Wing and Mr Edmund Weaver of Catthorp, the great Lincolnshire astronomer. He was coroner of Rutland 1727–42. There is a portrait of him by J. Vanderbank dated 1731 in the Hall of the Stationers Company, London. His son, also Tycho Wing (1726–76) was a mathematical instrument maker in partnership with Thomas Heath.

172 Everest: The Man and the Mountain

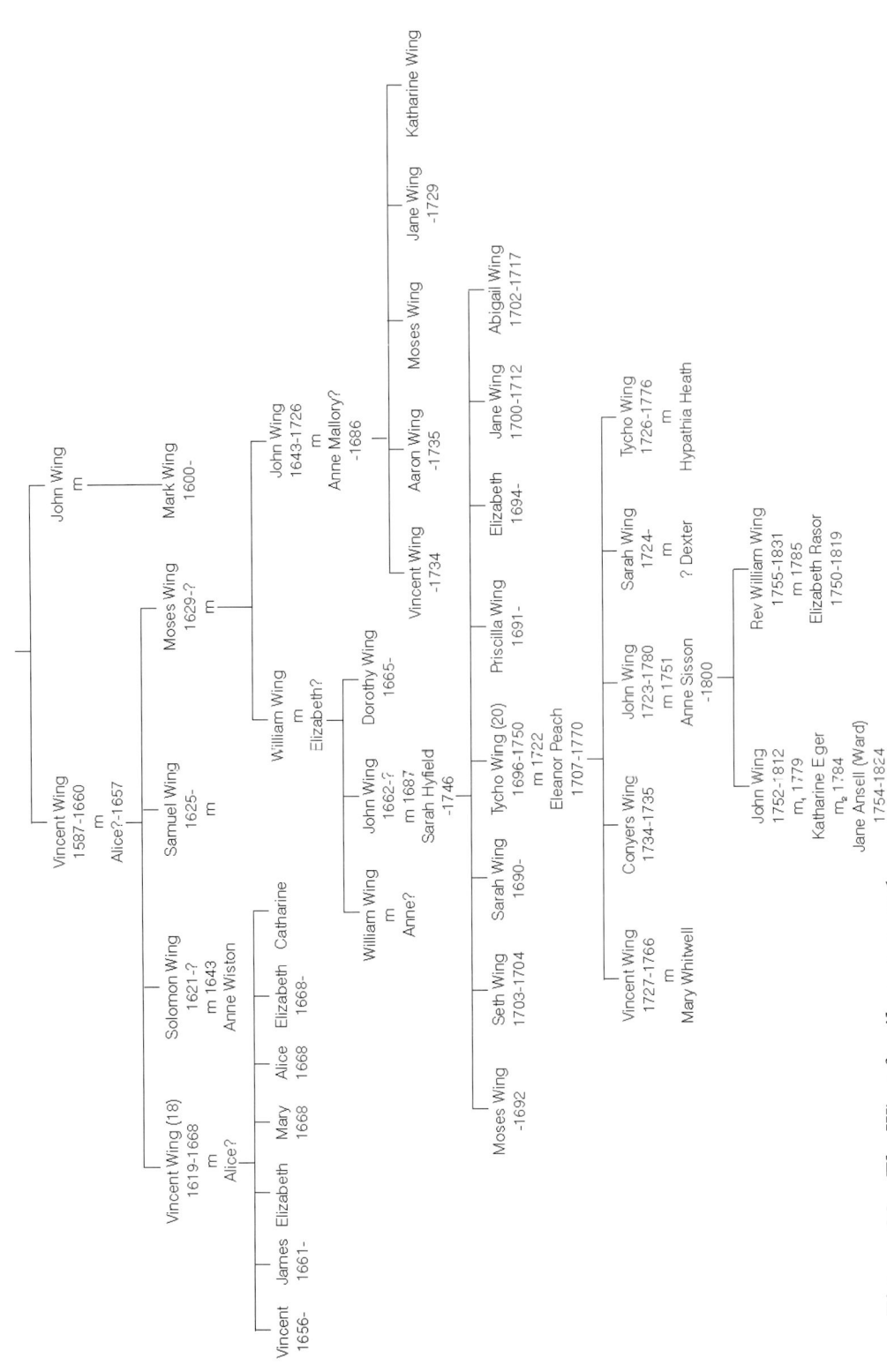

Figure 28 The Wing family tree, part 1.

Relatives and friends 173

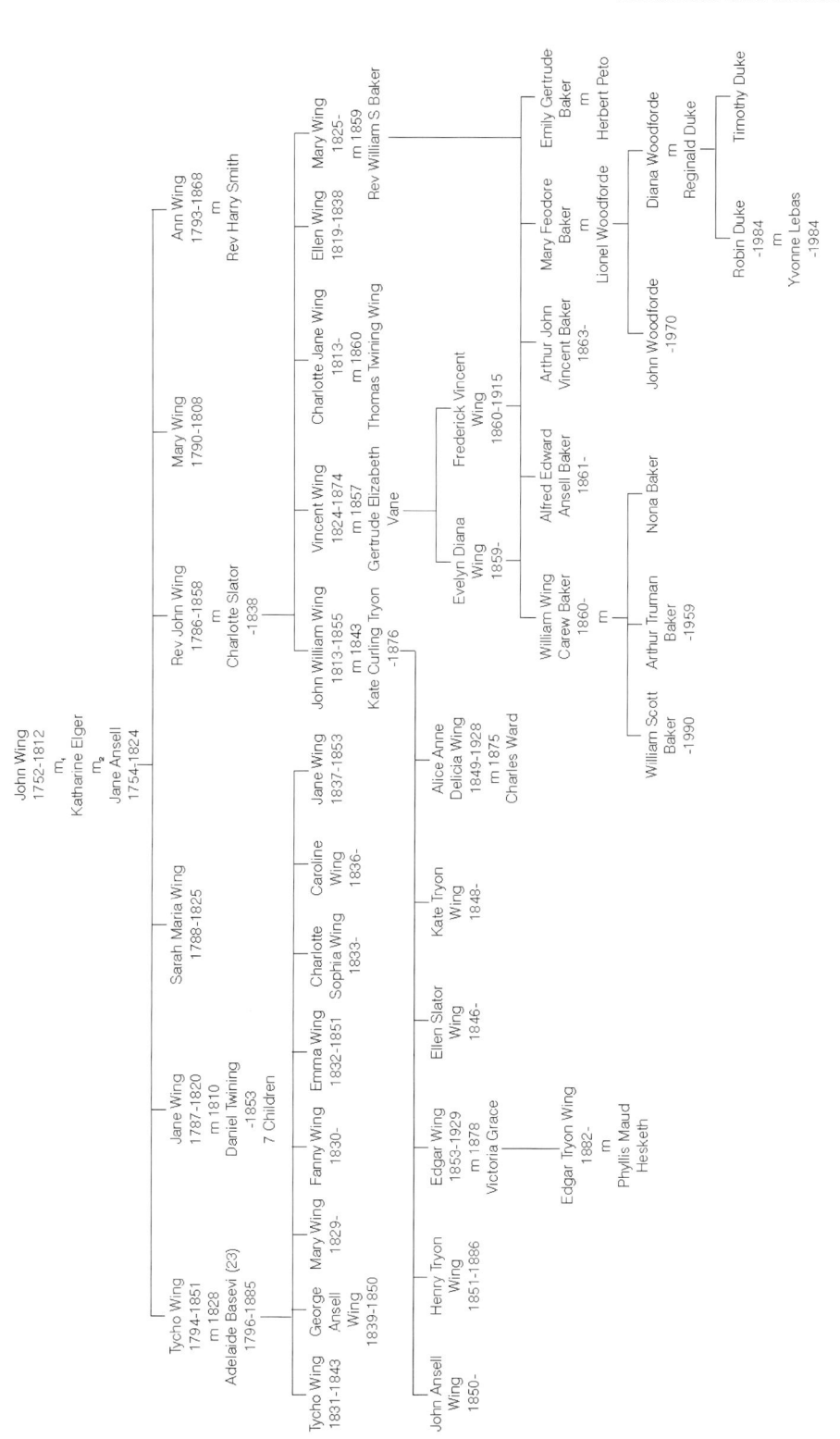

Figure 28 The Wing family tree, part 2.

{21} **Thomas Wing**

Born 3 July 1796

Married c. 1822 Mary Anne Paternoster {22}

Died 12 November 1850 at Brighton. Buried at St Andrew's (Old) Parish Church, Hove

He was the father of Emma, who became Lady Everest. Thomas was a barrister of Gray's Inn and from Hampstead. He died of phthisis at 21 Marine Parade, Brighton after a long illness. His death was witnessed by Mary Ann Judge – would this have been the married name of his daughter Mary Ann, who was born in February 1825?

{22} **Mary Anne Wing née Paternoster**

Born 1798

Married c. 1822 Thomas Wing {21}

Died 4 December 1880 at Sudbury. Buried with her husband {21} at Hove

She was the mother of Emma, Lady Everest.

{23} **Adelaide Basevi**

Born 1796

Married 28 March 1828 to Tycho Wing of Thorney Abbey at St James Church, Westminster.

Died 19 January 1885 at Tunbridge Wells

She was the daughter of George Basevi of 37 Brunswick Square, Brighton, and niece of Maria Basevi, wife of Isaac Disraeli of Bradenham, the father of Benjamin Disraeli. Her brother, James Palladio Basevi, went to India in 1853 and worked with the Great Trigonometrical Survey. Born in 1832, he died on service with the GTS in 1871 – probably from overexertion in the Himalayas where he worked at over 17 000 ft. He was very interested in pendulums and gravity measurements. Today there is a chiming clock in his memory gracing the entrance to the Geodetic and Research Branch building of the Survey of India headquarters in Dehra Dun.

{24} **Thomas Twining Wing**

Born 5 July 1826

Married (1) 1860 to Charlotte Jane Wing

Married (2) 1882 to Georgina A. Dewdney-Wintz

Died 2 October 1904 at Farnham

He was the brother of Emma, Lady Everest. He was one of the executors of George Everest's Will and described there to be 'of Bedford'. Thomas and Charlotte Jane were distantly related in that they had the same great grandparents through the male lines.

{25} **Captain Arthur Wing RN**

Born May 1828

Married 14 July 1868 to Elizabeth Hambly of Wadebridge, Cornwall

Died 3 December 1873

Brother to Emma Wing (Lady Everest). No offspring.

{26} **Henry Vincent Wing**

Born 19 October 1834

Married October 1878 to Jeney Mettenheimer

Buried Copford Green near Colchester

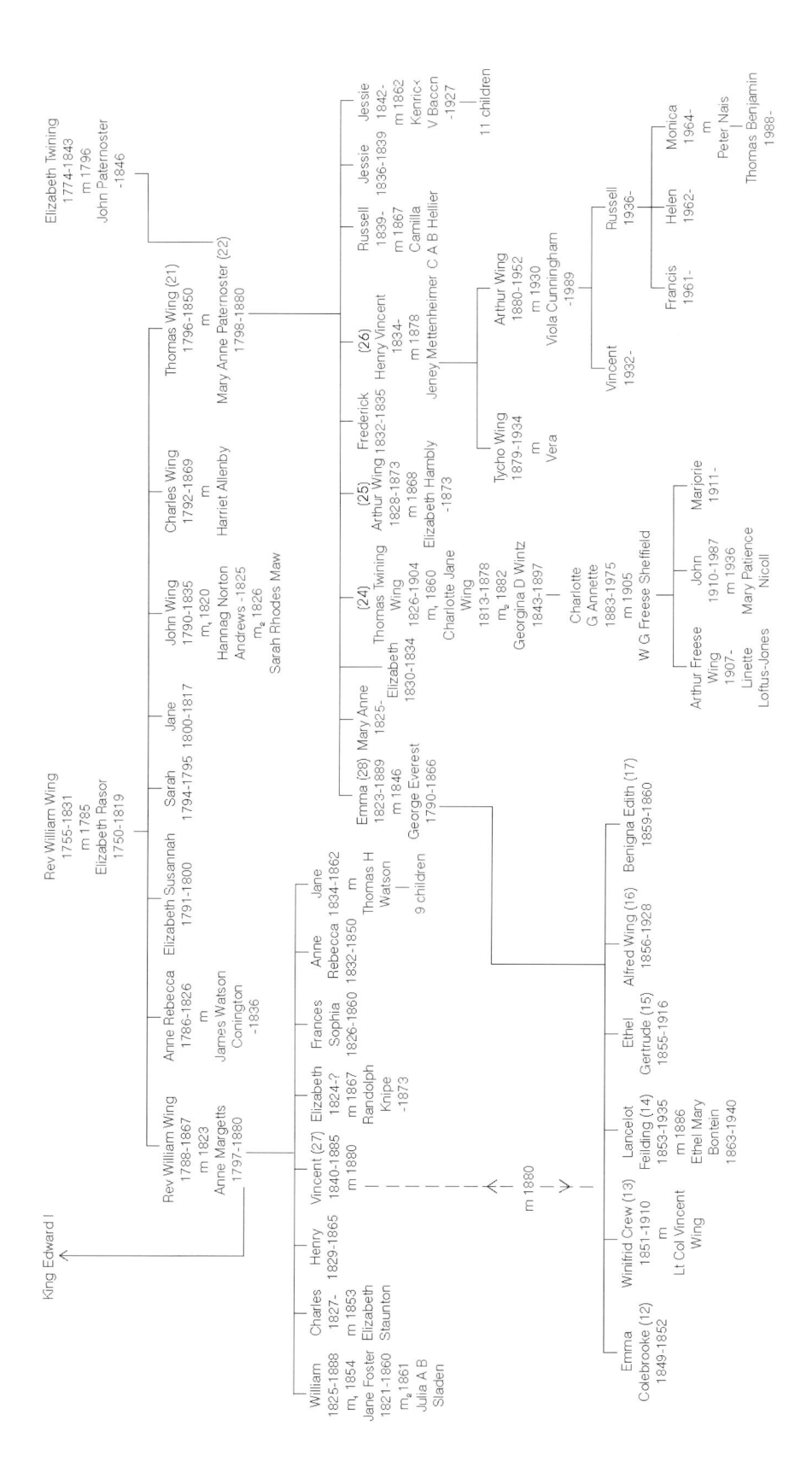

Figure 29 *The Wing family tree with Sir George and Lady Emma at the centre.*

{27} Lieutenant Colonel Vincent Wing

Born 25 May 1840 at Stibbington

Married 18 November 1880 at Christ Church, Lancaster Gate to Winifrid Crew Everest

Died 18 November 1885 at St Leonard's on Sea

Educated at Marlborough College, he joined the Royal Horse Artillery where he became Lieutenant on 21 December 1859, Captain 12 April 1872, Major 3 Oct, 1879, Lieutenant Colonel 19 Aug 1884. A horse trough was presented in his memory to Dorchester by his widow, where it is still to be found.

{28} Emma Wing (Lady Everest)

Born 12 July 1823

Married 17 November 1846 George Everest

Died 21 December 1889 at Upper George Street, Marylebone

She had four sisters – Mary Anne, Elizabeth, Jessie and Jessie; and five brothers – Thomas Twining, Arthur, Frederick, Henry Vincent and Russell. After her marriage she lived at 10 Westbourne Street, Paddington where she remained after her husband's death until at least 1875. By 1879 she was at 63 Queensborough Terrace but was no longer there in 1885. Sir George had a long lease on 10 Westbourne Street; it could have been for 30 years, and Lady Everest moved when the term was up.

{29} Professor Howard Everest Hinton FRS

Born 24 August 1912 in Mexico

Married 1937 to Margaret Rose Clark

Died 1977

He was Professor of Zoology, University of Bristol, from 1964 and Head of the Department of Zoology from 1970. The son of George Boole Hinton, he was educated in Mexico, the University of California and King's College Cambridge.

{30} Professor Sir Geoffrey Ingram Taylor FRS, OM

Born 7 March 1886

Married 1925 to Grace Stephanie Francis Ravenhill

Died 1975

He was educated at University College School, Cambridge University, and became a Fellow of Trinity College Cambridge 1910. At one time he was Yarrow Research Professor of the Royal Society. He did a variety of meteorological and aeronautical work for which he gained numerous awards. He worked on the first nuclear explosion. OM 1969 Kt. 1944 FRS 1919

{31} George Boole FRS

Born 2 November 1815

Married 11 September 1855 to Mary Everest

Died 8 December 1864

He was the son of a small tradesman in Lincoln. From the age of 16 he was teaching, first in Lincoln then Waddington. At 20 he opened a school of his own and all his spare time was spent studying Greek, Latin, French, German and Italian. His devotion to mathematics and logic came later. From 1849 he was Professor of Mathematics at Queen's College Cork. His wife Mary Everest was the daughter of the Rev. T.R. Everest and niece of Colonel George Everest.

Figure 30 Georgina Wintz, sister-in-law to Lady Emma Everest, and second wife of Thomas Twining Wing.

Other notable persons

{32} William Macbean Colebrooke (1787–1870)

He surveyed in Java during 1811–12. This included the Solo River, which Everest finished in 1815–16 at Colebrooke's suggestion.

{33} Emma Colebrooke (1.12.1799–?)

On 20 December 1820 Emma Sophia Colebrooke, a daughter of Robert Hyde Colebrooke, married William Macbean Colebrooke in Calcutta cathedral. [228 p. 391] There seems to have been some irregular kinship between Robert Colebrooke and John Garstin, who was Surveyor General of Bengal 1808–13. One of Garstin's six children, Charlotte, married James Charles Colebrooke Sutherland, on 26 August 1814 in Calcutta. He was a grandson of Sir George Colebrooke Bart, who was a Director of the East India Company; and chairman of it in 1769 and 1771.

{34} William George Feilding (1784–1868)

L.F. Everest had as a godfather, Colonel William George Feilding, after whom Lancelot obtained his second name. Feilding was 'a cousin of a former Earl of Denbigh'. [130 p. 12] From the *Dictionary of National Biography* it appears that the Feildings were from Feilding Hall in Warwickshire. (Also described as 'Of Newnham Paddox in Warwickshire' which was the seat of the Earl of Denbigh).

{35} John Bontein (1809–1878)

He married Elizabeth Mary Barnard and their only daughter was Ethel Mary who, in 1886, married Lancelot Feilding Everest {14}. He was an assistant in the Office of the Surveyor General from October 1838. Many of his later years were spent at Mussoorie.

Chapter 15

CHRONOLOGY

Everest, George: summary

Born	4 July 1790	Most probably at Greenwich
Baptised	27 January 1791	St Alphege Church, Greenwich
Married	17 November 1846	Emma Wing
Died	1 December, 1866	10 Westbourne Street, Hyde Park Gardens
Buried	8 December, 1866	St Andrew's (Old Parish) Church, Hove

George was the eldest son of William Tristram and Lucetta Mary Everest. His early education was at the Royal Military College Great Marlow and Royal Military Academy Woolwich.

1790	4 July	George Everest was born.
1804	11 July	Nominated as a gentleman cadet to the RMA.
1805	30 March	Joined RMA.
	30 November	Recommended for promotion and by RMA to East India Company.
1806		Entered the East India Company as a cadet and sailed to India.
	11 July	Arrived in India.
1812		Thought to have been at the siege of Kalinjar.
1813		Sent to Java.
1815	June–16 August	At Samarang in Java for reconnaissance survey of the Solo river. Then mapping interior areas.
1816	August	Reconnaissance of south coast harbours in Java.
1817		Involved in work on obstructions to the navigation of the outlets of the rivers Ichamati and Matabhanga near Calcutta.
	25 October	Appointed as a Chief Assistant to the Great Trigonometrical Survey of India. Effective from 1 January 1818. This marked Everest's entry into geodesy.
1818	End of year	Joined Lambton as Chief Assistant at Hyderabad.
1819		As a result of the route survey of his journey Everest planned improvements to the perambulator.

180   Everest: The Man and the Mountain

1820	8 May	Brother, John Everest, died.
	1 October	Sailed from Madras to the Cape of Good Hope on sick leave.
1821	End June	Obtained copy of LaCaille's report.
	31 August	Everest produced a report for his Directors and for Lambton on his investigations into the arc of LaCaille.
	31 December	Arrived back at Madras.
		Lambton produced his parameters for the figure of the earth.
1822	15 October	Everest parted from Lambton for the last time as he started work on the Bombay longitudinal series. Everest invented the vase light.
1823	13 January	First use of the wet and dry bulb thermometer in the field.
	20 January	Death of Lambton
	3 February	News reached Everest of Lambton's death.
	7 March	Appointed Superintendent of the Great Trigonometrical Survey under the Surveyor General.
	20 August	New attack of fever, and partial paralysis; very dangerously ill.
1824	End of May	Another serious attack of fever when near Hooshungabad.
	22 September	Still ill, without intermission since four months.
	13 December	Sironj base completed.
1825	11 November	Sailed for England on sick leave.
1826		Concept of Everest theodolite discussed with Simms.
	April-June	Wrote a memoir of 130 paragraphs on the GTS.
		Some time before mid 1827, went to Ireland.
1827	9 February	Elected member of Council of Astronomical Society.
	8 March	Elected a Fellow of the Royal Society.
1828	April	Visited Rome. Remarked on instruments in Naples Observatory. Also went to Venice, Vienna and Milan.
	November	Calculations of Arc started by Royal Observatory computers at Everest's home.
1829	February	At 50 Charlotte Street, Portland Place.
	March–June	Resident at 23 Beaumont Street, London.
	June	More equipment for compensating bars ordered from Troughton & Simms.
	5 July	Went to Ireland and visited Sir William Hamilton in Dublin and Colonel Thomas Colby on the Irish Survey.
	25 August	Nominated as Surveyor General of India in addition to his existing post as Superintendent of the GTS. Combined the two until his retirement.
	8 December	Recommended the appointment of Barrow.
1830	February	Completed the manuscript of his First Report.
	April	Tested compensating bars at Lords Cricket Ground. He published *An Account of the Measurement of an Arc of the*

		Meridian between the Parallels of 18°03' and 24°07'. 500 copies printed. Derived his first set of earth parameters.
	8 June	Left for India on the *Cornwall*.
	6 October	Arrived in Calcutta.
	8 October	Resumed his appointment to the GTS. Brought Argand lamps made by Simms and the first Everest theodolites as well as many other instruments.
1831		Introduced Chesterman's self-acting tape measure.
	23 November	Began measurement of the Calcutta base using compensating bars.
1832	28 January	Completion of Calcutta base.
		Survey of the Great Arc resumed and continued until December 1841.
	12 November	The great theodolite, as reconstructed by Barrow, reached Everest at Nojhli.
	December	Everest bought Park Estate from General Whish.
1833	10 May	Took up residence at Hathipaon – the Park Estate.
	15 October	Produced a differential perambulator.
1834	21 February	Official field office moved from the Park to Dehra Dun.
	31 August–5 September	Took an expedition into the Himalayas to see how far it was practical to extend series.
1835	January	First use of Everest's new 34 inch theodolite.
	18 February	Had no use in left thigh – an inflamed hip joint made him a cripple for life.
	End February	Began to walk again.
	May–October	Health suffered severely but would not go home. Confined to bed.
	October	Brother Robert stayed with him.
1837	2 September	Jervis appointed 'Provisional SG of India'
	10 November	Start of more attacks that periodically confined him to bed.
	1 December	Sironj base started.
1838	18 January	Sironj base completed by Waugh.
	14 July	Signing of the document by 38 eminent gentlemen.
	26 August	Jervis addressed the British Association.
	18 November	Everest keen to obtain a CB and put forward a case for its award.
		Fanny Parks visited Everest at the Park.
1839	October	Publication of Letters to the Duke of Sussex
	4 December	Commencement of observations at Kaliana.
1840		The 18 inch theodolite constructed by Mohsin Hussain.
	10 December	Everest back in field.

182   Everest: The Man and the Mountain

1841		Sought confirmation of ownership of the Park.
	25 August	Told Jervis that he was not going to retire in 1841.
1842	9 January	Completion of the Arc.
		Started disposing of the Park.
	November	Submitted his official resignation.
1843	3 May	The Directors of the East India Company recorded their appreciation of Everest's work and agreed to the appointment of Waugh as successor. Everest was to retire on the pension of a full Colonel.
	1 October	Everest left Hathipaon for the last time.
	1 December	Resignation was gazetted.
	16 December	Sailed on the *Bentinck*.
1844	9 September	Grant of Coat of Arms.
	November	Resident at 16 Bury Street, London.
1845	February	Elected a member of the Royal Institution and to the Royal Geographical Society.
	14 March	Again sent a Memorial to the East India Company about a CB.
	July–October	In USA.
1846	17 November	Married Emma Wing.
1847	June to September	Resident at Claybrooke Hall, Leicestershire.
	Early June	In a train crash on way from London to Hinckley.
		First observations made to the mountain.
		His *Account of the Measurement of Two Sections of the Meridional Arc of India Bounded by the Parallels of 24°07'15" and 29°30'48"* was published. It contained Everest's second set of parameters for the figure of the earth.
1849		Daughter, Emma Colebrooke Everest, born at Hove.
1850	12 November	Father-in-law, Thomas Wing, died at Brighton.
1851	13 September	Daughter, Winifrid Crew Everest, born.
1852	10 February	Daughter, Emma Colebrooke Everest, died.
		Computations indicated a highest peak of 29 002 ft. (XV).
1853	28 May	Son, Lancelot Feilding Everest, born.
1854	15 June	Brother, Thomas Roupell Everest, died at Wickwar.
		Daughter, Ethel Gertrude Everest, born.
1856	1 March	Waugh wrote to the Royal Geographical Society to inform it that he had named Peak XV as Mont Everest with height of 29 002 feet.
	3 August	Everest's son, Alfred Wing Everest, born.
	6 August	Major Thuillier announced the height of mountain to the Asiatic Society of Bengal.
	September	The Name of 'Mount Everest' reluctantly accepted by RGS.

Chronology

1857	10 January	Sister, Lucetta Mary, died.
1859	May	Became a visitor to the Royal Greenwich Observatory.
	3 September	Daughter, Benigna Edith Everest, born at Hove.
		Became a Manager of the Royal Institution.
1860	24 January	Daughter, Benigna Edith Everest, died at Paddington.
	October	Elected an Honorary member of the Asiatic Society of Bengal.
1861	26 February	Awarded the CB.
	13 March	Knighted.
1862		Chosen Vice President of RGS.
1863		Park Estate purchased by Col Thatcher
1863–5		He was on the Council of the Royal Society and also on the Council and a Vice President of the Royal Geographical Society.
1866	1 December	Died at 10 Westbourne Street, Hyde Park, London.
	8 December	Buried in Hove.

Later relevant events

1874	7 February	Brother, Robert Everest, died at Ascot.
1880	4 December	Mother-in-law, Mary Ann Wing, died at Sudbury.
1889	21 December	Lady Everest died at Upper George Street, St Marylebone.
1910	April	Daughter, Winifrid Crew Everest, died at Lyndhurst?
1916	21 March	Daughter, Ethel Gertrude Everest, died at Hever.
1930c		Everest's chronometer donated to the Royal Artillery Institution, Woolwich.
1932	July	Sketch of Sir George presented to the National Portrait Gallery. NPG Exbt. 2553.
1935	1 April	Last surviving child, Lancelot Feilding Everest, died.
1974	2 February	Everest's chronometer stolen from the Royal Artillery Museum, Woolwich.
1988	June	New portrait of Sir George Everest unveiled in the Royal Geographical Society; commissioned by the Survey of India veterans' organisation.
		Government of Upper Pradesh purchased the Park Estate.
1990	4 July	On bicentenary of his birth, a ceremony at his graveside; a special philatelic cover issued in UK.
		Director General of Tourism proposed renovations to the Park Estate.
	4–5 October	Bicentenary celebrations in Dehra Dun.
	4 October	Special stamp and philatelic cover issued in India; and cover with special postmark in UK.

184 Everest: The Man and the Mountain

Figure 31 *Special philatelic cover issued on the bicentenary of Everest's birth.* Reproduced by kind permission of A.G.Bradbury first day covers.

	4 October	Various presentations made to the Survey of India including a painting of Everest's Coat of Arms. The Survey of India unveiled a bust of Sir George Everest at their Dehra Dun Headquarters and a memento in the form of a replica of the 36 inch theodolite was produced.
	8 November	Commemorative conference at the Royal Geographical Society, London.
1991	25 March	BBC 2 TV Horizon programme on surveying in India.
	April–May	Granada TV series that included a programme on surveying in India.

Figure 32 Bust of Sir George Everest unveiled at the Dehra Dun HQ in 1990.

Part III

The measuring and naming of Mount Everest

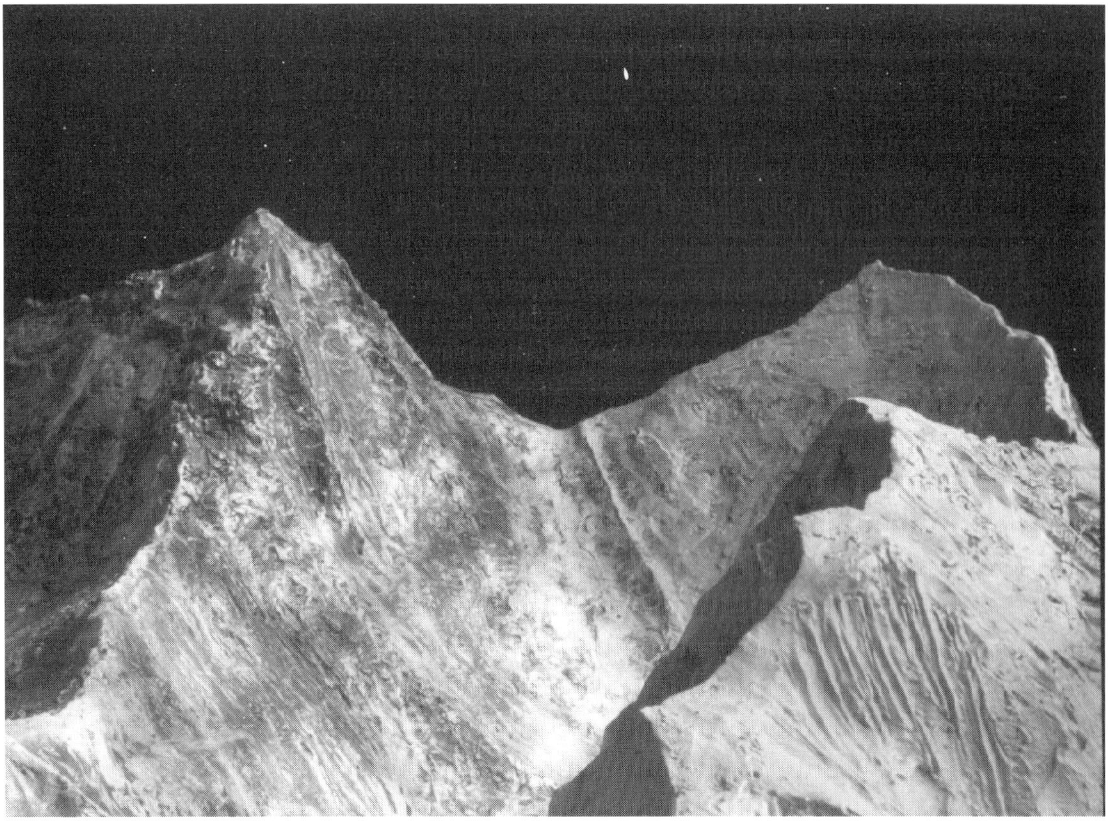

Chapter 16

THE MOUNTAIN: Measuring its height

Various Central Asian travellers of the early nineteenth century, such as Colonel Crawford and Captain Webb, tried observing the heights of some of the Himalayan peaks. Among their problems were the unknown heights of their observing positions and this led to inaccuracies of several thousand feet. As a result, their attempts were virtually worthless. [8 p. 367], [57 p. 454]

The heights of the massive peaks were obviously not going to be easy to determine. Not least of the other problems was that much of the eastern part of the Himalayas fell in Nepal to which no access was permitted. It was not until 1903 that an experienced British surveyor was allowed into Nepal and even then it was only as a result of the direct personal intervention of Lord Curzon. Until that time the only possibility for co-ordinating and heighting the peaks was by observations with theodolites set in the plains of Bengal and Oudh over 100 miles away. [54 pp. 42–4]

Which is the highest mountain?

In the seventeenth and eighteenth centuries Chimborazo (in Peru) at 20 702 feet had been considered the highest mountain and then in 1809 Dhaulagiri (in the Himalayas) came out at 26 860 feet. [194] Before 1850 Nanda Devi (Himalayas) had been considered a contender at 25 645 feet, then Waugh found Dhaulagiri to be better represented as 26 795 feet until briefly superseded in 1847 by Kanchenjunga (Himalayas) at 28 168 feet. [289 p. 345]

In 1845, Andrew Waugh felt that in his position as Surveyor General it was about time that he arranged for the measurement of the Himalayan peaks. Among the difficulties to be considered in such an undertaking was the fact that it was estimated that the Himalayan region contained some 40 000 peaks of which 25 per cent were under perpetual snow and probably less than 50 had a local name. [54 p. 42] One other main problem – that of atmospheric refraction – stopped, or at least slowed, his progress. Although Waugh was very much aware of the problem it was really brought home to him when one of his observers reported that the height of Dhaulagiri was apparently changing by 500 feet between morning and afternoon because of refraction. Such effects would naturally have a profound influence on any observations and could play havoc with all vertical angles. It is uncertain, however, how much was known at the time of Waugh about refraction and its effects on long rays such as those required to observe Mount Everest.

Probably the best text available then was Puissant's *Traité de géodésie* the third edition of which had been published in 1842. From that text it is obvious that much still had to be learnt on refraction. It was already acknowledged that it was a quantity that varied between the seasons and with the weather, but there is no comment on diurnal variations. This is somewhat strange since the diurnal variation of refraction was certainly well illustrated to George Everest in the early 1830s when he was able to make considerable use of the effect while fixing his Great Arc stations. He wrote to Boileau that the time of maximum refraction varied from midnight to sunrise but was generally about 3 a.m.

This was, however, very much the exploitation of refraction rather than the limitation or control of its effect as needed for heighting. In addition, Lambton in 1814, had reported that he had found refraction to vary between 1/4 and 1/20 of the contained arc. He then indicated that he intended to discover a law covering its variation and in particular he recommended the use of an hygrometer. As a result of these phenomena one might have expected that temperature and pressure would have been observed at the same time as vertical angles but this was not so during the period of the initial sights to what was to become known as Peak XV (and then later Mount Everest).

First observations

What was to be done? By the time that George Everest had retired in 1843 there was the beginnings of a chain of triangulation in the plains to the south of the foothills of the mountains. This was the North East longitudinal series and by 1851 it stretched from the north end of the Great Arc across the sub-continent to the Calcutta meridional series just east of the 88° meridian – some 10° extent in longitude. But, even so, points on this were over 100 miles from what has turned out to be the highest peak, and other major peaks.

It had originally been Everest's idea to carry the triangulation into the mountains but he had to abandon that idea when it proved impossible to penetrate the depths of Nepal. Thus once the west-to-east chain had crossed the elevations of Kumaon it then had to come down into the inhospitable plains and continue for some 800 miles along the foothills.

Although many peaks could be seen from these low-lying stations, in order to fix any one peak in position and height, it needed to be visible from at least two, and preferably more, known positions. The achievement of this was aggravated by the changing outline of any mountain as it was viewed from different aspects, and this made it almost impossible to use simple optical recognition to identify a particular peak. Two different observers might actually observe to the same peak but without knowing it. Hence each would give it a different notation and only when all the observations were computed could likely combinations be unscrambled.

The first recorded observation to this principal peak (Peak XV) seems to have been an isolated one on 13 and 14 November 1847 by J.W. Armstrong. While he was on the Gora series near the 83° meridian, he made an observation at Sawajpor, a few miles north of Muzaffarpur, and 200 miles from the great peak, of the vertical angle and distance to a point that he designated peak b. Note that by 'distance' it does not mean that he had equipment to measure directly. Distance in this sense was obtained by intersecting one point from two other points whose distance apart was known. With the observed angles it was then possible to solve the triangle so formed for the lengths of the unknown sides to the mountain. His result, with a 15 inch theodolite, converted to a height of 28 799 feet. [158 p 1], [232 p. 17]

During the same month Waugh and Lane were on triangulation near Darjeeling and observed various of the snow peaks. One of these they called γ and this later turned out to be identical to the peak b of Armstrong. [232 p. 82] In addition Kanchenjunga was found in 1847 to be at 28 146 feet.

In October 1849 it was recorded that 'One of Mr Armstrong's peaks appears to be the highest Himalayan mountain, and requires verification, having only a single determination of distance and one of altitude, the observations being taken at a great distance.' [263 (493) 243] Though the rays observed by Armstrong and Waugh were later rejected, they did contribute to the identification of the peak and to Nicolson's success. Lane's horizontal ray from Doom Dangi was used for computation of position but had no vertical angle to contribute a height. Verification of the results by Armstrong and Waugh did not come until December 1849 when James Nicolson made observations from several stations to a peak he called h. These rays over distances of less than 120 miles were much shorter than the previous attempts. He completed his work in January 1850 and presumably computed preliminary heights whilst he was at Monghyr during the rainy season. Even so, the Surveyor General was not in a hurry to disseminate any such values.

In December 1850 Waugh asked his chief computer Radhanath Sickdhar in Calcutta to revise the form of computing geographical positions of snow peaks at distances in excess of 100 miles and with azimuths up to 45° from north. For the next few years he was discussing refraction coefficients and the datum zero height (height value of the point from which the values of the mountains were to be calculated) which had to wait until Tennant completed tidal observations at Karachi at the end of 1855.

There is no official corroboration that 1852 was the first year that the Surveyor General knew for certain of the great height of Peak XV. As early as 1848 he had reported Armstrong's height for b and this would have been confirmed by Nicolson's field computations of 1850. [232 pp. 91–2]

To compound the problems, peaks that later turned out to be extremely high appeared just as ripples on the horizon, overshadowed, apparently, by nearer peaks which seemed to be much higher. In addition the geographical location of the peaks was just as uncertain as the heights and, since their positions related to sight lengths, and in turn to differences of height, it was just as critical to get the geographical position correct as it was the height.

Unscrambling the results

Originally, during field observations of 1847, the peak that was to become Mount Everest was designated as b by Armstrong, γ by Waugh and Lane and then, in 1849, h by Nicholson. The name was changed to Peak XV at headquarters by John Hennessey, the computer to whom Waugh expressed his great indebtedness for the work on the Himalayas.

When referring to work on the North East longitudinal series Waugh said that 'The lofty snow peaks situated north of Nepal are the most stupendous pinnacles of the globe. Their heights and relative positions should form permanent objects in the geodetic operations.' [158 p. 1] Waugh instructed his surveyors to observe from every station the height and position of every snow peak that was visible but not to attempt to identify them. This would be done when the plots were untangled in the office.

'You should be in the observatory before sunrise and all prepared to commence horizontal angles as soon as it is light. The vertical angles may be taken from 8 to 10 o'clock a.m.' Determination of heights above the geoid corresponding to the datum were complicated by refraction, as already mentioned, which in the case of sights to Mount Everest could have an average value of the order of 800 feet and be much greater on the longest lines. The effect was smallest during early afternoon and as a result the practice adopted from about 1850 was to observe vertical angles between 1300 and 1600 hrs. [232 p. 291]

To resolve the situation that materialised in the office when numerous observations to unidentified peaks came in from the field a four step approach was adopted. [54 p. 43]

1. The observation stations were projected on to a map and lines were drawn from each representing the directions to all observed peaks.
2. When direction lines from three or more stations met at a point it was tentatively assumed that the same peak had been observed from several points.
3. The distances to the possible peak from each observing station were then calculated and independent values of position found. Where several results agreed, the fix was continued with.
4. Using the observed vertical angles and computed distances, the heights for a peak were determined and again, if there were several in agreement, then the identification was accepted. Naturally there were many peaks that did not get the necessary agreements and were rejected.

A peak of great altitude

Seventy-nine Himalayan peaks visible from principal trigonometrical stations were intersected with the great theodolite. [210 p. 105] The largest triangle was of 1706 square miles with a maximum side of 151 miles. Thirty-one of the peaks were found to have names but the others had to make do with numbers.

By 1852 the chief computer became aware that there was a peak of great altitude but again because of refraction uncertainties it was not until 1856 that a reliable value was achieved. [56 p. 1]

Burrard [54 p. 43] suggested that the chief computer was sited in Calcutta and this was borne out by research done by Phillimore; but at the same time it was also proved that all computations relating to the peaks were done in Dehra Dun. Thus, while some authorities considered the chief computer, Radhanath Sickdhar, to have been the person who 'discovered' the highest mountain, Phillimore disputed this because of his location.

Both Airy [1 p. 328] and Howard-Bury [173 p. 10] credit Sickdhar:

and then one day the Bengali chief computer rushed into the room of the Surveyor-General Sir Andrew Waugh, breathlessly proclaiming 'Sir! I have discovered the highest mountain in the world!' [173 p. 10]

On the other hand Waugh specifically expressed thanks to Hennessey for his work on these computations. In any case it would seem more appropriate in such a situation to consider it as a team effort since the field observer J.O. Nicolson surely warrants a part of any praise that is being distributed. The height value had been determined from six independent observations from stations in the plains of India that were at an average elevation of 230 ft and distances of 108–118 miles from the peak. [57 p. 455], [158 p. 1] The geographical positions were determined from the same six observations and one additional station which together formed 11 triangles.

By February 1856 Hennessey's computing section had completed computations of the peak positions. [232 p. 91] As a result the positions were considered to be good to 0.25 seconds in latitude, 0.5 seconds in longitude and heights to 10 feet – although the heights later appeared to be all too low probably because of either deviation of the vertical (the angle between the tangent plane of the spheroid and that of the equipotential plane at the corresponding ground point) or refraction. Peak XV from among these was calculated at 29 002 feet.

Summary of early results

In summary the pertinent results up to that time were:

Armstrong (b) 13–14 November 1847 28 799 ft from a single observation. This value was derived from the nearer of the two stations of the Gora meridional series at 160 and 180 miles from which it was intersected. [232 p. 93]

Waugh and Lane (γ) 1847 from Darjeeling – not to be confused with the later observations from this vicinity in 1880.

Nicolson (h) November–December 1849. The results are listed below but the final height had to await various adjustments and knowledge of a reliable datum. He used a 24 inch theodolite at stations on the North East longitudinal series. [288 p. 111]

Despite the use of six stations the configuration of the rays was such that the maximum intersection angle was 36°, not by any means ideal.

Station	Distance (miles)	Date	Vert. angle	Height (ft)	Coeff. of refraction k
Jirol	118.661	27.11.1849	1°53′33.35″	28 991.6	0.073 525
Mirzapore	108.876	5/6.12.1849	2 11 16.66	29 005.3	0.073 651
Janjipati	108.362	8/9.12.1849	2 12 09.31	29 001.8	0.073 150
Ladnia	108.861	12.12.1849	2 11 25.52	28 998.6	0.074 613
Harpur	111.523	17/18.12.1849	2 06 24.98	29 026.1	0.072 655
Minai	113.761	17.01.1850	2 02 16.61	28 990.4	0.075 272
			Mean	29 002.3	

[158 p. 17]

Effect of refraction

The quotation of the coefficient of refraction to six decimal places was perhaps a stretch of the imagination but, with the extreme circumstances, the last of these decimal places could be significant since 0.000 01 was equivalent to about 1 foot in height. Gulatee indicated that Waugh had made extensive observations with reciprocal vertical angles between the Bengal Plains and near Himalayan peaks. How he reduced the results to obtain such coefficients is not recorded but they must contain considerable uncertainty. [158 p. 2]

Then the effect of refraction on long lines came to the fore. It could equate to 500 seconds, or over 1000 feet, on some of the longer rays to Peak XV, and could be aggravated by low elevations and diurnal changes. Refraction depends on the temperature, temperature gradient and pressure. It has been shown to be incorrect to consider always that simultaneous reciprocal vertical angles cancel out refraction not only in plains but also for long steep rays. Observations to Peak XV required corrections at the extreme of up to 1375 feet for refraction with possible angular variations on 100 mile rays of up to 200 seconds between a.m. and p.m. observations. [157 p. 6]

Later calculations were to suggest that the value of 29 002 feet was too low due to over-correction for refraction. As will be seen from the above table, Waugh used values of k ranging from 0.0727 to 0.0753 to obtain the celebrated results. Later, how-

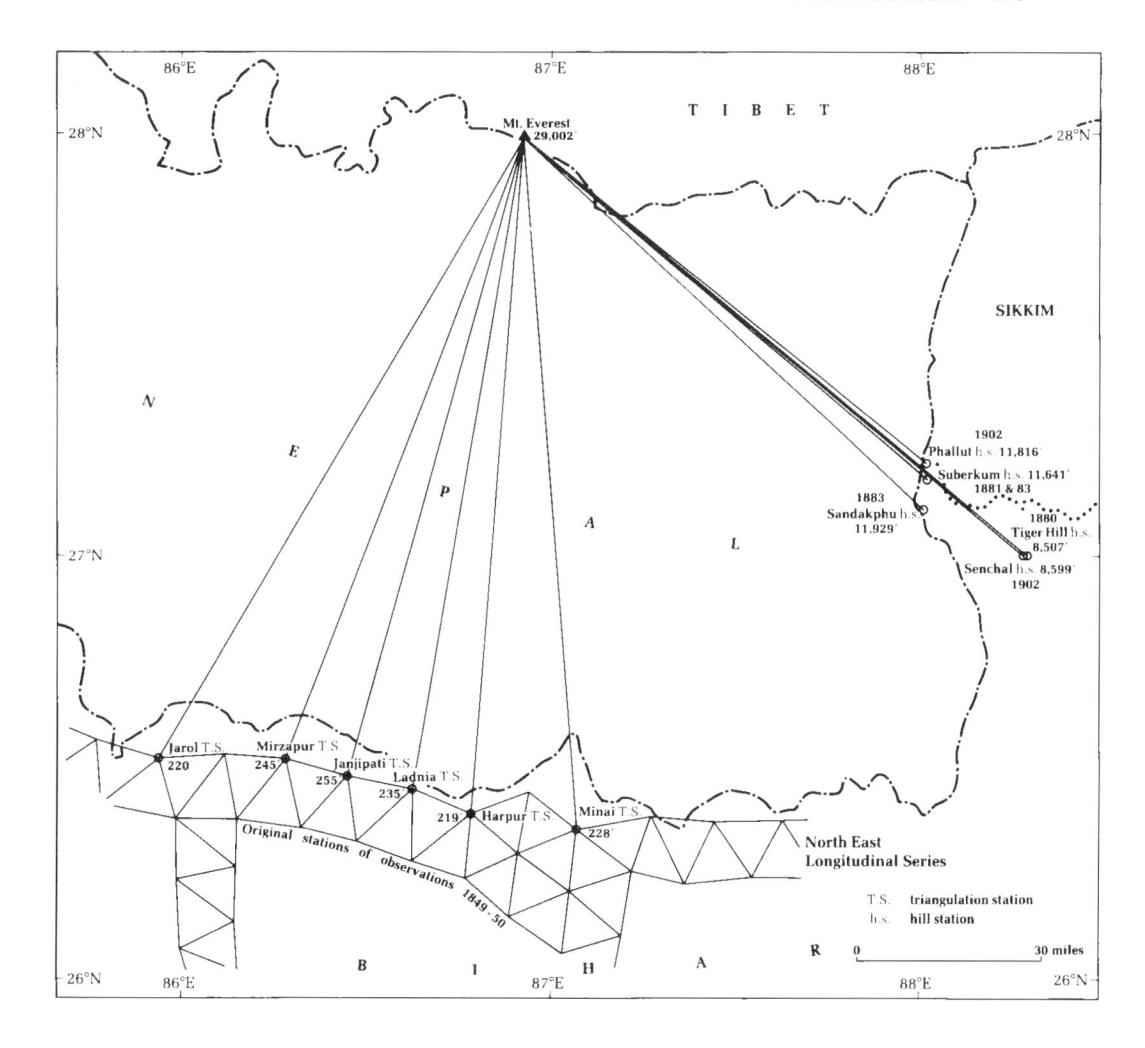

Figure 33 *Observations to Mount Everest at the time it was first coordinated and heighted together with the observations of 1880–1902.*

ever, these were considered to be too large by 15% and that a more appropriate figure of 0.064 should have been used. [179 p. 22]

Nevertheless de Graaff Hunter was full of praise for Waugh and the drive and enthusiasm that led him to get the necessary observations. [158 p. 17]

What is it that is being measured?

The question at that time was whether or not the '2' at the end of the height value should be omitted. If left it might appear to imply an enhanced accuracy but if it were dropped it would then look as if the value was only known to the nearest 1000 feet. [175 p. 287]

Is 29 002 an exact height? De Graaff Hunter used an analogy.

> The height of the Eiffel Tower is stated to be 984 ft. But what about the legs of the tower and foundations – should not they be included? Then beneath the tower is a complex of radio installations. The point is that not only top but bottom also needs defining before quoting a height. In the case of a tower it would be possible to decide what the bottom was and get access to it. In the case of the Eiffel Tower however one would need to remember that it changes by some 3 inches seasonally with temperature. With mountains the base is certainly not visible. To use the term mean sea level is only a concept since the sea is not under mountains. Spirit levelling would avoid the difficulty or to a lesser accuracy the use of trigonometric heighting techniques. [175]

Problems affecting the accuracy

There are three main factors that generally affect the accuracy of the height determination of a mountain:
- atmospheric refraction,
- geoidal separation,
- deflection of the vertical.

In the case of Mount Everest, however, the effects were compounded by:
- the vast lengths of sights,
- the great difference of height from the Plains to the peak,
- the huge mass of the mountain range,
- the distance of Mount Everest from the sea
- the impossibility of occupying both ends of the line. [8 p. 371]

It had also to be remembered that the 29 002 feet was a spheroidal height i.e. a value related to a specific figure of the earth. Whilst India used the Everest spheroid[*] determined in 1830 with a datum at Kalianpur and separation there of zero, it was so arranged that the geoidal separation[†] varied from +50 feet at Karachi to –45 feet at False Point. [158 p. 5] Thus to obtain sensible height comparisons it was really essential to use geoid values or those pertaining to a line of potential coinciding with mean sea level. It was thought that the difference under Peak XV could be as much as 150 feet and Gulatee argued that because of different countries having different spheroids it was also sensible to adopt geoid height values for their universality. [178], [157 pp. 5–6]

---

[*] spheroid equals the mathematical shape most nearly fitting that of the earth. There are many versions, of which two were determined by George Everest.

[†] distance between the spheroid and sea level surface at a given point.

Another problem was that of datum. Several reference surfaces were available, some real some imaginary. The most important datums were the reference spheroid and the geoid (mean sea level surface). However there was a lack of knowledge about the separation between the geoid and spheroid. The particular spheroid used is almost incidental other than that it forms the mathematical surface that best fits the geoid in a particular region.

Geoidal height is understood to be that above sea level – freed from tidal effects – but the sea was 450 miles from Peak XV. Practically this can be overcome in fairly flat country by spirit levelling but it fails in hilly country. [175 p. 290]

A theodolite when levelled indicates the normal to the geoid so observed vertical angles are really geoidal angles. Height formulae generally make the assumptions that observed angles are spheroidal and that the geoid is a sphere. Deflection of the vertical is the difference between the normal to the geoid and normal to the spheroid at a point and requires a combination of astronomical and trigonometric observations to sort it out. As seen in relation to George Everest's work in South Africa (Chapter 4) there is a tendency for large mountain masses to pull the plumb line slightly out of the vertical towards the mass. The effect is for vertical angles near mountains to be too small. [157 p. 6]

Whilst heights are required above the geoid and can be found from the trigonometrical heights by application of corrections for the deviation of the vertical and for geoidal separation, at the time of the early heighting observations there was a complete lack of information to allow an application of these corrections.

As Gulatee [157] said, in the light of more recent knowledge, the value of 29 002 feet contains several sources of error. Besides refraction difficulties, no measures were made for temperature or pressure, nor were observations taken at the time of minimum refraction. Even when refraction was allowed for in the computations the coefficient used varied from 0.07 to 0.08 which would seem to be too large for a ray that extends from a few hundred feet above sea level at one end to 29 000 feet at the other terminal. Gulatee considered that the error due to this fact could be as much as 200 feet.

He also considered the distances to have been far too long for accurate results (albeit the shortest that were possible at the time), and that no account was taken of deflection of the plumb line. The result might then be described

> either as a preliminary geoidal height or a rough height above the Everest spheroid so placed as to touch the geoid under the North Bihar Plains. This is not our present definition of the Everest spheroid. [157 p. 8]

Announcement

In a communication to the Proceedings of the Royal Geographical Society for May 1857 Lieutenant Colonel A.S. Waugh wrote from Dehra Dun on 1 March 1856.

> You are aware that the computations of positions and elevations of all the principal

peaks of the stupendous Himalaya, comprising 18¾ degrees of longitude, from Assam to the Safed Kho, have been provisionally completed,… [289 pp. 345–6]

Previous to publication, however, it is essential that the computations should be scrupulously revised and every refinement of correction introduced…

…I am now in possession of the final values for the peak designated XV… We have for some years known that this mountain is higher than any other hitherto measured in India, and most probably it is the highest in the world. [232 p. 91]

Waugh at this stage decided to name the peak Mount Everest (see Chapter 17). Although it was in 1855–6 that this new, unnamed peak was announced to be 29 002 feet, it was not until 1860 that Waugh felt justified in concluding that this was the highest peak in Asia. In 1856 another unnamed peak, K2, had been discovered at 28 250 feet and other higher ones could still have materialised. As with Peak XV (Mount Everest), no undisputed local name has ever been found for K2 and it has kept the designation given to it by Colonel Montgomerie in his field book. [57 p. 455] By 1860 the three most elevated peaks known in the world were Peak XV, Peak K2 and Kanchenjunga, but Waugh could not have known then that no higher ones would be discovered. The final value of the co-ordinates of Peak XV were given as latitude 27° 59' 16.748" longitude 86° 58' 05.852" and height 29 002 feet.

The pertinent values

The geographical values were from seven observations that varied by only 0.5 seconds in latitude and 0.4 seconds in longitude. Note, however, that the longitude was in terms of the old value for the longitude of Madras Observatory – a value that was to be later corrected by some 3'. [289 p. 347] Waugh considered that it would be interesting to record the independent results of all the observations of the mountain and of three others.

		Mean
Everest	28 990, 29 026, 28 999, 29 002, 29 005, 28 992	29 002
Choomalari	23 946, 23 941	23 946
Kunchenginga*	28 151, 50, 63, 47, 80, 42, 72, 60, 40	28 156
Dhaulagiri	26 815, 60, 43, 06, 61	26 826

* alt. sp. Kanchenjunga

The geographical positions of the other three mountains were given as :

Choomalari = I	27° 49' 41.5" N	89° 18' 43.1" E	from 2 control stations
Kunchenginga= IX	27 42 09.4	88 11 26.3	from 9 control stations
Dhaulagiri = XLII	28 41 48.0	83 32 08.6	from 8 control stations

On the same date, 1 March 1856, that Waugh had addressed his letter to the Royal Geographical Society, he felt justified in promulgating similar information in India and this he did in a letter to his deputy Major Thuillier in Calcutta. Waugh's letter continued:

I ... append an attested statement on the geographical positions and elevations of ...Mount Everest ... You are at liberty to make use of these results in anticipation of my forthcoming report ...greatly indebted to assistant J. Hennessey... for the revised computations. ... Mount Everest from 7 stations = 29 002 feet. [No precaution was overlooked including 0.2 feet] ... for correction required to the height of distant mountains to reduce to the ellipsoid when difference of latitude is great. [287]

In March 1858 the Surveyor General issued a complete list of the peaks I–LXXX with their local names where these were known.

Not a prominent peak

From many aspects Everest is by no means a prominent feature. In 1921 it was described as

Mount Everest, for its size is a singularly shy and retiring mountain. It hides itself away behind other mountains. On the north side, in Tibet, it does indeed stand up proudly and alone, a true monarch among mountains. But it stands in a very sparsely inhabited part of Tibet, and very few people ever go to Tibet. From the Indian side only its tip appears among a mighty array of peaks which being nearer look higher. Consequently for a long time no one suspected Mount Everest of being the supreme mountain not only of the Himalaya but of the world. At the time when Hooker was making his Himalayan journeys – that was in 1849 – Kanchenjunga was believed to be the highest. [173 p. 9]

In the last quarter of the nineteenth century Hari Ram and Gunderson Singh acquired information on routes in Nepal and possibly reached within 15 miles of Makalu and perhaps 18 or 20 miles of Mount Everest. Up to 1881 this seems to have been the nearest which any trained individual, even Asiatic, had penetrated towards the mountain and this remained so until 1921 [134 p. 87–95], although Captain J.B.L. Noel made a reconnaissance of the area in 1913 [173 p. 14] 'Mount Everest cannot be approached by a European within eighty miles, owing to the persistency with which the Nepalese Government keep Englishmen out of their country. Sandakphu, thirty eight miles from Darjeeling... commands the finest view of Everest that is anywhere obtainable from British Territory'. [137 p. 442]

More on refraction

Around 1870 General J.T. Walker was investigating refraction in the Punjab Plains and met the phenomena of inverse – or negative – refraction. This makes the observer feel that he is in a saucer at dawn, later seeing objects such as survey targets sinking below the horizon as do ships at sea – except that they are at least moving!

Between 1881 and 1902 five observations were made from the Darjeeling Hills of Sikkim where refraction was more predictable. One station was occupied twice [57 p. 455]. The coefficient of refraction was taken then as 0.05.

Sir Sidney Burrard was the next to consider the heights of Himalayan peaks. Because of his concern for refraction he specifically set up experimental observations near Dehra Dun between stations at Nojhli (887 feet), Mussoorie (6930 feet) and Nag Tiba (9915 feet) to snow peaks 50, 60 and 100 miles away. [179 p. 22]

The results for Mount Everest computed by Burrard in 1905 ranged from 29 134 feet to 29 151 feet with a mean of 29 141 feet. The stations from which these were derived were at elevations between 8500 feet and 11 900 feet and at distances from 85 miles to 108 miles. He also obtained the same value by recomputation of the original data with a different coefficient of refraction. [158 p. 2] For k he used 0.0645 for rays from the Bihar plains and 0.5 for the much higher observing stations in the Darjeeling Hills.

Despite the agreement between his results, Burrard did not claim any finality because deflection of the plumb line had to be omitted through lack of data relating to it. As de Graaff Hunter put it [179 pp. 22–3] 'Indeed the era of the geoid had not yet arrived'. Burrard gave the error sources in his determination of the height as perhaps as much as 200 feet. [292 p. 12] Much later, Fellowes gave his assessment as approaching 400 feet. [134 p. 86]

Nevertheless Burrard's 29 141 feet was a value that stuck for many years. It appeared on American maps and an eminent mountaineering author stated in 1949 that 'the true height of Mount Everest is 29 141 feet'. He then went on to give an erroneous explanation of the difference from 29 002. From 1907 Dr J. de Graaff Hunter further investigated the refraction problems and their dependence on height, temperature, and pressure. In 1913 he published formulae for atmospheric refraction and in 1922 deduced that the local spheroidal height of Mount Everest should be 29 149 feet. He had made an effort to reduce earlier observations by using extrapolations of refraction but it gave a somewhat confused picture. It was neither above the Everest spheroid nor above the geoid. He allowed a geoid height of 70 feet to give the elevation of Mount Everest above sea level as 29 079 feet, (sometimes quoted as 29 080 feet). [157 p. 9]

In 1929 Bomford gave the estimate for geoidal height as less by 160 feet than the spheroidal height. [57 p. 457] (However, de Graaff Hunter [179 p. 23] says that Bomford used a geoidal rise of 100 feet.)

In 1913 Hunter noticed that diurnal refraction varied as temperature with minimum refraction at maximum temperature. He also noticed that the change was smaller on the longer of two rays from one station. He then went into the theory of refractive effects. Burrard's 1905 values agreed closely with a theoretical formula given by Gulatee but which was originated by Hunter.

Geoidal height

De Graaff Hunter's 1922 value of 29 079 feet was the first attempt at a geoidal height. He had used all known data relating to deflection of the vertical. Refraction was derived by a formula for which the temperature was required and when assessing old observations this was derived from values for the same season and hour as at the time of original observation. The basic difficulty with refraction was that for rays of 115 miles an error of 0.000 01 in k was equivalent to about a foot in height. Although all the observations were reliable, the big problem was: where is sea level in relation to the mountains? [57 p. 456]

Hunter in 1928 found refraction as 0.063 for the Bihar plains and 0.053 for the Darjeeling Hills. But although it appeared that the refraction problem had been cracked it was to be 1952–4 before adequate observations were available to be able to say the same for deflection of the vertical. [179 p. 23] All the various results for the mountain height were given by Gulatee thus:

Station observed from	Distance (miles)	Height (ft)	Date	Vert. angle	Height (ft)	k
Suberkum	87.636	11 641	4.10.1881	1° 35' 13"	29 141	0.0463
Tiger Hill	107.952	8 507	30.05.1880	1 21 44	29 140	0.0471
Suberkum	87.636	11 641	16/26.5.1883	1 35 11	29 137	0.0463
Sandakphu	89.666	11 929	28.05.1883	1 29 21	29 142	0.0457
Phalut	85.553	11 816	16.03.1902	1 37 50.2	29 151	0.0450
Senchal	108.703	8 599	23.02.1902	1 20 23.4	<u>29 134</u>	0.0513
					29 141	

[158 p. 17]

Some people imagine that if a climber were to reach the top of the mountain with the most sophisticated equipment (this was de Graaff Hunter writing in 1928) then all problems about height would disappear. Not so. He then went into the various possible methods for determining the height: barometric pressure/hypsometer/mercury/aneroid, spirit levelling; vertical angles. Barometric methods, he concluded, were too inaccurate and spirit levelling was quite impractical. [175]

Various accepted values

Briefly, from 1892 to 1903 the charts of triangulation showed 28 994 feet but this then reverted when it was realised that such change was premature (because of small modifications in the heights of observing stations). [158 p. 6] Freshfield commented in 1904 that 'In one of the last maps from the survey office at Calcutta to the RGS Map Room, the height of Peak XV is reduced by 8 feet, from 29 002 to 28 994'. [143 p. 363] (Note that he did not use the name Mount Everest.) The change had originally been made to account for an 8 feet (on average) reduction in the accepted heights for the six observing stations in the plains that had been used in the 1849 observations.

Thus the General Reports for 1892 to 1903 contained charts of the triangulation with 28 994 feet but in 1903 the old value was restored when it was realised that the change required might be more complicated than at first seemed. At the same time the heights of three of the observing stations used in the 1880–1902 work had been reduced by about 16 feet each.

The geographical position determined for the mountain varied slightly over the years. The original position for Everest was 27° 59' 16.748" 86° 58' 05.852".

Note that this longitude was referred to the old value of Madras Observatory of 80° 17' 21" and required a correction of 3' 25.5" to convert it to the new value later adopted by the Admiralty and Royal Astronomical Society. [289 p. 347]

In 1905 the co-ordinates were changed to 27° 59' 16.22" 86° 55' 39.91".

In 1953 the values were given as 27° 59' 15.85" 86° 55' 39.51".

This last pair of values arose from the 1952–4 observations under the guidance of Gulatee where additional horizontal angles to the peak resulted in a movement of some 40 feet south and 40 feet west of the formerly accepted position. The effect of such a movement towards the observing stations would be to lower the elevation by about 5 feet. [179 pp. 8, 24], [107 p. 8]

Hunter introduced the uncertainty due to plumb line deflection as this was needed to convert observed angles from geoidal vertical to spheroidal vertical. [175 p. 297] Although not enough values were known in India it was assumed that they were small at the Plains stations for Everest observations.

From best estimates, Hunter found the spheroid height as 29 149 ft with a probable error of 4.6 ft. The height related to the geoid he thought to be less by about 69 ft. An increase in field observations after 1910 was sufficient to allow a tentative estimate of 29 080 ft for the height above the geoid. Thus he finished with 29 080 p.e. 30 ft. If the later value of –109 feet for the geoid separation were used then the result would be in even closer accord with other values at 29 040 feet. [175]

Burrard, with his spheroidal value of 29 141 feet was considered to have been incorrect in his choice of datum. In effect it was even more vague than that of Waugh. It corresponded to a geoid height of 29 032 feet. [158 p. 16], [175 p. 292] Then Bomford in 1929 proposed an increase in the geoidal separation to 100 feet and hence reduction in geoid height to 29 050 feet. [179] All very confusing for the sake of a few feet!

Expeditions to Mount Everest

During the nineteenth century Mount Everest could not be approached by a European within 80 miles owing to the attitude of the Nepalese Government. [137 p. 442] It was not until 1903 that Captain Wood was allowed into Nepal to visit Kaulia and he was able to unravel some of the problems of names and peaks but he did not produce any new value for the height of Mount Everest. [54 p. 44]

Sir Francis Younghusband KCSI, KCIE in his preface to [173] said:

> so far as I know, the first man to propose a definite expedition to Mount Everest was the then Capt Bruce, who, when he and I were together in Chitral in 1893, proposed to me that we should make a glorious termination to a journey from Chinese Turkestan across Tibet by ascending Mount Everest.

This suggestion formed the basis of the first land attempts on the mountain.

The Mountain 203

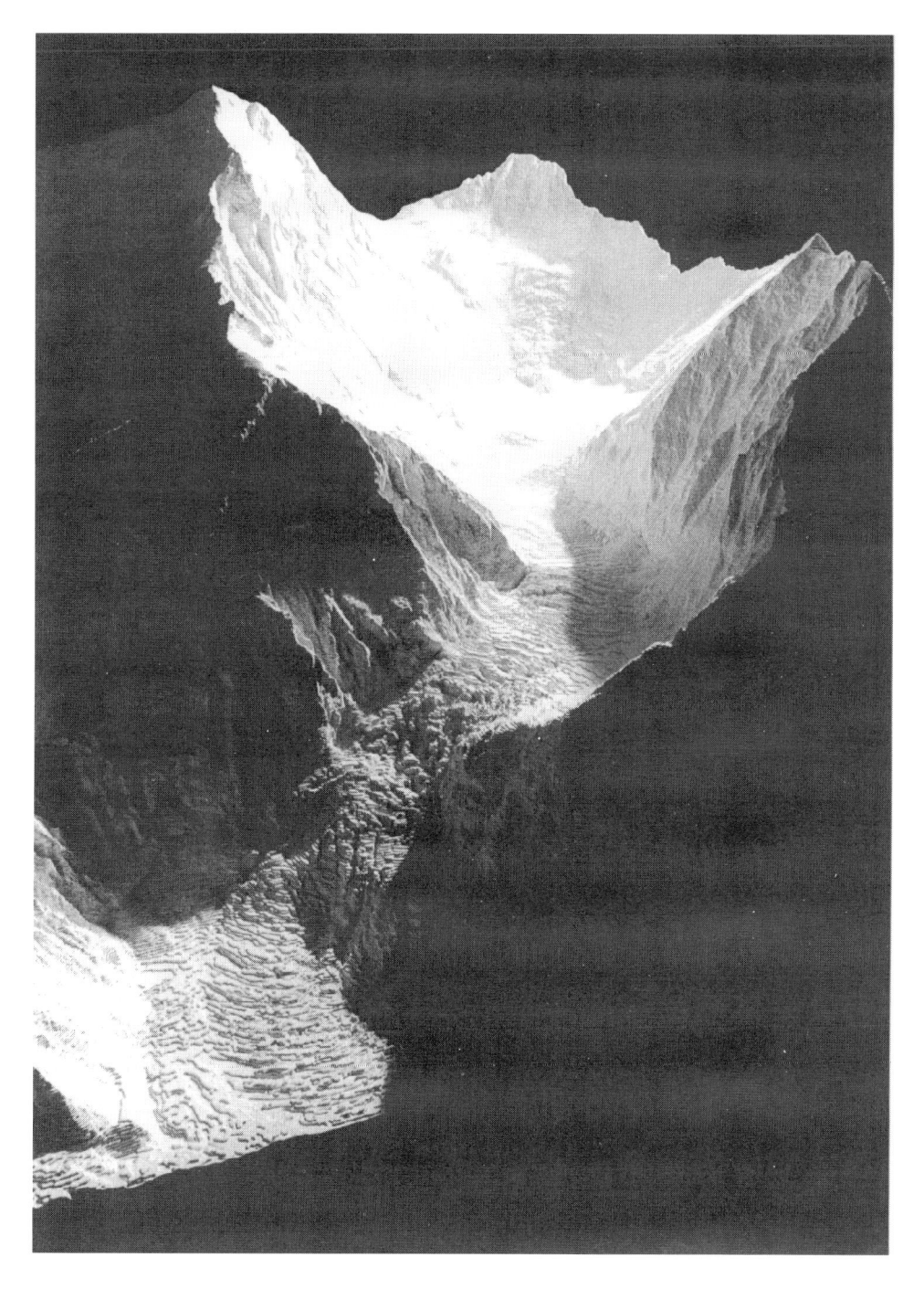

Figure 34 Part of the scale model of Mount Everest constructed from the aerial photography of Dr Bradford Washburn.

Then in 1921 a Survey Detachment under Brevet-Major E.O. Wheeler MC, RE set out with the aim to make a general survey at 1 inch to 4 miles of the whole unmapped area to be covered later by the Expedition; a detail survey at 1 inch to 1 mile of the neighbourhood of Mount Everest and a revision of the 1 inch to 4 mile sheets of Sikkim. [55 p. 319]

For the detail survey use was made of the Canadian pattern of photo-survey equipment that was particularly applicable to mountainous areas. It was a form of plane table by photography using the same sort of control net as for normal plane tabling but with the alidade and plane table replaced by a theodolite and camera. Larger areas could be covered in a given time and the equipment was more portable in the climbing conditions met in such areas. [173 p. 320]

On 26 March 1922 there was the first attempt on the mountain by General Bruce, then on 7 June 1922 a second attempt on which seven porters were killed and a third attempt on 7 June 1924 which saw the loss of Mallory and Irvine.

The first attempt to fly over the mountain was in 1933. [292 pp. 12–19] The plan for such a flight was conceived in March 1932 and by August the same year sanction for it was received from the Government of Nepal. The first test flight reached 35 000 feet in January 1933. At that height the temperature fell to −60°C outside and −40°C inside the cockpit. The first flight over the mountain was on 3 April 1933 followed by a second on 19 April 1933. [134 p. xviii]. While the position of Mount Everest in latitude and longitude was not too critical to the aviator, its height was, because if miscalculated by a few hundred feet it would affect the flying. [134 p. 83]

Both oblique hand-held photography and vertical, mounted camera exposures were made. Developments around that time enabled exposure times to be improved from 1/25 second to 1/60 second. Infra red photography was also just in its infancy and available for use. [134 pp. 50, 67] It must be remembered, however, that no amount of photography, over-flying or climbing would produce accurate results for the height of the mountain.

By 1928 Hunter was advising that any change in the accepted value for the height should await the availability of results from shorter rays.

In 1930 Joseph Rock, following comments by Theodore Roosevelt Jr, used a compass and barometer to suggest that in Sichuan, China, there was a higher mountain Minya Konka (now Gongga Shan) at 9250m (30 250 feet) that was 400m higher than Everest. Chinese measures later found it to be only 7550m. [74]

Later, Rock also suggested that Anye Machin (now Maqen Gangri), east of Kun Lun was over 9000m (30 000 feet). This idea had been started in 1923 by Brigadier General George Pereira. In 1929 Rock got within 60 miles of it and with only compass and hypsometer gave a value of 29 529 ft. [194]

In 1934 Mason appealed for the uniform adoption of 29 002 feet as the official height [214 p. 155]. He stressed that all authorities up to then had said that the time was not ripe for a change. The difficulty was that if the value for Everest were to be changed then one must also change the heights of various features in the neighbour-

Figure 35 Aerial view of the Mount Everest region.

hood. In addition, even estimated heights achieved by climbers relate to 29 002 ft. [214]

In 1949 an American explorer, Leonard Clark, returned to Anye Machin with a crude theodolite borrowed from the Chinese highway authority and measured the height as 9041m (29 661 feet). This was higher than Everest by 193m but Chinese measures in 1970 gave only 6282m (20 610 feet). [74]

Knowledge of the relation between the geoid and spheroid was essential if the geoidal height was to be obtained accurately. This was initially chosen in 1926 as zero at Kalianpur so that geoid and spheroid could be coincident under all 10 Indian bases measured to that time. In 1927, with the introduction of the International spheroid, the separation at Kalianpur was changed to 31 feet – with the spheroid above geoid. [158 p. 5] Brigadier Guy Bomford in 1951 counselled against any change until more data were available for the corrections that were required.

The opening of Nepal

From about 1950 Nepal was opened and the situation changed. The result was that the Survey of India was able to execute good topographic triangulation in 1952–3 that passed within 40 miles of Mount Everest at elevations from 8000 to 15 000 feet.

A meridional arc reached from the longitudinal series on the line Ladnia-Harpur to Namche Bazar only 17.5 miles from Mount Everest. [179 p.23] Four Laplace stations and three baselines were incorporated and astrolabe observations for deflection components were taken at many of the stations. These indicated a rise of about 109 ft from the geoid. All this considerably reduced the uncertainty in deflection values since they were now only extrapolated over 17.5 miles rather than 90 miles.

From all these observations Gulatee calculated the deflections sufficient to be able to produce a chart of them with a vertical interval of 5 feet.

Modern observations

During 1952–4 a new determination was made using modern instrumentation. In 1952–3 this was a Wild T2 theodolite and in 1953–4 a Geodetic Tavistock.

From Gulatee's computations there was a scatter of 16 feet in the results and the weighted mean geoidal height was 29 028 ± 0.8 feet at the season of minimum snow. This agreed closely with the 29 032 feet found in 1952–4 by applying the geoidal rise then known, to Burrard's value of 29 141. The 0.8 feet was from internal evidence only and the actual figure was thought to be possibly up to 10 feet. [158 pp. 13–14]

This produced a new official value of 29 028 feet ± 10 feet but there were still too many indeterminates. The official height remained as 29 002 feet until 1954 except for the few years from 1890 when a correction of 8 feet was temporarily admitted from spirit levelling. [232 pp. 93–4] If the 109 feet geoidal rise were applied to nearby Makalu the value found was 27 824 feet. [158 p. 15]

Gulatee then analysed the two old values of 29 002 feet and 29 141 feet. Of the former he said that it was a vague value. The datum surface was ignored and a faulty value of refraction used. Various sources of error apparently compensated one another to give a result quite close to the new value. [158 p. 17]

From observations at 12 stations during 1952–54, all within 50 miles of the mountain, and all lying above 8500 ft elevation, the results for the geoidal height of Mount Everest were calculated to lie between 29 022.8 feet and 29 038.7 feet. [158 p. 23]

It was shown that at Lower Rauje the deflection value was 71" which is said to be the largest in the world with respect to the International spheroid. If it had been related to the Everest spheroid it would have been even greater.

The geoidal height at Mount Everest of +92.3 feet compared with that of –17 feet at Ladnia was equivalent to 1 foot separation per mile of distance! [158 pp. 10–12]

In 1975 a Chinese expedition made Everest 8848.13 ± 0.35m (29 029.24 ± 1.1 feet) [8 p. 367]. Chinese mountaineers had put a light alloy beacon on the summit of the mountain and were able to allow also for the snow depth. The beacon survived for at least three years. They had nine stations averaging 5800 m (19 000 feet) elevation and from 8.5 km to 21.2 km (5.3 miles to 13.2 miles) from the mountain. The triangulation net that allowed this closeness included five Laplace stations and fifteen astronomical stations. Fuller details are given in [8 p. 380], [72], [63].

The use of the Global Positioning System

Then on 7 March 1987 an item in the *New York Times* announced that the astronomer Professor George Wallerstein from the University of Washington had a value for K2 of 8859m instead of 8611m (29 064 feet instead of 28 250 feet). This he had found during measures in the summer of 1986 using satellite signals. This would have made it the highest mountain by 11m. [74]

Wallerstein, however, had repeatedly cautioned that his observations were of a preliminary nature. He had side-stepped the need to bring heights from sea level, with all its inherent problems, by using a doppler receiver and satellite signals. On 8 June 1986 Wallerstein and others had taken a receiver to a point near the base of K2. They took an altitude fix then used a theodolite to triangulate the altitudes of several surrounding landmarks that had last been surveyed in 1937 by Michael Spender.

Wallerstein found later that Spender's values had all been about 900 feet lower than his. Spender had used the summit of K2, taken as 28 250 feet, as his datum. Thus Wallerstein concluded that K2 must be 900 feet higher than thought.

In August 1987 Professor Desio, a geologist from the University of Milan, used the Global Positioning System (GPS) in the vicinity of both K2 and Everest. He referred the results to the International ellipsoid World Geodetic System (WGS) 84 and this gave the ellipsoidal heights as Everest 8833m (28 979 feet) and K2 8579m (28 146 feet). Converted to orthometric heights these are 8872m (29 107 feet) and

8616m (28 267 feet) against the previous 8848m (29 029 feet) and 8611m (28 250 feet) where the 8872m was said to be ± 20–30m and the 8611m ± 7–17m.

Desio considered that there were two possible sources of doubt – the effect of the ionosphere and the local height of the geoid relative to the WGS 84 ellipsoid. [74]

In the early 1980s a study by the Nepalese Government changed the heights of 72 peaks but Everest (Sagarmatha) was not amongst them. Well-known peaks such as Kanchenjunga, Dhaulagiri and Gaurisankar were changed. Kanchenjunga went from 8598m to 8586m. Dhaulagiri lost only 1m while Gyalzen peak lost 554m and Nala Kankar lost almost 900m. [28]

In September 1992 the height of Everest was remeasured by a joint Italian-Chinese exercise and valued at 8846.10m in relation to mean sea level in the Bay of Bengal. The change in elevations quoted variously as 2.03 m or 6 ft 7 inches lower than previously accepted. Is it really realistic to quote the result to the nearest 0.01 metre (or 0.4 inch)? Even allowing for the sophistication of today's electronic equipment perhaps only the nearest 0.5 m can have any significance.

Simultaneous observations were taken from both Nepal and China to a survey target with laser reflectors and a GPS receiver on the summit. Four further GPS systems were operated from two valleys at about 5300m elevation and their relative three-dimensional positions found from the Navstar network. Other equipment included the later electronic theodolite and mekometer distance measuring instrument.

The idea was to obtain results related to both the Indian Ocean and the Yellow Sea and hence two quite independent evaluations. The orthometric height announced on 23 April 1993 was that related to the Bay of Bengal; that computed from the Yellow Sea is awaited with interest. How good will the agreement be?

In May 1998 an American expedition used Trimble global positioning equipment (GPS) to establish a station near the summit of the mountain to monitor its vertical movement. A hole was drilled in the face of the summit at Bishop Ledge, a steel bolt inserted and GPS equipment attached to it. This will serve as a benchmark for future measurements. Together with other bolts along the route to the summit the aim is to try and confirm the estimate that the mountain is rising by about 10 mm per year due to changes brought about by plate tectonics under the earth's surface. The expectation is that the actual height will be somewhat less than the quoted figure of 29 028 ft by perhaps 30–60 ft although this would conflict with all previous measures including the GPS ones of 1987. Information from the devices is logged and transmitted by satellite telephone to the Massachusetts Institute of Technology (MIT) for analysis. The definitive results from this latest heighting are awaited with interest.

Two prophesies from 1884 by W.W. Graham are a worthy postscript to this chapter.

> I believe that Mount Everest will have to take a lower rank. From… two peaks NW of Everest we were all agreed, considerably higher. [150 p. 69]

> …lay Mount Everest, and I pointed it out to Boss, who had never seen it, as the

highest mountain in the world. That it cannot be he replied – those are higher, pointing to two peaks which towered far above the second and 100 miles further north. I was astonished, but we were all agreed that, in one judgement, the unknown peaks, one rock and one snow, were loftier. (Note. The Boss referred to was Emil Boss who at that time had climbed to a higher point than anybody else.) [149 pp. 440–1]

As Bradford Washburn said in the BBC Horizon film on *Measuring the Roof of the World*, because of the movement of the continents, India is gradually going under the Himalaya massif so that little by little Everest is rising every year and will continue to do so for a long time to come. So not only are there the problems of access, instrumentation, refraction and datum definition, but also the geology of the region adds to the doubt about the height of this incredible peak.

Summary of principal results

		Height (ft)	(m)	
1847	Armstrong	28 799	8778	Isolated observation
1849	Nicholson	29 002.3	8840	From 6 stations in the plains
1892–1903		28 994	8837	Change in heights of control stations
1905	Burrard	29 141	8882	Recomputation. Mean of six values
1922	De Graaff Hunter	29 149	8885	Adjustment to Burrard 1905 value
1922	De Graaff Hunter	29 079	8863	Allowing for geoid height and refraction adjustment – sometimes given as 29 080
1929	Bomford	29 050	8854	New value for geoid height
1933	The Times Atlas	29 141	8882	
1952–4	Gulatee	29 028	8848	Geoidal height
1975	Chinese	29 029.24	8848	Beacon on summit
1987	Desio	29 107	8872	Orthometric height
(1986	Wallerstein	29 064	8859	For K2)
1992	Italian/Chinese	29 022	8846.10	Expedition using GPS

Chapter 17

THE MOUNTAIN: The issue of its name

As we have seen, the earliest scientific attempts to determine the heights of Himalayan peaks were in 1810 by Colonel Crawford and 1812 by Captain Webb. Unfortunately neither knew the heights of their observing stations so the results for the peaks were hopelessly astray.

The highest peak in the world was successfully observed scientifically in the late 1840s by a number of observers from different locations. Because of the difficulty in identifying individual peaks at ranges of over 100 miles one peak began life with the designations γ from observations near Darjeeling; b from observations by Armstrong working on the Cora meridional series and h by Nicholson working on the North East longitudinal series.

Refraction was a major problem and occupied Waugh and Radhanath Sickdhar for some years before coefficient values were accepted. When it was then ascertained that γ, b and h all referred to the same peak it became designated in the computing office as Peak XV. But as soon as it was determined that this was probably the highest peak in the world the question arose as to what it should be called. What was the local name? What authority was there to draw upon?

Announcement

When Waugh wrote on 1 March 1856 to his deputy Major Thuillier indicating that the calculations relating to all the mountain peaks were complete and had been checked and corrected, he indicated that Peak XV was in all probability the highest in the world and as such could not be left without a name. For any normal feature it would be extremely likely that a local name could readily be found but this was no ordinary feature: it was one peak in the midst of thousands. Waugh commented:

> I was taught by my respected chief and predecessor Colonel Geo Everest, to assign to every geographical object its true local or native appellation…But here is a mountain, most probably the highest in the world, without any local name that we can discover, or whose native appellation, if it have any, will not very likely be ascertained before

we are allowed to penetrate into Nepal and to approach close to this stupendous snowy mass … the privilege, as well as the duty, devolves on me (Waugh) to assign a name to this lofty pinnacle…

In virtue of this privilege, in testimony of my affectionate respect for a revered chief, in conformity with what I believe to be the wish of all Members of the scientific department, over which I have the honour to preside, and to perpetuate the memory of that illustrious master of accurate geographical research, I have determined to name this noble peak of the Himalayas 'Mont Everest' [289 p. 346]

From the circumstances of its discovery, from 100 miles away, it was impossible to make close enquiries as to the local name. Waugh pressed for permission for a survey party to enter Nepal, but this was denied. [232 p. 94] Lack of access exacerbated the problem of finding a local name and finally Waugh felt entitled to propose one of his own. He chose 'Mont' as more appropriate for a single definite peak as opposed to an extended massif but changed it to Mount a year later. The name 'Everest' was cordially accepted by the Secretary of State.

At a meeting of the Asiatic Society of Bengal on 6 August 1856 Major Thuillier announced on behalf of Colonel Waugh, the Surveyor General, the discovery of a mountain in the Himalayan Range that placed it higher than any previously ascertained value. Initially Dhaulagiri (Dewalagiri) had been thought highest at 26 860 ft then in 1847 Kanchinjinga (Kanchenjunga) came in at 28 168 ft. Now this mountain was at 29 002 ft. [14]

Other possible names

The proposed name of Mont Everest was challenged initially on 29 August 1856 by B.H. Hodgson who considered that there was a local name. Whilst worldwide geography could not be overlooked, and the highest mountain could not be left nameless, a controversy started that rumbles on still. The only disadvantage of the name Everest was that it created a precedent. [56 p. 16] A meeting of the Asiatic Society of Bengal in September 1856 agreed to accept Waugh's suggested name although those assembled were not too happy about it. [15] Thus the name Everest was given to the mountain which George Everest himself almost certainly never saw.

The reaction of Sven Hedin, a Swedish explorer writing in 1926, was that 'Everest … was a conscientious officer, able but not outstanding. By sheer accident without a trace of want of breath, he has become undying.' According to Hedin, the name 'Everest' had served to confer undying fame on a mediocrity, but he was more or less a lone voice particularly in that context. [163]

Hedin had asked who discovered Mount Everest. ('Discovering' Everest really meant the discovery of its height.) [56 p. 5] There was no one individual since there were many observers on the GTS and several computers involved. In any case, at the time it was thought there could be other higher peaks, so it was not of any particular note. [56 p. 6]

Burrard defended Everest as 'the dominant figure of the Great Trigonometrical Survey. He had been the first observer to foresee the necessity of greater accuracy of observation'. He had also been the first geodesist to foresee the effects of mountain attraction upon geodetic measurements. He foresaw that the Indian jungles could not be triangulated unless theodolites could be raised 30–40 ft and lamps still more. He was a brilliant mathematician. Surely no mean claim to fame. His colleagues gave his name to the mountain to commemorate his scientific work in the same way that parts of the Moon are named Copernicus and Tycho. [56 p. 8]

When Everest heard of the rare honour that was bestowed upon him he responded by saying that the very kind way in which Waugh had spoken of him was far beyond his merits. He certainly never contemplated having the mountain named after himself. He confessed an objection in that his name was not pronounceable by a native in India. It could not be written in either Persian or Hindi and the natives could not pronounce it.

In the choice of name of Mount Everest no question of nationalism has ever been heard in India. In 1865 Europeans and Indians of the Survey were unanimous in feeling that the scientific success of their Department would never have been achieved without the genius and inspiration of Everest. [56 p. 7]

Early maps

The background to finding a local name can be traced to the early eighteenth century. In 1711 two Chinese Lamas instructed by Jesuit Fathers made a survey of Tibet under orders of the Peking Government. [56 pp. 3, 12] From their results the Jesuit Fathers compiled a map of Tibet in 1717 and on the southern border they showed a mountain range 40 miles long named Tchoumou Lancma.

In 1733 D'Anville published in Paris a map of Tibet based on the Lamas' survey and Jesuit plot. [56 p. 3] This map was used throughout the 18th century but was out of date by the 19th century. Hedin wrote 'one loses oneself in guesses when trying to identify the representations on D'Anville's map'.

Hedin, writing in 1926, went on to say 'this correct Tibetan name Tchomo Lungma appears as Tchoumou Lancma on maps which were prepared from native information by French Jesuits in Peking in 1717, and printed by D'Anville in Paris in 1733'. [163] He concluded that the highest peak claimed by the English for 1852 was on French maps 119 years before and that the real Tibetan name for Everest of Tchomo Lungma was known to the Jesuits in Peking 190 years before. He gave the position of Everest as only 60 miles from the position of an imaginary point on the Tchoumou Lancma range, given by D'Anville in 1733. (D'Anville, however, named a long range, not any individual mountain.) [163]

Sven Hedin added that:

In 1921 the Mount Everest expedition under Col Howard-Bury found that the Tibetans had the name Tchomo Lungma for Mount Everest... If the similarity is only coincidental then ... the geographical positions of the name on latest English and old French maps should be noted. The modern position of Mount Everest is given as 27° 59' 16" N 86° 55' 40" E [157 p. 1] whilst on D'Anville's map it is 27° 20'. The modern longitude east of Ferro is 104° 55' and D'Anville had 103° 50'. This is surprisingly accurate... [163]

Burrard held that neither the Lamas nor Jesuits knew any more than that the whole area was mountainous – as did the Tibetans – and that nothing was known about Mount Everest until it was observed by theodolite in 1849. [56 pp. 3–4].

Objections

At the same meeting of the Asiatic Society of Bengal there was a contribution from Brian Hodgson, who had lived in Nepal as Political Officer for 20 years. He wrote to the Royal Geographical Society from Darjeeling on 27 October 1856 to say that, although he sympathised with Waugh, he must, in justice to his Nepalese friends, state that the mountain did not lack a native name. He said it should be Devadhunga, Holy Hill, Bhairavthan or Mons Sacer. In support of this he referred to an earlier paper in the Royal Geographical Society Journal in which a rough sketch showed Deodhunga in a position that tallied with the site of Mount Everest.

> and it were indeed a strange circumstance, if so remarkable a natural object had escaped the notice of the people of the country and thus remained unnamed. Nor would it have been very creditable to me after 20 years' residence in Nepal, had I been unable to identify that object. [289 p. 349]

He stated that both Mount Everest and Devadhunga were 'about 100 miles NE of Kathmandu; both are midway between Gosainthan and Kangchan; and both identifiable with the so-called Kutighat or great Gate, which annually for half the year is closed by Winter upon the Eastern highway of Nepalese commerce and intercourse with Tibet and China'. [289 p. 348]

Hodgson wrote similarly to the Royal Asiatic Society regarding his version of Devadhunga and at that time the members of that Society agreed with him thus:

> Your letter of the 27th October, together with your observation on the incongruity of assigning a European name to Indian localities already provided with native appellations, was received and read at our last meeting of the 17th instant, and I have the pleasure to inform you that the members present unanimously expressed their concurrence with your view of the case. [282 p. 88]

Interest aroused

In October 1856 Waugh wrote to Charles Lane thus:

> You are aware of the great interest which has been aroused in the case of Mount

Everest. If you have an opportunity of investigating the real native name of that stupendous pinnacle...it will be a great acquisition... You yourself observed Mount Everest from Doom Dangi and Banderjoola near the origin of the Assam Series, but you took no vertical angles. It would be extremely valuable to rectify that omission and ... your verificatory observations will be very highly appreciated... [263 (718) A-1], [232 p. 86]

There was a limit to how long Waugh could stall telling the world of this eminent peak and on 5 August 1857 he wrote to Major Thuillier along the following lines:

In August last year you communicated the results to the Asiatic Society of Bengal. This sparked Mr Hodgson to equate Everest with Deodanga, Bhairavathan, Bhairavlangur, and Gnalthamthangla.

The arguments adduced there were so palpably conjectural, resting on hearsay evidence alone, that I thought it needless to refute them,... The true geographical latitude and longitude and height of Deodanga are unknown to Mr Hodgson or even its true bearing and distance from any locality which can be recognised as a fixed point of departure...

But then he has put it subsequently in a somewhat unfair light to the Royal Asiatic Society, as well as to have his conclusions on a point of great ambiguity promulgated as certainties... Thus I [Waugh] laid it before a Geographical Committee to investigate, by Departmental Order of 22 April 1857.

The sketch map published by him [Hodgson] in the Journal of the Asiatic Society December 1848, gives his idea of the configuration of that part of the Himalayas; a more erroneous impression of the formation of the country was never formed; he represents a solitary mountain occupying a vast tract... this single mountain, however, is entirely imaginary. The range presents the appearance of a 'sierra' with innumerable peaks and groups of peaks. Among these nine have been fixed by the General Trigonometrical Survey of India and are marked XII to XXI... Besides these nine, several others are more or less partially visible, which we were unable to identify; and those who have any experience in conducting geodesical operations in the Himalayas can harbour no doubt that many other peaks do exist which have been concealed from our view by intermediate ranges... [288 pp. 102–5]

It was contended that the Surveyor General in fact did Nepal a favour by not taking Gaurisankar from a peak they all knew by that name and putting it on another. [56 p. 5] Waugh observed that

You will perceive the gist of the question is not whether the mountain should be called Mont Everest, or by its true native name (which is a principle not disputed by any one), but whether it can be called Deodangha without risk of error, in the absence of satisfactory proof that this is really its native name. [282 pp. 89–92]

What Waugh named was the pinnacle itself, not the general mountain mass and for this reason initially called it Mont Everest instead of Mount Everest.

The Schlagintweits

In the period 1854–57 three brothers, Hermann, Adolphe and Robert de Schlagintweit undertook a scientific mission to India and Central Asia at the instance of the King of Prussia, and with the concurrence of Lord Dalhousie and the Court of Directors of the East India Company. The Schlagintweits were interested in a wide range of scientific activities, and since the discovery of Mount Everest had just been announced, Hermann Schlagintweit felt it appropriate to try to take some observations to it. They passed through the area, but General Walker considered that they probably never distinguished Everest.

The Schlagintweits had made contact with Hodgson who had told them that the mountain had the names: Devadhunga, Bhairab-than, and Bhairab-langur (Nepalese); Gualham, Tangla, and Gualham-tangla (Tibetan). Hodgson had also used Nyanam. [282 p. 89]

In 1855 Hermann Schlagintweit visited Katmandu and drew a panorama from the hill named Kaulia and obtained the name Gaurisankar for the highest peak (the bright, or white, bride of Siva). [137 p. 454] This repudiated Hodgson's name of Devadhunga. [54 p. 43] The three Schlagintweit brothers did not appreciate the steps necessary to identify correctly a peak among so many. Rather than direction, distance and height all they took was direction. When their results were published it became obvious that they had mistaken Makalu for Everest.

Writing in 1904 S.G. Burrard said 'There is no doubt now that Schlagintweit was misled in his identification of Mount Everest … it was Makalu that he drew as Everest, both in his panorama of the snows from Falut and in his picture, which is preserved at the India Office.' [54 pp. 43–4]

As a guest of India, Hermann had no intention of criticising the Survey of India and indeed expressed his thanks for the co-operation given to him by Waugh. From Falut in Sikkim he saw in the direction of Everest what he took to be the mountain and took observations to it. Unfortunately from that position Mount Everest is really almost hidden and appears to be quite a secondary point. A similar situation occurred when he observed from Kaulia near Katmandu in the direction towards Everest; he had again failed to test for distance and was in fact observing Gauri-Sankar.

It was most unfortunate that he did not spare the time to consult Waugh or his colleagues before travelling on. It could have saved many years of controversy. When he returned to Berlin the added mistake was made in the drawing office of treating Makalu, Everest and Gauri-Sankar as one peak. When published in 1866 with the appellation Gauri-Sankar, this name was adopted in Europe. [57 pp 461–2]

A committee

In 1858 the Surveyor General appointed a committee to sort out the problem. The members of the committee were: Lieutenant Tennant, Lieutenant Montgomerie, J.

Hennessey, W. Scott and J.W. Armstrong, each eminent in their own area. All members came to the conclusion that the identity of Devadhunga with Everest was not only doubtful, but far from probable. [282 p. 89]; that Hodgson was mistaken in his direction and the Schlagintweits' panoramas did not include a new peak at all. Nevertheless each had their supporters. [56 p. 5] The comments of the members of the committee were, in brief, as follows.

W.H. Scott: 'no evidence to establish satisfactorily the identity of Mount Everest with Deodanga or Bhairavathan…' He quoted from a letter by Major Ramsey, Resident of Nepal, to Major Thuillier, 11 June 1855.

> You are doubtless aware that no European has ever travelled in the interior of this country, and that all the information we possess is derived from the reports of persons who are totally devoid of scientific knowledge, and are accustomed in their comparisons of distances to trust to vague estimates formed by parties who have travelled through the different districts.

He included evidence to repudiate that Everest could be seen from the Nepal valley and Sikkim.

J.B.N. Hennessey: 'no evidence at all…' Again he gave evidence of its invisibility from various stated points.

J.W. Armstrong: 'purely conjectural…' Again he gave evidence to show 'it is not visible. … enunciated without personal observation, and based upon the vague information of untrained travellers'.

Lieutenant J. F. Tennant: again he gave proof of non-visibility '…hardly admissible as evidence of anything … liable to great error … untrustworthy … No evidence to show Everest and Deodanga are identical.' [288 pp. 106–11]

In 1862 the Schlagintweits published volume 2, *Hypsometry,* of their *Results of a Scientific Mission to India and High Asia.* This gave results of the expedition of 1857 and included the name Gaurisankar for the highest peak. Details were given of how Hermann deduced this.

> I saw it first from … a distance exceeding 84 miles. The Hindu name I found, to be Gaurisankar, Gauri = white or fair, a name of Purvati the wife of Shiva; Sankar, or Sankara, one of the forms assumed by Shiva. Gaurisankar is the term in use among the Hindu Pandits of Nepal. The name given to it by the Tibetans, and that by which it is generally known in the northernmost parts of Nepal, is Chingopamari… [282 p. 89]

Walker summarised the position to date:

> From the information given by the Schlagintweits in a panorama drawn from Kaulia and supplemented with horizontal and vertical theodolite angles it was possible to calculate the bearings, distances and heights to each peak. This was done by Tennant and Walker with the results; Everest from Kaulia azimuth 82° 08'; Distance 108.6 miles; Apparent elevation 1° 32'. Details that both Everest and Makafu are nearly in same line from Falut but whilst the latter is 1200 ft lower it has the greater apparent elevation, and thus obscures Everest. [282]

From this data it was obvious that several mistakes had been made in identification. The position of Everest on the chart was an impossible one and whilst it had an apparent elevation above peak XXI of 10', if it were visible at all it would be at –25'. It was most likely that he had mistaken Makalu for Everest and Sihsur for Makalu. [282 pp. 90–2]

General Walker asserted that what the Schlagintweits drew and heard called Gaurisankar could not be the 'great Nepal peak'; he found mistakes in their atlas. After some arguments, Freshfield went only as far as to admit that the Schlagintweit work 'requires confirmation' and then struggled to find other supporting evidence. [137 p. 458]. Certain European maps of the area (1866) confused Everest with Gaurisankar. This emanated from Berlin and the Schlagintweit data. [57 p. 462]

No names substantiated

Burrard asserted that the various claims by Hedin should be impartially judged. He said the only interest that then attached to Everest was due to its great height. [56 p. 4] When discovered, the Survey had the problem of a peak of great geographical importance that was without a name. It was the responsibility of the Survey to prove the authenticity of any local name before using it on a map.

Burrard analysed the situation thus:
- 1849 Survey of India fixed a high peak by intersections from low-lying jungle plains of India,
- 1852 computations indicated that there was a very high peak,
- 1852–65 the names Devadhunga and Gaurisankar were suggested but not accepted.

Writing in 1904, S. G. Burrard concluded that 'All subsequent information goes to show that there is no peak in Nepal called Devadhunga. Mr Hodgson's sincerity has never been doubted, and it is believed now that the name Devadhunga is a mythological term for the whole snowy range.' [54 p. 43] Referring to Hodgson, Waugh said 'the sketch map published by him … a more erroneous impression was never formed … this single mountain is entirely imaginary.' His name of Devadhunga was unknown to the Himalayan people. [209 p. 459]

From 1852 to 1865 much thought was given to the name. However all the supposed native names suggested were proved to be based on mistakes. After exhaustive enquiry Andrew Waugh, with concurrence of Sir Roderick Murchison, President of the Royal Geographical Society, accepted the name Everest and height of 29 002 feet in 1865. [56 p. 1]

1884

In 1884 D.W. Freshfield entered the arena with the assertion that the local name of the highest mountain was Gaurisankar (or Devadhunga). He said 'M. Reclus and the editor of Petermann's *Mitteilungen* followed by a large number of continental geographers, adhere to the high-sounding Gaurisankar in place of Mount Everest or Everest, and I propose to use any influence I may possess over the literature or orography in the same direction.' [137 p. 452].

Freshfield decried the evidence produced by Waugh and his Geographical Committee to repudiate Hodgson's claims of 1856. [137 p. 453] He then quoted a comment by General Walker 'The great mountain masses of which it is the highest pinnacle are known to Nepalese and Thibetans by various designations, of which Devadhunga, "the home of the gods" may well be accepted as most in harmony with the religious instincts of the people of the country,'

He went on to say that:

I incline, therefore, to believe that even though the name Gaurisankar on the Phalut panorama (by Schlagintweit) should prove, as seems probable, to be wrongly placed, the peak recognised on the spot by Schlagintweit as the highest, and computed by him at 29 196 feet, was the 29 002 feet summit of the surveyors…

I cannot … bring myself to share General Walker's firm belief that owing to the distance and configuration of the chain it is extremely improbable that the great peak is seen… from the confines of the Khatmandu valley… [139 pp. 179–80]

Freshfield also referred to Sir Joseph Hooker's book *Himalayan Journals* of 1854 where he described the area from four points as 'A white mountain mass of stupendous elevation, at 80 miles distance from Tonglo, called by my native people Tsungau, in about lat. 27° 49' and long. 86° 24' – perhaps the one measured by some of Colonel Waugh's party, and reported upwards of 28 000 feet…'

He went on to suggest 'if the longitude given, 86° 24', is thirty minutes out, Sir Joseph Hooker's estimate of its distance shows that the '2' is a misprint and that '54' should be read'.[139 p. 183]

Freshfield rambled on at length but said 'Practically, perhaps, the matter at issue may seem a very small one – a mere matter of convenience and tastes. But ethnologically and historically it has considerable interest.' [139 p. 185]

1886

In 1886 Freshfield summarised the state to date. He was critical that Waugh had called upon his own experts rather than independent travellers to testify. He indicated that, whilst they said the 29 002 ft peak was not visible from Khathmandu, that was not the point since Hodgson had used the term 'confines of the Khathmandu district'. He further disputed the boundary of Sikkim as putting doubt on that original statement also, and referred to an evasive reply. He then went on to refute Walker's criticism of Schlagintweit's sketches and commented on the possibility of Peak XVIII covering sight of 29 002 ft and suggested that there were places from where this would not happen. Further discussion was related to the variation of refraction and whether an appropriate formula was used. He accepted that the German atlas must have been subject to much carelessness in preparation. [139 pp. 176–7]

He then commented

'does it become as obvious, as Gen Walker supposes that the peak the Schlagintweits saw cannot have been Everest? I do not think so. [139 p. 181]

I cannot admit that General Walker has as yet finally overthrown Hermann Schlagintweit's identification, supported, as it is to some extent, ...by other evidence [139 p. 182]

He further quibbled over Walker's reference to pinnacle as opposed to range of mountains, and considered it quite invalid. He then expounded on the mystery of Waugh using the term 'mont'. At least he agreed with the principle put forward by Waugh and Walker that local and national appellations should prevail. Thus the whole argument really turned on the existence, or not, of a local name.

1888

In 1888, when Hermann Schlagintweit was dead, his younger brother Emil published an article in German in which he suggested the compromise of 'Gaurisankar-Everest'. It was left to General Walker to point out that these were two separate mountains and so such a combination was inadmissible. [57 p. 462]

Burrard commented on Waddell's publication of 1899 [279] in which he came up with the name Jamokankar (or Chamokankar).

let us suppose... that the mountain called Jamokankar is identical with our Mount Everest. What then? Will it be incumbent upon us to abandon the name of Everest and to adopt that of Jamokankar? I think not.

It will, I think, be lamentable if former advocates of the name Gaurisankar, seeing their cause is doomed, continue the struggle under this new flag of Jamokankar. Already, to our regret, has Mr Freshfield, a life-long defender of the claims of Gaurisankar, declared in favour of the Tibetan name. [54 p.45]

1901

In 1901 Dr S. Ruge wrote in Petermann's *Mitteilungen* on the distinction between Mont and Mount Everest. He asserted that Waugh had initially used Mont and that it should be retained. 'It would be petty pedantry to degrade it into an ordinary Mount.' Freshfield responded immediately both to the German publication and the *Alpine Journal*. He took issue with the use of Mont and set out to ridicule its use. Then he added the point that the Tibetan name found in 1885 by the Pundit surveyor Chandra Das was Jomo-kang-kar, the Lord, or Lady, of the Snows. [140 p 34–5]

1903

In 1903 Freshfield took up the subject again by first quoting from Waddell [279] 'it is physically impossible to see Everest either from Khatmandu, or the Kauli or Kakani peaks, whence H. Schlagintweit believed he saw it, and got his local name, Gaurisankar…' To this the response was 'His assertions are convincing at first sight; but they do not bear examination … But what can be seen from the city itself (i.e. Khathmandu) never formed any part of my argument.'

He then quoted from Boeck [71] which contained photographs from the hill that Schlagintweit had visited many years before. Freshfield purported to be able to identify the peak of 29 002 feet on these and in addition Boeck was apparently given the name Gaurisankar for the peak.

In particular Freshfield stressed the fact that Everest could be seen from Sandakphu [141 p. 297] In 1903, Freshfield accused the Survey of India of falsifying results to avoid finding a native name. This resulted in Colonel Gore, as Surveyor General, getting permission for Captain Wood to go to Kaulia and verify by theodolite whether Mount Everest could be seen from there. It could not, except from the summit of Kaulia – and then only as a small, low, insignificant peak. By using a second station he found that Everest and Gaurisankar differed in position, height, and shape. His report published in 1904 finally refuted the Gaurisankar idea.

In 1903 Captain Henry Wood visited Nepal by order of Lord Curzon. He proved by trigonometrical observations that Gaurisankar and Everest were different peaks 36 miles apart, and that the peak singled out by Hodgson and Schlagintweit (Peak XX) was lower at 23 440 ft and was the famous Gaurisankar of the Nepalese. [54 p. 441] Wood confirmed there was no Nepalese name for Peak XV. [56 p. 5]

Lord Curzon commented on the report by Captain Wood thus: 'Everest from Kaulia is an insignificant point just visible in a gap in the main range … The name Gaurisankar is given by officials of Katmandu to Survey Peak XX. The name is not known to any of the Nepalese hill men.' Peak XX is 36 miles from the highest peak, Peak XV, but forms part of what, according to the principle adopted by European orographers, would be considered the same group. "Peak XV is 108 … miles distant from Katmandu … and is not visible from the city … it was, as I suspected, wrongly identified by Dr Boeck.' [143 pp. 362–3]

Thus, whilst Schlagintweit was correct to give the great peak visible from Katmandu and Kaulia the name Gaurisankar, he was incorrect to equate it with Peak XV. Among the various mistakes made were bearing errors of up to 2°, considering as the same, two peaks that were in fact 47 miles apart, and, by Dr Boeck, a mistake of 32° in direction that swung his whole area of interest. [54 p. 44]

1904

In 1904 Freshfield was able to refer to a series of photographs taken in 1903 by Mr Hayden when he was at Khamba Jong in Tibet. They formed a complete panorama, albeit not all from the same viewpoint, including Kangchenjunga and 'the peak known to the Survey as XV, or Mount Everest, and to Tibetans as Chomokankar'. [143 p. 361]

In 1904 Colonel Waddell and Sarah Chandra Das heard the name Chomo Kankar applied to Everest by Tibetans in Tibet; although Burrard refuted this suggestion. [56 p. 10] He said that, when naming the 29 002 ft peak, 'Waugh was unaware that the Tibetans who lived to the north of the great mountain already had several names for it that were more apt and considerably more mellifluous, foremost among them Chomolungma, which translates Goddess Mother of the Land'. [194 p 178]

Mount Everest was first observed from the north by the Lhasa expedition of Colonel Ryder. No name was found for it but the fallacy of Everest and Gaurisankar being the same was exposed. [56 p. 8]

1907

When Natha Singh of the Survey of India was surveying on the southern side of Everest he heard the name Chholungbu used by Tibetans as they brought salt across the border for sale in Nepal. Natha Singh was also sometimes quoted as having heard it as Chomo Lungma but this is incorrect. [56 p. 11]

1909

General Bruce was told of the name Chomo Lungmo by Sherpa Bhotias in Nepal.

1921

In 1921 the Dalai Lama gave Sir Charles Bell permission for the first Mount Everest expedition to go into Tibet. In this permission the peak was described as

> To the west of the five treasuries of great snow is the southern district where birds are kept (Cha-ma-lungma). The word Lungma means valley and not mountain; this reference to Mount Everest as a valley shows that the peak of Mount Everest itself is not known to the Government of Tibet by any name, but that the mountain is identified as being near a bird sanctuary. [57 p. 460]

Colonel Howard-Bury, leader of the first Mount Everest expedition, was told of the name Chomo Uri by his drivers, and also Chomo Lungma. [56] Thus five Tibetan names had been applied to the highest peak during the early twentieth century:

1904	Chomo Kankar	Colonel Waddell and Sarah Chandra Das
1907	Chholungbu	Surveyor Natha Singh
1909	Chomo Lungmo	General Bruce
1921	Chomo Uri	Colonel Howard-Bury
1921	Chomo Lungma	Colonel Howard-Bury [56 p. 10]

Returning to the fray in 1922 Freshfield jubilantly referred to the Himalaya expedition of 1921 and said that it had succeeded where the endeavours of the Survey of India during the past 66 years had singularly failed. It had ascertained, he said, on official authority, the local and native name of the highest mountain in the world. In the Tibetan passport granted to the 1921 party it referred to the neighbourhood of

Chomo Lungma – The Mother Goddess of the Country. 'the real Tibetan name of Mount Everest definitely ascertained during first expedition of 1921 to be Chomolungma … was actually known to Europeans as early as 1733'. [221], [56 p. 13] Freshfield stated that the 1921 Himalayan expedition had succeeded in ascertaining the local name of the mountain. (He referred to the entry in the Tibetan passport of the party.) At the same time he said he did not want to reopen the old controversy! He tried to compromise by suggesting that Mount Everest be retained for the loftiest pinnacle of the Chomo Lungma range. The neighbouring range to be the Makalu range; thus dividing a 50 mile range in two. [144 p. 192]

Since the 1921 expedition, various spurious names have crept in, such as Changtse, Lhotse, Nuptse, which are not terms that would be used locally, since they mean north peak, south peak, west peak, etc., which are quite inappropriate to the Tibetan. [56 p. 17] Odell, however, highlighted the fact that Burrard had published in 1931 that the word translated in the passport as Chomo Lungma should have been Cha-ma-lung. The pertinent passage had been translated by Sir Charles Bell as 'To the west of the Five Treasuries of Great Snow (in the jurisdiction of White Glass Fort, near Rocky Valley Inner Monastery) is the Bird Country of the South (Lho-Cha-ma-lung). Now the Tibetan words Lho-Cha-malung were said by Sir Charles to be short for Lho-Cha-dzi-ma-lung-pa, which meant nothing more surprising than "the southern district where the birds are kept"!' [222 p. 128]

Sir Charles Bell's comments based on conversations with Tibetan officials in Lhasa suggested that the evidence for Chomo Lungma was no better than that for Chomo Kankar. In commenting on an entry in [249] by H. Ruttledge, he came to the conclusion that yet again there was evidence of no local name. This related to the translation on the passport granted to the fourth Mount Everest Expedition by the Tibetan Government. [222]

There is a further mention by Colonel Sir Edward Tandy in his report of 1926–7 that:

> The Nepalese only give specific names to a few snow-covered peaks of remarkable aspect, but each group of snowy peaks was called a Himal, or 'Abode of Snow' and received a name. Thus Mount Everest dominated the Maha Langur Himal.
>
> …
>
> It would seem, therefore, to be quite clear that no native name for Mount Everest itself is in existence on either the Tibetan or Nepalese side.

In 1952 a newspaper cutting of 7 June reported that the Chinese had renamed Everest. A circular issued by the Chinese Ministry of the Interior, quoted by the *Peking People's Daily*, sharply criticised those who had 'blindly adopted' the name of Mount Everest, which commemorated, an 'imperialist colonial administrator'. 'Our highest mountain' said the circular, 'is now to be named Chumulongma – sacred mother of the waters – the name it was given on a map published in 1717 during the reign of the Emperor Kang Hsi'. It was pointed out in China over a year ago that the use of the name Mount Everest amounted to an acceptance of 'cultural aggression'.

The modern Hindi name for the mountain is Sagarmatha (Head of the Ocean).

Summary of main claims

August 1856	Everest	Waugh
October 1856	Devadhunga	Hodgson
1857	Bhairavathan	Hodgson
1857	Bhairavlangur	Hodgson
1857	Gnalthamthangla	Hodgson
1857	Deodanga	Hodgson
1862	Gaurisankar	Schlagintweit
1885	Jomokangkar	Chandra Das
1899	Jamokankar (Chamokankar)	Waddell
1904	Chomo Kankar	Waddell and Chandra Das
1907	Chholungbu	Natha Singh
1909	Chomo Lungmo	Bruce
1921	Chomo Uri	Howard-Bury
1921	Chomo Lungma	Howard-Bury
1952	Chumulongma	Chinese
1975	Qomolangma Feng	Chinese
1990	Sagarmatha	Hindi

What of the future?

The time has surely come to resist more alternatives and accept 'Everest' as the English language version. If other countries in the locality wish to use a different name in a local tongue, then there can be no objection so long as it is an alternative and not to the exclusion of Everest.

POSTSCRIPT

*T*he preceding pages have described the life and career of a remarkable man. Various of his many qualities have been mentioned but it is appropriate to summarise them here.

He can be described as a genius for invention, as shown in his developments to theodolites, beacon lamps, the perambulator and air pump for filling barometers. He had an obvious mechanical ability both for the in-depth understanding of a complicated instrument and for the design of his own models. At the same time he was a first rate mathematician, mostly self-taught in the higher mathematics required for a clear understanding of geodesy and the figure of the earth. He was able to hold his own with the top experts of the time. But above all he was a truly professional surveyor.

He was single minded and determined to achieve his goal of completing the task on which Lambton embarked in 1800 of a meridian arc from the southern tip of India into the foothills of the Himalayas. To achieve this he had on numerous occasions to display courage and stubbornness beyond the call of duty. Where many others would have given up and succumbed to the ravages of the climate and terrain, Everest persevered until his object was successfully achieved.

He was not afraid to back his own views and in this he was obstinate in the extreme. It was his development of the beacon lamp and his use of it contrary to the standard approach of Lambton that undoubtedly saved many of his followers from the ravages of more fever than was essential. Possibly because of his impulsive nature or the effects of his frequent bouts of fever, he often spent much more time and heat on trivial matters than was warranted.

He was a difficult man to work under, if for no other reason than that nothing but the best was good enough for him, and he required rigorous exactness.

> "That which is used for a basis of other operations ought to be itself as free from error as instrumental means and human care can make it…"

> "The angle book must on no account whatever be suffered to fall into arrears or be scratched or daubed, or slovenly written …"

"The sun never should rise and set twice on an unexamined angle book."

"Where … errors combine instead of compensating we learn … the true value of prudence and a rigorous attention to accuracy in principle as well as practice…"

As a fiery correspondent Everest did not mince his words.

"You are certainly most irregular. Who but a half crazy person would have chosen a time when it was blowing great guns to burn his lights in utter defiance of my orders, you certainly did this …"

"Much ought to have been accomplished … but … Capt Shortrede made one of the most utter failures on record …"

(Nevertheless, where praise was due, he would dispense it.)

Protocol meant a great deal to him, and this was illustrated when he wanted to go into Gwalior. After lengthy acrimonious correspondence with the Resident of Gwalior he found no officials to greet him at Dholpur on the Gwalior border and wrote furiously to the Supreme Government complaining of the Resident's neglect. Everest remained obstinately at Dholpur for two weeks, refusing to cross until a State official should appear with ample provisions.

His devotion to the task he had set himself is ably illustrated by his own comment about his time in the Doab during season 1833–4.

My hours for meals were irregular, with those for sleep. Every personal comfort was thoroughly abandoned … I never had one leisure hour. [231 p. 434]

Waugh, writing in 1850, said 'Blacker, with the exception of Col Everest, was the ablest and most scientific man that ever presided over this expensive department'. [286], [229 p. 424]. Phillimore's thoughts were that 'through Everest's energy and genius the shape and character of the Great Trigonometrical Survey has now been determined, and the pattern set for its steady extension and for the construction of worthy topographic surveys and maps'.[231 p. 320]

Above all Everest was a man of vision. Lambton in his last report wrote 'It would indeed be gratifying to me if I could but entertain a distant hope that a work which I began should at some future day be extended over British India.' [210 p. 71] Everest certainly contributed enormously towards the achievement of this wish and said himself 'That the whole of India will be eventually covered with triangles may be looked for as a result almost as certain as any future event can be…'

The Ode from Horace that Everest made his children learn by heart very much mirrored his outlook on life. It refers to he who values most the middle way, is hopeful when the going is difficult since bad luck will come to an end, yet has spirit and fortitude. In effect it ties up with the strong belief he had in the golden mean.

Most know his name because of the mountain. Mention must be made of the modern mapping of that area by Dr Bradford Washburn of the Boston Museum of Science, USA. Using photogrammetric techniques, he organised a mapping project at 1:50 000 scale that has had superb results. Not only are there now fine, accurate and highly detailed maps of the area of 102 km² around Mount Everest, but he also had constructed a magnificent relief model at 1:2500 scale with 5 metre contours (see Figure 34). The completed model is some 12 ft × 15 ft in dimension with a height of over 5 feet. It is certainly the most ambitious and most successful attempt at representing the mountain in this way. [205]

The bicentenary of the birth of Sir George spawned two television films relating to the Survey of India, with mention of him and some of his achievements. Unfortunately these were very general in their content and really did not say nearly enough about the man and his achievements. There is material in abundance that could be used to produce a full length documentary on his life but the opportunity for this has probably now been overtaken by these other related, but not specific, portrayals.

It is hoped that this publication will in some small way redress the situation and give a detailed insight into a most remarkable man: one who to date has been almost unsung and who, if it had not been for the mountain, would probably have been long forgotten.

Appendixes

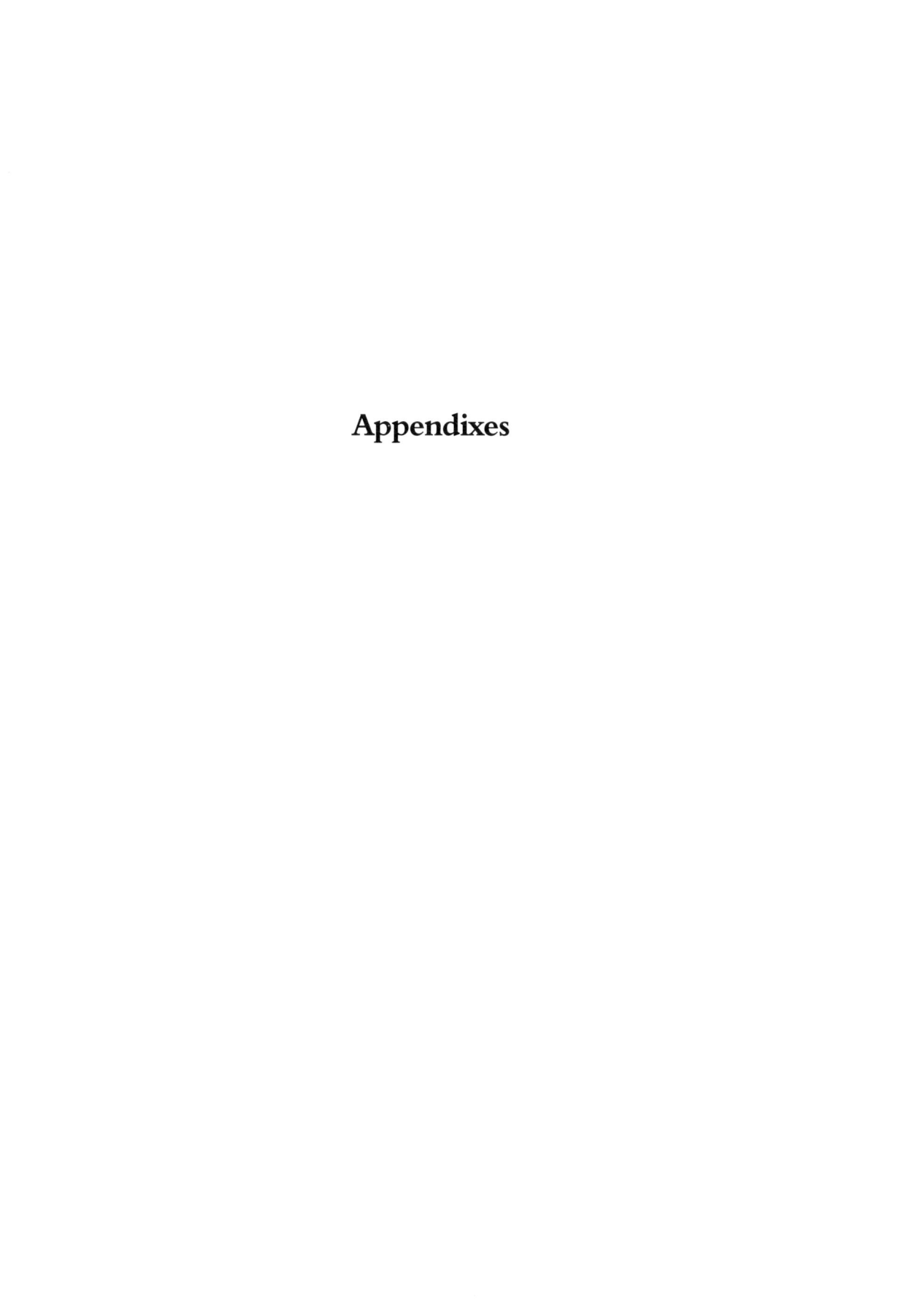

Appendix 1

INSTRUMENTATION used in the Survey of India

*E*verest first went to India early in the nineteenth century following a century that had seen considerable survey activity in Europe - not least of which was the arc measure from which the metre resulted. As a consequence of this activity there were considerable developments in the instrumentation that was used. In particular the quadrant was giving way to the theodolite and wooden measuring bars were replaced by chains and then by compensating bars.

During the 37 years that Everest was connected with survey in India there were further changes in instrumentation. Since Everest played a major role in the promotion of some of these it is appropriate to devote a section to the instrumentation used in India during his period.

Everest was very much a field man when office paper work allowed, so he had a deep interest in getting the best possible results out of the equipment at his disposal. He was observer, repairer, innovator, developer all in one, and had a detailed understanding of the construction of every item he used.

The following are notes on the various instruments that were used in India during the first half of the nineteenth century with rather more detail where the item was developed in some way by Everest.

Barometer

The barometers most commonly used in heighting operations were called mountain barometers. When the mountain barometers were available they were very susceptible to damage and this led Hodgson in 1822 to devise a means of constructing suitable instruments locally. With Lieutenant Herbert he had purchased such tubes as he could find in Calcutta and turned them into barometers of sufficient accuracy for general practical use. [263 (198) 1]

Mountain barometers were portable, and could be supported in a tripod. When the tripod was closed it formed a case around the tube. The key to safe carrying of such a fragile instrument was to carry it with the cistern, or reservoir for the mercury, inverted and then to turn the screws until the mercury almost touched the top of the tube, thus preventing oscillation of the mercury.

Each barometer would have one thermometer attached to it and also a free one, and generally two sets of units were required to determine a reliable height. The reduction of the observations were then carried out by either the formula of Laplace or that of Bessel. The difference between these was that Laplace took account of the readings of the barometer and thermometer only while Bessel included the hygrometer.

In the late 1830s Everest developed a barometer pump designed to enable the glass tubes to be filled in vacuo without fear of breakage. He had six of these built by Mohsin Hussain. [263 (344) 38–45] The authorities allowed Everest to take one sample home to England when he retired so that it might be used as a pattern for others. [229 p. 138]

Chain

The use of chains in the Survey of India fell into two categories: for baseline measurement and for more routine work such as revenue surveys. In total there would appear to have been at least five different lengths of chain used at some time or other: 33 ft, 50 ft, 66 ft, 70 ft and 100 ft.

For the revenue surveys, not only during the early years of the nineteenth century but even to mid-century, the surveyors used chains of 33 feet length, with 16 links, called annas, each 2 feet 0.75 inch long, such that forty square chains equalled one acre or goonta. To check the chain regularly, the measurer carried a wooden staff of 8 feet 3 inches length, which would check 4 links or one quarter of the whole chain at a time. [231 p. 238] These would be of a cruder pattern than those used for baselines.

On the early baseline measures, for example that at Dod-Ballapur in 1802–3, on the Mysore survey, the chain was of 66 links, each 1 foot long. This chain was compared with two 'known' lengths with distinctly different results. Against a line measured with a 4 foot ruler as 20 feet 0.325 inches the chain measured 66 feet 0.96 inches at 86°F. When compared against a distance by brass ruler the chain was found to be 66 feet 0.56 inches. A difference of 0.4 inch would never have been tolerated in later years by surveyors such as Everest since over the whole baseline it was equivalent to some 20 feet!

The chain used by Lambton was '…of blistered steel, constructed by Mr Ramsden, and is precisely alike, in every respect, with that used by General Roy in measuring his base of verification on Romney Marsh. It consists of 40 links of 2½ feet each, measuring, in the whole, 100 feet at 68°F. It has two brass register heads, with a scale of six inches to each'. [228 p. 253].

In some revenue work the chain was replaced either by a rope jureeb or by a bamboo rod. Whilst the accuracy was sufficient for the purpose, the former was open to much abuse. Occasions were noted where the rope was up to 15% in error and this could be either plus or minus according to the circumstances, i.e. whichever was the most profitable!

The bamboo rod of 12 feet length and 3 inches diameter was considered to be more reliable than the chain since the chain was found to stretch anything up to 18 inches after only a short time in use and would be frequently checked against the bamboo. When actually in use in revenue surveys one inch was allowed on each chain length for the unevenness of the ground. [229 pp.162–3] Great care was needed to get any reasonable accuracy.

All the baselines measured between 1800 and 1830 – when Everest brought the compensating bars to India – were measured by chain. The only line on which Everest used a chain was the base at Sironj in 1824. The early lines were measured thus: a line of tripods was set out to take five coffers each of 20 feet length. These were aligned and put to an even slope by adjusting screws. The chain was then laid in the coffers such that its rear end could be fastened to a post and a weight suspended from the front end. The scales at each end were then read. Twenty men were used to lift the chain carefully out of the coffers and replace it on the next bay. The temperature was recorded at each bay.

Chronometer

In the early nineteenth century the Surveyor General was supplied with a gold chronometer by the East India Company to make the necessary astronomical observations to correct his surveys. It was considered so valuable that he was obliged to supply a special receipt of its arrival. [228 p. 65] It had in fact been purchased from the Colebrooke estate for Rs. 1600. [263 (81) 107] At the time a new theodolite cost about Rs. 420 and a new sextant about Rs. 210 so the chronometer was extremely expensive by comparison. [228 pp. 226, 330]

In 1810 a chronometer was supplied to Sackville to enable young officers to be instructed in astronomy. However it was soon reported that it had stopped functioning after

> ... its twentieth revolution and is no longer in use. I regularly and carefully wound it on every day at the same time. On the march I had it carried in a bearer's hand by my side, and in my tent never allowed it to be put out of my sight. It had previously gone very irregularly... [263 (82) 138]

In 1825, as Everest was about to go to England on sick leave, he reported that:

> The Great Trigonometrical Survey is very ill supplied with chronometers, there being but two, one of which ... is absolutely worthless,...[263 (171) 403]

Although time was a very important ingredient in the surveyor's equipment it was still troublesome and expensive. If the chronometer stopped, which it often did for no apparent reason, then it took great trouble to reset it correctly.

Circle, repeating (astronomical)

The repeating circle, sometimes called the repeating altitude and azimuth instrument, or the repeating theodolite was particularly used for the astronomical observations but could also be used similarly to a theodolite.

Lambton had purchased one of these instruments after his great theodolite was damaged in 1808. By Cary, it arrived in 1810, had an 18 inch diameter horizontal circle and was tripod mounted.

> By disengaging the end of the lever, the Telescope and its supporting pillars may be made to revolve independently of the horizontal circle and by fixing the end of the lever again, the Telescope with its pillars move together with the Horizontal Circle; to accomplish which the horizontal circle has a double axis, one within the other. [263 (266) 105].

In his Report of 1839 Everest referred to these instruments as suffering three defects: the graduations were not lines at 5 minute intervals but dots at 15 minute intervals; one axis rotated within another and this led to instability and eccentricity; and the telescope was of too small a power.

In an effort to improve the reliability and stability of his equipment Everest resorted to various exercises of cannibalism among the instruments available to him. The 18 inch theodolite of 1840 was made up by Mohsin Hussain, almost all of it, from pieces of other instruments and local material. [229 p. 143]

Circumferentor

In the early days of Everest's survey work the circumferentor was still in use for detail survey. Used in conjunction with a plane table and perambulator it remained popular into the 1820s. On the Revenue surveys there were as many circumferentors as plane tables and they were an essential accessory in this form of survey. [229 p. 213] Often the circumferentor would be used to survey the boundary and the results transferred to the plane table before the filling in of the detail.

The circumferentor was gradually superseded by the prismatic survey compass that was invented by Henry Kater about 1812 and marketed by Gilbert & Son. Kater had seen service under Lambton from 1803 to 1806 when his health broke down, so he would have been particularly familiar with the requirements of survey in such remote areas.

Compass – surveying, pocket and prismatic

In the early years of the nineteenth century there was considerable use of the pocket compass on exploratory, route and small-scale surveys. Traverse by perambulator and compass was common. The prismatic compass was invented by Kater in 1812. It

Appendix 1   Instrumentation   235

consisted essentially of a compass, graduated limb, vernier scale and four sighting vanes. Readings were possible to a few minutes of arc. [272 pp. 53–4] It was the sort of instrument that local youths could be taught to use successfully. [229 p. 389].

For many there was a preference to use a good compass instead of a circumferentor in traverses where vertical angles were not required. The new Kater compass by Gilbert had a silver metal ring of eight inches diameter attached to the fly. The circle of this was divided to 20 minutes of arc.

Compensating bars (See also Chapter 6)

One of Everest's worries when carrying out the Great Arc observations before going on sick leave in 1825 was accuracy. He was well aware that any inaccuracy in the baseline measurements would reflect through the whole arc and into any subsequent calculations relating to the figure of the earth. However, up to that time the bases had been measured by chains in coffers with variable results.

So he was very excited by the new idea developed by Colonel Frederick Thomas Colby, and used in Ireland, of measuring bars that compensated for temperature changes. They were first used in Ireland in 1827 and when Everest heard of them he arranged to visit Colby and see them in operation.

Figure 36   *The Colby compensating bar equipment as used on baselines in India from 1832.*

The theory behind the use of such bars was straightforward; however, in practice there were second-order effects to be considered as is so often the case in scientific constructions. The two bars of brass and iron had respective expansion coefficients of about 0.000 010 485 and 0.000 006 9844 for 1° F or approximately in the ratio of 2 to 3.

The two bars, each 10 feet 1.4 inch (10.117 feet) long, were set parallel and securely fixed together at their centres. Towards each end a tongue of metal was mounted more or less perpendicular to the bars and such that it pivoted on each bar. Exterior to the bars there was a mark on each of the two tongues which were so arranged that as the bars expanded or contracted so these marks remained a fixed distant of 10 feet apart.

In Figure 38 the pivots are at A and B and the mark at C, A and B were 1.8 inch (0.1500 feet) apart and the tongue was a total length of 6.2 inches (0.5167 ft). By similar triangles, if the brass expanded by roughly 1.5 times that of the iron then the length BC would need to be about 3.6 inches (0.3000 ft) to define a point of no movement. A similar arrangement existed at the other end of the compound system.

Great care and patience was required to locate the point C, and its companion C', at the other end of the bars by successive approximations. At that time the expansion coefficients were not too well known and this made the calibration of the sets of bars against a standard 10 foot scale very tedious. The whole system was placed in a lined wooden box and set on rollers within the box. Only the tongues bearing the points C and C' protruded and these were observed by microscopes.

Figure 37 One end of a compensating bar showing the linkage and the micrometer.

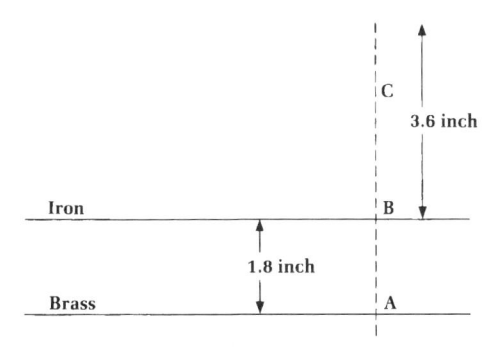

Figure 38 *Diagram illustrating the principle of the compensating bars.*

Heliotrope

The heliotrope (as in the flower, turning towards the sun) was invented in 1820 by Gauss and announced in London at the Astronomical Society in 1822. (It is possibly more appropriately called a heliograph.) Such was the reflecting power of this instrument that at 10 miles distance it could be too bright for the telescope, and it was used quite easily at 66 miles. [145] The heliotrope was among the new instruments that Everest acquired while on sick leave. Unfortunately he did not take nearly enough of them to India in 1830 and the surveyors had to make up many of their own from whatever was available in the local bazaar. Some were also made by the Mathematical Instrument Maker Henry Barrow (see Appendix 2). Once made, time was then spent training the *khalasis* (local staff) to operate them correctly so that the sun's reflection was always pointing towards the observing station. [231 p. 87]

Everest insisted that heliotropes be used for daylight observations on primary triangles but doubted whether they were essential for minor triangles. He saw some disadvantages in their use, such as the need to have accurate knowledge of the direction of the ray, and the difficulty of use in cloudy weather. In the centre of the mirrors was a small unsilvered aperture of about 0.1 inch diameter. For correct alignment a ring with cross-wires was placed at a distance of about 3 feet. The signalman would look through the unsilvered aperture and move the cross-wires until they intersected the distant station. Then the mirror was rotated until the sun's rays fell on the cross-wires.

Heliotropes of 9 inches diameter could be used up to 100 miles. For shorter distances a much smaller aperture was used: of the order of 0.1 inch of aperture for every mile of distance required. [272 pp. 477–9]

Hygrometer

In 1814 Lambton wrote about the problems of refraction and the lack of complete understanding of it. He attributed much of the variation to the 'different degrees of moisture in the atmosphere at different times. …The hygrometer will be a necessary instrument to indicate the degree of moisture.' [228 p. 259]

In 1807 Kater had published a 'Description of a very sensible Hygrometer' in which a species of grass proved particularly sensitive to humidity. However, while Lambton was probably aware of the article, he no doubt wanted something a little more scientific to solve his problem.

It was Henry Voysey who, in 1818, first suggested the idea of comparing wet and dry bulb thermometers as a measure of humidity. Everest said that he did not recall Lambton ever having an hygrometer other than one of the Kater sort which soon became useless. Voysey's idea was to apply a piece of wet muslin over the bulb of a thermometer during the time when vertical angles were being observed. Unfortunately he does not appear to have put this idea into practice and it was left to Everest, in 1823, when he was in the field with Voysey, to introduce the practice. From the first observation on 13 January 1823 at Netoli, Everest had the technique continued at all possible opportunities. [263 (171) 284]

Lamps

The use of lamps of various sorts as a target for triangulation observations was a considerable landmark in the Great Trigonometrical Survey. At least four different versions were in operation at one time or another while Everest was in India. He was instrumental in their introduction on a regular basis and developed various versions. In particular their use enabled night observations and this could take advantage of the greater effects of refraction and thus 'raise' targets that would otherwise not be visible.

(a) Argand or reverberatory lamp

An argand lamp was a hollow cylindrical burner in which the admission of air from inside and out increased the luminosity. It was named after Aimé Argand, a Swiss physicist. Argand lamps were made by Simms to Everest's design during the period 1826–30. They comprised an argand burner with glass chimney, fed from a reservoir of oil, and placed in the focus of a parabolic mirror.

They were particularly effective except when conditions were smoky or misty. On one occasion such a lamp was used at over 40 miles. Initially they could not be used when there was any air movement but Everest devised a means of enclosing the whole lamp in a wooden case with a tin chimney and a circular glass aperture fitted into the front door of it. When this was in place the light was perfect even in a torrential downpour with violent winds. [231 pp. 88–9]

The light disc of this lamp was 12 inches in diameter and when the atmosphere was clear this could be formed into a luminous point but in some conditions it would become an ill-defined disc that vibrated like a sheet of fire. The only solution then was to await better atmospheric conditions, and this might well be late at night. [263 (325) 179–235]

(b) Blue lights

As early as 1803 Lambton was using blue lights on his survey across the Indian Peninsula. [228 p. 239] Large blue lights were fired at Savendroog and observed at two other stations that were 135 miles apart. Lambton and Kater were doing the observing but because of inclement weather the lights were only just visible. The technique was supposedly to observe the instantaneous moment of extinction but large numbers of observations were required to get a reasonable result. [228 p. 259] Blue lights were very powerful up to ranges of some 60 miles. They were also useful at shorter distances in hazy weather.

In 1824 Everest was having difficulties obtaining sufficient blue lights. He had supplies of the various component compounds and was lucky enough to find someone who knew how to make them into the required form. The size of the problem will be appreciated when it is realised that 22 lights were required at each station, and each weighed 3 lbs, so a camel could only carry 160 of them.

Everest used blue lights only on lines over about 25 miles; below that distance, the vase light was effective. Blue lights could only be used properly in calm weather and the composition could be so inflammable as often to be quite unmanageable. There was a constant dropping of a lava-like substance from them and they gave off a dense smoky gas that reacted with any metallic substance. [231 p. 87]

The recipe for making these lights was:

Consider a composition of 739 parts then the components are –

gunduk	= sulphur	136
neel	= indigo	20
shora	= nitre	544
shungruff	= sulphuret of mercury	3
hurtal	= sulphuret of arsenic	32
ood	= gum benzolin	2
kaphoor	= camphor	2

They were made into a cylindrical shape 14½ inches long and 2¼ inches diameter. All was wrapped in 3 layers of paper, 2 layers of coarse cloth and 1 layer of sheep's bladder." [263 (171) 271–6]

With such a recipe the lights were expensive and so could not be kept burning continuously. They were fired at regular intervals and generally cut into lengths that lasted for about four minutes.

(c) Vase lights

The vase light was invented by Everest in 1822. [229 p. 235]

The vase light consists of a common earthen dish about 10 inches in diameter, and filled with cotton seeds and common oil. This is placed on the mark, and, to prevent the flame being blown aside, a large earthen pot, in the side of which an aperture has been cut, is inverted over the dish. An aperture is also cut in the top to allow the smoke to escape... The materials for this light are procurable in nearly every village. [272 pp.476, 477]

(d) Drummond lights

These were designed by Lieutenant Thomas Drummond (1797-1840) of the Ordnance Survey but, although Everest took six to India, he found them too cumbersome and costly as well as difficult to manage. They used the effect of oxygen played through an alcohol flame on a ball of lime. [231 p. 88]

Level

Everest had little interest in levelling as his prime occupation was in the determination of angles. Levels were common on the revenue surveys but descriptions of them are scant. When Everest required elevations they were invariably determined by vertical angles. Levelling staves were generally of 10 feet length divided into feet, inches and tenths of inches and with a movable target vane. [228 p. 203]

Pendulum

Although there were pendulums in India during Everest's time, he did not make use of them. This was probably due more to lack of time than lack of inclination. He had taken part in pendulum experiments while on leave and took two of the Kater variety back to India with him in 1830 but did not use them. Before shipment they were on display in London for three weeks for inspection. [231 p. 105]

Perambulator

Route location by perambulator and compass was a common method in the early nineteenth century. For work in such rough terrain the perambulator needed to be built very stoutly, but this was often not the case, and the perambulator simply fell to pieces as it was jolted along. To allow for slippage of the counter over rough surfaces, a turn was subtracted. This might be 1 in 30 or even as much as 3 in 30 in hilly country. Everest was interested in the perambulator. As early as 1819, after his route survey from Chunar to Hyderabad, he wrote

> I have a plan for improving perambulators which I think you will say is not without its merit. It is this. Let the common construction adopted by Mr Berge [apprentice to Ramsden] be the ground work, but instead of a mahogany wheel at 9 ft 2 in circumference I propose one of cast iron or brass of 10.56 feet in circumference, the

axle of the wheel works as hitherto and by means of teeth and pinions turns the long rod with the endless screw, which works into a wheel of 50 teeth (α); upon the same arbor with α is a pinion of 5 teeth which works into β of 50 teeth; and on the same arbor, or axle, with β is a pinion of 5 teeth working into γ of 50 teeth which turns upon the same axis as α. Thus a revolution of α is 10.56 × 50 feet = 528 = 1/10 of a mile, of β is a mile and of γ is 10 miles. [263 (154) 59]

While still in Calcutta after his return from England, Everest devised an ingenious series of dials and differential wheels. These he adapted from the double-wheel perambulators of Gilbert that had long been said to have been wrongly constructed. He had his ideas made up by Mohsin when they were established at Mussoorie. Although the new instrument was praised by those who used it, there were various suggestions for changes in the graduations and Everest considered it quite possible to have several versions. In acceding to this, he emphasised that it was not really his invention since it was based on the pattern in use previously in Madras since 1780.

Everest did not lose interest in the perambulator, even after retiring. In 1861 he wrote to Dr Shaw with yet another suggested arrangement. He thought the wheel should be of 8 ft 11.39 inch circumference (1/59th part of 528 ft). There would be two toothed dials. The back dial would have 60 teeth and its face divided to 60 parts. The front dial would have 59 teeth and be divided to 100 parts. When the front dial had made one revolution, the back dial would have gone all but one tooth or 1/10th of 5280 = 528 feet. [121]

Plane table

Use of the plane table for mapping both for civilian and military purposes has a long history. As far as India is concerned the earliest record of its use is for a rapid sketch around Salem in 1793 although the first recorded use for topographical survey is in Mysore in 1799. This was much earlier than in many other national organisations, including the Ordnance Survey. With the setting up of the Madras Military Institution in 1804 plane tabling was an essential part of the training. It was introduced at the Institution through Anthony Troyer who had learnt of the art in Austria.

Over the years its design changed little from the graduated wooden frame with compass and sight rule. Its size varied from 12 inches square to something larger than 24 by 18 inches. It was considered to be '…not inferior in accuracy, and of superior facility in use, to any other instrument. …[it] is a very portable and durable instrument, subject to less casualties than any other'. [263 (127) 67–8]

Sextant

Sextants were used over many years for astronomical observations. The main problem with them was the need to protect the artificial horizon from wind and insects. [228 p. 229] Most surveyors had either a sextant or a reflecting circle – which was on the same principle but with a full circle instead of just an arc.

Figure 39 *Plane tabling in Kulu region in 1904, but appropriate to Everest's time.*

The size of sextants varied from 8 to 15 inches radius with readings down to 10 seconds by vernier or nonius. Generally there were telescopes of different powers available. Care had to be taken over the effects of temperature which could cause variations of as much as 7 minutes of arc in a 10 minute time period. [231 pp. 178, 217]

Standard bars (See also Chapter 7)

Until Everest brought a calibrated standard bar to India in 1830, the standard of measure in the country was very haphazard. Lambton had used a chain 'set off from Ramsden's bar at 62 °F'. Another chain had been set off at a temperature of 50 °F. Unfortunately Lambton found the relative lengths of these two chains varied continually and checked them against his 3 foot brass scale that had been 'laid off by Cary from the scale of Alexander Aubert'. [231 p. 46]

In 1813, at Bellary, Lambton tested his chains against a three foot brass scale. He built a low brick wall with inset brass studs, the first five at 2½ feet and the remainder at 10 feet intervals. Using beam compasses, a length of 2½ feet was transferred to the initial studs to give a 10 foot length, which was then set over the full 100 feet. The chain was found to be 0.034 inch longer. A similar system was used at Hyderabad in 1814. In June 1821 the same technique was used at Hyderabad to check the standard chain against Cary's 3 foot brass scale; then again in 1825. [263 (172) 295]

With the passage of time the joints of the chain kept as standard seized up and affected the overall length. Because of this lack of a suitable standard, so much doubt lay on the precise measures of Lambton that Everest remeasured the Sironj base when he had the compensating bars.

Two 10 foot standard bars and two six inch brass scales were made for Everest by Troughton and Simms. Before being transported to India they were compared and certified. The bar and scale labelled A went to India in 1830, those labelled B not until February 1833. This delay was to allow them to be compared with the Tower Standard. Then in 1834 and 1835 it was possible to compare the A and B bars at Dehra Dun.

On his retirement in 1843 Everest took the B bar and scale back to England and deposited them at the Ordnance Survey for comparison with the English national standards. [263 (323) 35–46; (452) 59–60, 67–8]

Tables

In the pre-calculator era much resort had to be made to tables. Although Charles Babbage had begun work on his calculating engines neither his nor any others were operational during Everest's time in India. Hence all calculations were by logarithmic tables although various aids such as traverse tables were available. It has to be remembered, however, that all tables were compiled manually, so there were often inherent errors. Long after he retired Everest commented on errors in the logarithmic tables that had affected the results of his Great Arc computations. Any surveyor who remembers working with logarithmic tables will recall those by Shortrede but they were only compiled around the time Everest was retiring. It was, however, the same Robert Shortrede as was employed on the Survey of India under George Everest. His logarithmic tables were published in 1844 and his version of traverse tables in 1864. The other well-known seven figure mathematical tables by Chambers were only published in 1844. Everest produced tables of his own for geodetic calculations and a set of these is at present in the library of the Royal Society.

Theodolites

(a) Everest theodolite

The first mention of the Everest theodolite was:

> I have devoted some consideration to the improvement of the common theodolite which are both more cumbersome and expensive than they need be and after frequent examination of all the best devices I could meet with in the shape of the various makers in London, Mr Simms has at my suggestion designed an instrument which contains all the useful parts of the old construction, is quite free from

superfluous apparatus and is cheaper by one-fourth. I beg to suggest the propriety of keeping one of the 7 indented for at the India House as a model and sending two to each of the Presidencies.

The model has only 5 inch diameter but the principle is so perfectly applicable to all instruments for secondary triangles that I should respectfully recommend the propriety of adopting this as the Honorable East India Company's form for all small theodolites not exceeding 12 inches diameter and preserving on all future occasions the strictest uniformity. [90 fol. 458–64]

The first six of this pattern left for India with their inventor in 1830. The Everest pattern theodolite had two opposing vertical arcs instead of a complete vertical circle; the low height of the telescope standards prevented any transit of the telescope, and thus limited angles of elevation to 30° and of depression to 45°. [231 p. 144] Everest theodolites were particularly useful on the revenue surveys and sought after for many years as the ideal instrument.

The Everest theodolite was manufactured by Troughton and Simms and became very popular. The horizontal circle carried four arms, on three of which were verniers and the fourth had a clamp to fasten the index to the horizontal limb. This could be adjusted by a tangent screw. It was possible to repeat a measurement where the circle, fixed to a centre, was movable within the tripod support and adjustable with a clamp and tangent screw. The pivoted telescope was carried on a wide, flat horizontal bar that carried a spirit level. The vertical arcs were attached to the telescope. This

Figure 40   *The Everest pattern theodolite.*

was read by two verniers that also carried a spirit bubble on the arm connecting them. [272 p. 70]

(b) Great theodolite

Made by Cary, the great theodolite arrived in India in 1802 (See Chapter 9). When packed for transportation it weighed 1011 pounds and was valued at £650. The azimuth circle was 36 inches in diameter and the vertical 18 inches. Each circle had two microscopes. This instrument should not be confused with the 3 foot theodolite of Everest.

The great theodolite was much more suited to primary than secondary triangulation. In 1808 it met with an accident when it was being lifted to the top of a pagoda in Tanjore. A support rope gave way, the box broke and the limb hit against the side of the building and rendered it useless. [263 (198) 109]. The particular damage was to the tangent screw and its clamp, and the circle was considerably distorted.

Over a period of six weeks it was restored by Lambton. He took it to Trichinopoly, where he shut himself up in a tent, into which only the head artificers were allowed to enter. He then completely dismantled it, and gradually brought the circle back to its correct shape. '... in a manner surpassing his most sanguine expectation... but the circumstances of the case were never, I believe, officially brought to the notice of Government...' [263 (198) 109]

Figure 41 The 36 inch theodolite first used in 1802.

Because of this accident Everest introduced the idea of regularly changing zero to counteract the distortion of the horizontal circle. 'If, therefore, the zero could be changed a sufficient number of times, it was a fair assumption that the errors would be annihilated ... I have arbitrarily assumed ... nine times for the whole semi-circumference, by which means every twenty degrees have successfully come under the micrometers.' [96 pp. 47–9]

Before the accident it was felt that three or four observations with only one zero change were sufficient to get acceptable angles but now nine or twelve zero changes were necessary and these gave eighteen or twenty-four observations to mean out. [263 (171) 322]

In 1825 the instrument met with another accident. On 10 February a violent storm blew up with strong winds, hail and rain. All the tents were blown down and the great theodolite, which was already mounted on its stand for observing, was knocked over by the falling tent. One of its lower screws was broken.

By early 1825 Everest was not at all happy with it because of the wear and tear of time and constant use. The tripod was warped, many screws were out of order and the dots that marked the divisions were almost illegible through constant cleaning and the effect of dust. [263 (171) 316]

Everest had it put into store at Saugor and except for one use of it by Olliver in 1826 on the Calcutta longitudinal series, it was still in the store when Everest returned from his sick leave in 1830. Everest had wanted to take it to England when he went in 1825 so that it might be properly repaired but he was not allowed to do so. Instead he had it renovated by Barrow in Calcutta and then kept it in use through his remaining years in India. Its first foray into the field after reconstruction was at Nojhli on 12 November 1835 after a two-month journey on the Jellinghi from Calcutta. It was 1860 before it was pensioned off to the Survey museum. [229 p. 259]

(c) General theodolites

A variety of models of theodolite were used in India over the period from 1800 to 1843. Many were of poor quality and unwieldy. Some, made to special order by the likes of Lambton and Everest, were of better quality.

(d) 3 foot theodolite

On Everest's return to India in 1830 he took a wide range of theodolites with him. At the top end was the 3 foot (actually only 34 inches) diameter (not to be confused with Lambton's great theodolite) then two 18 inch, four 12 inch, twelve 7 inch and four 5 inch. Almost all were by Troughton and Simms.

On the 3 foot theodolite the horizontal circle had been hand-divided by Troughton and was read by five flying microscopes. The vertical circle was of 18 inches diameter and fixed to the telescope. This had a focal length of 39 inches and magnifications from 49 to 98. The whole weighed 16 cwt when crated.

Figure 42 Theodolite constructed for George Everest in 1830.

(e) 18 inch theodolites

The two 18 inch theodolites were also built to Everest's specifications but there were some variations in the graduations of the circles. One exhibited large periodic errors which were not in the other. However, by judicious observing techniques, it gave results as accurate as the other.

Both horizontal circles had three fixed microscopes 120° apart and the telescopes were of focal length 20 inches and magnification 25 to 50. To cap all of these there was the 18 inch instrument constructed by Mohsin Hussain in 1840 almost entirely from local materials. It was used on the Gurwani meridional series near the 82° E meridian line. [231 p. 144] This is some 275 miles east of Delhi and passes through Allahabad.

Zenith sector

Lambton had a zenith sector by Ramsden before he had his great theodolite. It had an arc of radius 5 feet and covered 9° either side of the zenith. Its micrometer was graduated to single seconds on which fractions of a second were readily estimated. Fourteen coolies were needed to carry it. [263 (63) 63]

Lambton first used it in 1801 at Bangalore, and Everest was the last to use it in 1825 at Kalianpur. In 1823 Everest had to be lowered into the seat at this instrument to take the observations because he was too ill to move himself [229 p. 244]. Today it rests with Everest's 3 foot theodolite in the Victoria Memorial Museum in Calcutta. [228 p. 252]

In 1823 another zenith sector, by Dollond, was purchased from Lambton's estate by the Surveyor General and was set up in his small observatory in Calcutta. It had a 6 foot focal length telescope with 3 inch aperture. There is no record of Everest ever using it. [229 pp. 259–60]

Figure 43 The 5 foot zenith sector used by Lambton and Everest for astronomical observations.

Appendix 2

INSTRUMENT MAKERS

Surveyors provide their own instruments

In the early days of the Survey of India the policy of the East India Company required surveyors employed by them to supply their own instruments. The Company held a small stock of special items, and in 1802 records state that '... a few good ... chronometers, sextants and theodolites should be sent out ... kept in store or deposited in this Office, to be delivered out occasionally as Surveys might be ordered'. Those instruments that had been sent out were by inferior makers and thus not very serviceable. In the early nineteenth century the Company used Mr M. Berge, successor to Ramsden, as their instrument supplier, and when asked about the low quality of his instruments he was scathing of remarks made because he considered that he used the best workmen.

By 1804 military conflicts prevented the Surveyor General from obtaining either a perambulator or compass. He had only two Ramsden theodolites in the Office and they were almost useless 'there is not a Mathematical Instrument Maker in Calcutta to repair them'. In 1808 one surveyor, whose theodolite was worn out, persuaded the Surveyor General to assist in his purchase of a new one because 'the Military Board would not grant the theodolite and other articles they had in store'. [228]

In 1809 a special request was made for the Government to purchase astronomical instruments because of their high cost. At the same time, however, the luckless surveyor still had to pay for perambulators, compasses and similar items. 'When surveying instruments are issued from the Arsenal of Fort William, or any of the subordinate magazines, ... the instruments shall be ... paid for on delivery...'

Surveyors were generally expected to effect their own repairs. In 1813 Hodgson wrote 'Here I was detained a day to repair my Wheel, the axle of which had worked loose. I was obliged to send to a distant village for workmen...' [228 pp. 221–8] The situation of a surveyor who broke a screw on his own personal theodolite highlights the state of affairs in India. There was a 'great disinclination of Government to replace anything ... broken on the public service, the survey must necessarily be at a stand till I be supplied with another theodolite'.

In 1808 when Lambton had an accident with the great theodolite it was he who

> ... shut himself up in a tent, into which no person was allowed to penetrate save the head artificers. He then took the instrument entirely to pieces, and, having cut on a large flat plank a circle of the exact size that he wanted, he gradually, by means of wedges and screws and pullies, drew the limb out so as to fit into the circumference; and thus in the course of six weeks he had brought it back nearly to its original form. [96 p. 46]

After the accident 'It seems that for a period of several years whatever repairs were required for the several instruments employed in the operations of the Survey were effected by native artificers attached to the nearest depots for Ordnance Stores'. [265(II) p. 53]

Although there were watch makers in Calcutta – David Mills had been there since 1793 – they do not appear to have dabbled much in the survey instrument field. [229 p. 213n] In 1822, however, it is recorded that Everest sent certain parts of the great theodolite to the jeweller George Gordon & Co of Madras for repair. [263 (171) 46]

In 1815 the Madras Government specified that military surveyors had to provide for themselves: one theodolite, one circumferentor with compass, one 100 foot brass chain, set of drawing instruments, etc. [229 p. 211] Requests for loans or replacements of instruments remained few and far between. On leaving the service it was sometimes possible to sell one's instruments to the Government but even then a good price was seldom attainable. Often much undervalued figures were accepted to assist with the cost of passage home.

In 1825 Everest had suggested that all the more important instruments, many of which had had upwards of 26 years of service without overhaul, should be sent to England for repair. Unfortunately the Government did not agree. [229 p. 260]. The turning point came when Everest was forced to go to England on sick leave from 1825 to 1830. He used the opportunity to contact all the leading instrument makers, study the latest developments and commission a large number of instruments. On Everest's return to India in 1830 the great theodolite was still in the store at Saugor. Everest brought it out of hiding and got Barrow to renovate it. As a result, it was subsequently used regularly on the principal triangulation until 1860.

Appointment of Henry Barrow

On the advice of Everest the Directors of the East India Company appointed a Mathematical Instrument Maker to go back to India with him. This was Henry Barrow. William Richardson of the Royal Observatory introduced Barrow to Everest in January 1829. Everest wrote in November 1830:

Appendix 2 Instrument Makers 251

I found him an intelligent, clever person, one of the principal workers of the trade. His character stood exceedingly high with the most eminent opticians in London for ability, punctuality, and unimpeachable integrity, and from the profits of his employment by Troughton, Dolland, Jones, Watkins and others, he cleared between 300 and 400£ per year … I had no scruples whatever in mentioning Mr Barrow, and my proposal was readily acceded to by the Court of Directors on his producing testimonials highly creditable to himself.

There are few persons in England who so situated would have been induced to quit their native country on a doubtful speculation but Mr Barrow had a wife and 6 children, and he felt, I presume, what every man of innate talent must more or less feel when he finds himself kept down below his natural level in Society by the want of capital to set up for himself besides which there was a considerable leaven of enthusiasm in his character for he had by way of amusement erected a small observatory on the roof of his house at his own cost and stocked it with instruments of his own making.

Barrow travelled to India with Everest and was found accommodation sufficient for himself, his wife and six children, and a workshop in Calcutta. This was at number 7/6 Theatre Street, which was initially leased for three years but actually remained in use for much longer. The workshop consisted of six rooms: a central one 38 × 22 feet running more or less north-south; on each of the east and west of this there were two rooms 18 × 16 feet; and to the north, one of 18 × 7 feet. This last room was well protected from any direct sun and so would be excellent for a dividing room.

… he would … find it necessary to place together in one room, a lathe, a vice bench and an apparatus for dividing, but in the division of instruments it is especially necessary that the dividing room should be kept quite separate and distinct, under lock and key, and never entered but by the Artist himself or someone, on whose discretion he particularly confides. A similar first floor served as living accommodation.

The division by hand of an instrument too large for a dividing engine, is an operation requiring great patience and much time, it is requisite that the temperature of the apartment should be kept quite equable during the operation and therefore the aspect must indispensably be to the North, so that light may be admitted without heat…

…Now Mr Barrow, though there is no dividing engine in India, will be called very shortly to put a new limb to the old large theodolite of the Great Trigonometrical Survey (36 inch diameter) and to divide it by hand, which he cannot do without such an apartment… [263 (265) 78–81]]

Barrow was initially allocated some ten staff of various sorts and by May 1831 this was supplemented with a further six, but it was, nevertheless, difficult to find skilled people. There were times when equipment sent in for repair was put to one side because of the lack of expert staff. [229 p. 126]

Barrow was the same age as Everest, so he was making his first venture overseas at the age of 40. He was somewhat forthright in his language and resented any interference from Everest. Not until this time had anyone dared to question Everest's ideas and comments.

Barrow refused the services of Mohsin and wanted everything left entirely to his own hand. He specifically declined to teach his craft to others. [263 (302) 38] By early 1832 Everest was writing to Barrow in strong language:

> The style in which you presumed yesterday to accost me was such as to be totally incompatible with our relative situations … For many days past I have perceived an inclination on your part to be offensive … After imputing to me an assertion which I never made, you rudely and flatly contradicted me twice … The tone and manner which you assumed were insulting and overbearing in the extreme, and totally at variance with that respect … which I will insist on preserving inviolate … If you compel me to have recourse to … higher authority, you will have yourself only to blame. [263 (266) 147–9]

Barrow's behaviour persuaded Everest that he must set out firm rules for the manner in which the workshop was run and the conduct of its staff. This entailed the submission of regular daily reports. He sent a copy of his rules to the Government, who, whilst supporting his assertion of authority, doubted whether some of his rules would improve efficiency. [229 p. 122]

In February 1832 Everest was writing to Casement:

> … relative to an unpleasantness which has taken place in my Department … Mr Barrow's impertinence to me (vide statement A) I am well disposed to treat it as unworthy of my notice provided I can be assured that it will not be repeated; the answer however, of that person to my letter does not seem to me to hold any such prospect, for though Mr Barrow knows very well how to behave himself when he pleases, yet when he is disposed to be saucy, a person with a more aggravating tongue, or more inclined to be forward, familiar and vulgar, is rarely to be met with. Until of late he has never ventured to practise any rudeness towards me, on the contrary his behaviour has been as decorous as I would desire and I submit the necessity of putting a stop to a bad habit of the kind, before it gains strength by continuance.
>
> … with reference to Mr Barrow's refusal to obey my orders … I have no desire whatever to interfere, and it is manifest that my object must always be to insist on the strictest obedience to the mathematical instrument maker, on the part of his workmen, but circumstances have come to my knowledge of late, which persuade me that it is injudicious to subject these people to the uncontrolled caprice of a person who possessing not the slightest acquaintance either with their language or manners, can only communicate with them through his Baboo…
>
> Statement A.
> On the afternoon of Tuesday 21st February I went over at 4 o'clock to see the progress of the work at the Shop of the Hon'ble Company's Mathematical Instrument Maker, Captain Wilcox was with me. … these cups seem to me to be of so simple a construction, that I should think it might be possible to get them done out of the house by contract…

Appendix 2 Instrument Makers

Mr Barrow replied to this not uncivilly at first ... as he talked he gradually got angry without any cause that I could divine and assumed an insolent and dictatorial tone towards me ... in a very surly tone 'Yes, if natives can do these kind of things, what's the use of my being sent as Mathematical Instrument Maker?...' [263 (283) 39–43]

The friction continued despite assurances by Barrow that he would comply with Everest's stipulations. In June 1832 Everest took exception to the amount of time spent by the workshop on items for the Arsenal whilst apparently neglecting work on the altitude and azimuth instrument that was urgently required.

Everest contrasted Barrow's attitude with that of Troughton. When the new instruments were being designed in England, Troughton and Everest would exchange ideas and when agreement on the best approach was reached, Troughton carried out the decision whether or not it was his own preferred plan. Barrow was referred to as '... no celebrity ... an able mechanic ... his opinion ... is of no weight whatever. Astronomers and geodesists are the best judges of the instruments which they are to use'. [263 (286) 43–9]

Whilst Everest's remedy would have been to place Barrow directly under the control of de Penning, the chief computer in Calcutta, the Government did not agree. Nevertheless, it appears that Everest sent many of his workshop orders to de Penning for onward transmission to Barrow. 'I enclose an order to Mr Barrow... If he should still be disobedient, a further statement ... must be communicated to me forthwith, that I may lay the matter before Government, as I have done with his former acts of refractoriness...' [263 (316) 87–9]

By 1837 Everest found that he needed an instrument maker in the field with him and requested that Barrow should join him. In particular he was to make various alterations to the two astronomical circles that had been brought out in 1830 and were now found to be quite unstable. The workshop in Calcutta was then to be left under the control of de Penning. Barrow left Calcutta on 11 August 1837 and reached Dehra Dun 13 September 1837.

Half of the team were put at Barrow's disposal when he arrived at Kaliana since skilled workmen were very loathe to go from Calcutta to join the field groups. De Penning tried persuasion in the form of additional salary but only when this was a considerable increase did a few agree to go. When Everest heard of the terms he was wild.

I am deeply ... mortified at the results of ... discussions with the workmen, who ... have shown themselves most unworthy servants ... Pretensions so extravagant ... cannot be listened to for an instant ... I suggest ... that they be instantly dismissed ...[263 (342)57–61; (304) 199]

Government agreed, and the whole of Mr Barrow's previous establishment were dismissed. Barrow set to work at Kaliana in October 1837 but three months later he was writing that it was impossible to carry out the work in Kaliana because of lack of the appropriate facilities. If he were in Calcutta with his workshop there would be no problem. When Everest returned to Kaliana on 9 March 1838 he was disappointed to find that, in his absence, progress had been very poor. There had been various setbacks in the complicated work, due possibly to the rudimentary conditions that Barrow had to work under. This made Barrow very touchy. Direct supervision seemed the only answer.

Repair of the astronomical circles

The two astronomical circles each had a vertical circle of 3 feet diameter and azimuth circle of 2 feet diameter. Troughton and Simms started work on their construction in 1829. Since this firm had constructed similar instruments before, Everest felt that any intrusion by him would not improve the design.

The instruments reached India in 1832 but it was 1837 before there was a chance to give them a fair trial. At Kaliana, Everest put each of them on a substantial masonry pillar since he felt the wooden tripod supplied was not solid enough. A sandstone slab of 33 inches diameter capped the masonry, and since all was on a very sound foundation no vibration was expected. As a result of tests Everest felt it to be absolutely necessary to construct for each instrument a pair of new columns, a pair of new outriggers and friction rollers, a new axis for the azimuth action, a new table and a new azimuth circle. [109 p. 142] In fact, almost a new instrument!

At that time Everest had his quarters at the Park, Mussoorie (see Chapter 12), and he took both the instruments and Barrow there for some six months by which time all that remained to be done was the graduation of the circles. This Barrow refused to do despite previously agreeing to the manner of division (copying them from a previously graduated instrument) suggested by Everest and later described in detail in [110].

No doubt looking for a way out of an *impasse*, Barrow requested six months' leave from 29 August 1838 but received the short response that the work should be finished first. This immediately raised the temperature of the relationship and Everest recorded:

> About three days after this, he again accosted me in the workshop in a tone of increased anger ... I believe he said ... something to impugn my plan of dividing ... but his passion seemed ... to have so far got the better of his reason, that I neither clearly understood what he meant to say, nor do I believe he understood himself... [263 (342) 252–3]

Everest laid down several rules that Barrow should follow. These included

> Whenever you may desire to communicate your opinions to me ... on professional points, you will be pleased to notify your wish ... in writing ... you are prohibited to commence any verbal discussion with me in the workshop, or elsewhere, ... unless invited by me to do so... [263 (348) 298–9]

By 5 November there appeared to be no progress and Everest commented:

> If Mr Barrow cannot, or will not, divide instruments, he is not required in India. There

Appendix 2 Instrument Makers

is nothing whatever, save this, in which my sub assistant Said Mohsin, is not his equal, and ... [in] many points ... his superior, being in far better practice as a workman. [263 (342) 252–3]

Everest then left it up to Barrow to use whatever method he thought appropriate to divide the circles but there was no progress. In January 1839 Barrow reported that the circles were in working order although still containing the vibrations that had been the original cause of the trouble. He suggested that as the work had been under Everest's direct supervision then any inadequacies rested with Everest. Naturally this was not well received and in return Everest put all the blame on Barrow.

Everest wrote in January 1839:

Be pleased ... instead of wasting time in committing your impertinence to paper, to employ your whole energies in endeavouring to put the instruments into working order, which I fully see is a job requiring all the skill you are master of [263 (350) 8–10]

Further petty criticisms flowed, with Everest even resorting to reporting as Superintendent of the Trigonometrical Survey to himself as Surveyor General who then passed on the instructions to the Mathematical Instrument Maker. In January 1839 he wrote to Barrow: 'On receipt of your letters ... I handed them to the Superintendent of the G.T. Survey, a copy of whose report sent to my office is enclosed for your ... guidance.' [263 (350) 18]

Figure 44 The horizontal circle of the 24 inch theodolite.

Figure 45 *The vertical circle of the 24 inch theodolite.*

In one instance Everest questioned Ramsden's ability, to which Barrow riposted 'I think it would have been quite as well if you had the same distrust of your mechanical powers before you had altered the 2-feet circles by Troughton, who as a mathematical instrument maker stood preeminent'.

The reply was 'though you are at liberty to think just what you please, yet an official letter is not the proper channel wherein to give vent to impertinent comments on me or my proceedings...' [263 (302) 152; (350) 15]

In consequence of the refusal by Barrow to divide the horizontal circles, he was sent back to Calcutta in February 1839 to report to the Town Major and left the service on the day of his arrival there, 19 April 1839. The comment from Everest after Barrow had left the service was that 'The whole of this affair caused me a great deal of uneasiness, for I had every inclination to be friends and oblige Mr Barrow... I had a high value for his skill as a workman, and was quite ready to conciliate him by any sacrifice consistent with propriety.' [263 (344) 108–84] Whether he actually resigned or was dismissed is uncertain. The term 'discharged' is used. [229 p. 124], [231 p. 419]

In 1840 Everest described Barrow thus:

Though Barrow was a skilful workman, yet, latterly at least, he was really not so expert as Said Mohsin and … as to the practical part of conveying instruments on the march, and the various wants of parties in the field, … or doing anything out of the mere routine of his own particular business – Mr Barrow was about as completely helpless as need be. [263 (402) 82–95, 133–49]

On his return to England, Barrow built scientific instruments in a workshop in London. Surprisingly enough, two theodolites that he built in 1845 were destined for the Survey of India. In 1849 he became a Fellow of the Royal Astronomical Society.

As Everest had no access to a dividing engine of any sort, he arranged a system where the graduations required on a blank circle were copied from an already graduated circle. The new circle was to be of cast iron inlaid with gold to receive the graduations. At every tenth degree a circular disc of silver was let in to allow the large figures to stand out. An elaborate piece of machinery was designed and built such that the circle to be graduated was fixed to a substantial masonry pillar. The circle to be copied was allowed to revolve above and carried with it a cutting tool and clamp. It was thus necessary to be able to read off the angular spaces passed over by the rotation of the upper circle and for this there were four reading microscopes with micrometers attached to the lower circle. The graduation implement was of the form used by Mr Hindley of York.

Everest had all the necessary items made and a special room built at the Park to house it: a small room 12 feet square with a parallel verandah all round. In the centre of the room was a masonry pillar. Openings which faced the meridian direction as well as to east and west were sufficiently large to allow the instrument to be brought in. In these doors there were round windows to allow lamp light to enter from the verandah but not to be so close as to affect the temperature. The light was from two reverberatory lamps with Argand burners. Further refinements, such as double doors were not completed by the time it was necessary to start dividing.

Graduation of the first circle started on 6 August 1839 with various disadvantages. The method was new, no person there had ever used a cutting tool or even seen others using one. The weather was unfavourable: constant mist and heavy rain with a raw wind. Hence several precautions were overlooked. One particular problem was of being able frequently to check the relative positions of three particular radii: that through the centre of the divided circle; that determined by the microscopes; and that passing through the point of the dividing tool. All should emanate from the centre of rotation but they could vary for many reasons. By the time the first circle was completed it was 24 August, and Everest had encountered various difficulties, either because of his eagerness or his failure to consider some possible effect, but even so it was a magnificent effort. The graduations were down to 5 minute divisions.

By the time for graduation of the second circle the building had been fully completed and a modified method of division was used based on the experience gained with the first one. [109], [110]

Mohsin Hussain

In 1824 Syed Mir Mohsin Hussain was put in charge of the instruments held in Calcutta. Hussain had started out in the workshop of George Gordon in Madras but in 1819 was taken on by Blacker (then Quartermaster General in Madras) to repair instruments. After Blacker became Surveyor General in 1823 he sent Mohsin to Calcutta. When Hodgson became Surveyor General in 1826, he taught Mohsin how to take astronomical observations and found him to be a steady observer. [229 pp. 313, 485]

In 1830 Everest came upon Mohsin Hussain in Calcutta. At the time he was doing little of note, but Everest soon took a liking to his manner and work and saw the use of Hussain as a safeguard against the unpredictable nature of Barrow. [263 (402) 78–95, 133–49] When Everest moved headquarters from Calcutta to Dehra Dun in 1832 he took Mohsin Hussain with him. [229 p.129] At his Park Estate, Everest built a two-room workshop and engaged a team of workmen to be under Hussain. This team accompanied Everest into the field each season and were invaluable to him.

When Everest recommended Hussain for an appointment as sub-assistant in 1836 his glowing reference mentioned the work by Hussain on the apparatus for comparing the chains and standard bars, in 1832. On the remodelling of the 18 inch theodolites of Cary, Barrow only played a small part, yet it was his name that appeared on the instruments with no mention of the major part played by the native artist.

The remaining work on the circles was then entrusted to Mohsin Hussain who completed the task successfully by October 1839. [229 p. 98] Everest was particularly pleased with a 'stop' that Hussain devised for the circles '…an exquisite contrivance … as remarkable for its neatness and simplicity as for its efficiency; furnished with a graduated micrometer head, of which one revolution is equivalent to the difference between the graduating lines of the whole degree and the five minute spaces'. Hussain had both constructed the copying apparatus and cut the divisions and for this was awarded an honorarium. It had taken two and a half years of labour to put the two instruments in working order.

During the stay of Barrow at Dehra, the workshop at Calcutta, although nominally under the direction of de Penning, was, after protests from de Penning, entrusted to Ernest Gray in November 1838 for a period of six months, with a six month extension. On Barrow's departure, although Everest wanted Hussain to be given the post, Government insisted it should be Gray. Everest took exception to this by indicating that there was nothing that he would want Mr Gray to do for the GTS while Hussain was still around. [263 (344) 72–4] Gray found himself severely handicapped through lack of accommodation and tools and the fact that he had to divide his time between his watch maker's business and the Survey. Without further incentive he would not give up his watch making business and so was not retained after the twelve month contract.

Government, in its wisdom, then appointed Alexander Boileau to take charge of the Calcutta office and he retained the position until 1843. After much persuasion the Directors agreed to Mohsin Hussain being Mathematical Instrument Maker in succession to Barrow but they left his salary as it was.

Mohsin Hussain was with Waugh's party at Kalianpur and on the Bidar base measurement so that any problem with instrument repairs could be effected on the spot and not be subjected to the usual extensive delays. [229 p. 55] Otherwise even the loss of a screw would mean that the other instrument at Kaliana would also be rendered useless for the same period since simultaneous observations were being taken. Such was the necessity for Hussain's attendance that it was arranged for him to travel by Dak, which was by relays of ponies or bearers, and was the fastest mode of transport then available. [229 pp. 124, 170]

In total, Hussain was at three baseline measurements repairing any of the delicate parts of the bar systems that became deranged. He repaired the large theodolite before Everest's first use of it in 1835, [263 (289) 222–5], made the reverberatory lamps efficient, constructed lifting cranes for the instruments, built several barometer pumps to Everest's own design and was in general of inestimable use. 'He has both genius and originality, his conduct is marked by the highest probity, and… he is one of the few … on whose word I could place entire reliance.' [231 p. 458] While Everest was still in India, Hussain was allowed to remain at the field headquarters but as soon as Everest retired Hussain had to return to Calcutta. [229 p. 131]

Appendix 3

AN ATLAS OF INDIA

Early maps

Whenever a new territory was entered, there was an immediate requirement for maps of the area. These were not simply for finding one's way around but for a host of other uses: military operations, tax gathering, statistics on commerce, agriculture, mineral wealth and many others. In the terminology of the twentieth century, maps were the basis of the gathering of all forms of geographic information.

From the 1760s, maps were made of various parts of India with one of the first exponents being James Rennell who was Surveyor General in Bengal. He naturally introduced mapping techniques akin to those pertaining in Europe at that time and in 1780 produced *A Bengal Atlas*.

As might be expected, funding for the growing requirements was not generous, so the mapping was piecemeal and usually by untrained staff. Instead of waiting for an overall triangulation to be completed before mapping, the Company endeavoured to match up the scattered surveys, and this, of course, was contrary to all standard survey practices. They considered that a well-drawn, neatly presented map must of necessity be a quality product in all its aspects – which it often was not. An example of this was work produced by Thomas Jervis – about whom much is said in Chapter 9 – of the Konkan area. Because 'it looked very attractive' it was considered to be good, so he was given a large sum of money. Only on detailed inspection by surveyors, rather than administrators, was it found to be very inconsistent and not reconcilable with maps of surrounding territory.

To exacerbate the situation, several authorities in each province dabbled in mapping for their own requirement. All efforts to bring material together failed and no early map of the whole country was possible.

Prime aim of the Surveyor General

One outcome of this was the rolling into one post of the three provincial surveyor generalships as they existed in 1814. The requirements of the post were laid down and one prime aim was the compilation of a general map of India. [81] This followed from the concept laid down by the Directors as the main duty of a Surveyor General of India. He was obliged to compile large-scale maps of all parts of India from the best available material. He was to maintain these and was to be the sole authority for such maps and had to ensure that copies were only made for authorised persons. [229, p. 274] This last tenet continued to be exercised for much of the next 170 years!

In 1822 the Directors of the Honourable East India Company decided to start the compilation of an Atlas of India that would cover the whole country at a quarter of an inch to the mile. The compilation occupied many years but sheets were engraved as they became available and formed the accepted map until the turn of the century. [229 p. 2]

The prompt for this move came when the Directors found that ideas put forward by both Mackenzie and Hodgson closely followed those in England of Aaron Arrowsmith. In fact it was in 1822 that Arrowsmith produced both an Atlas of South India and a single sheet Sketch of the Outline and Principal Rivers of India. [81]

The Atlas

If surveys that were then in progress were to be incorporated in the Atlas it was necessary to ensure that they were all carried out on the same principles and to the same standards. The large-scale series of maps at 4 miles to 1 inch were to be bounded by lines of latitude and longitude, and in parallel with this there was to be a small-scale map of India. Unfortunately there was a lack of draughtsmen to keep up with the new material that was being continuously collected. As the need for a complete Atlas became increasingly necessary, the solution to the lack of cartographers had to be met by sending the material back to London.

Valentine Blacker was the man in the hot seat at the time of the introduction of this idea although it had previously been discussed with Colonel Mackenzie, who was Surveyor General in Calcutta. Blacker set out to assess the worth of all existing material and considered that the Atlas must be linked directly to Lambton's triangulation work. This was rather contrary to the much earlier suggestion by Rennell who had considered the appointment of a special astronomer to determine the positions of selected points throughout the country as a basis for a rapid mapping scheme. [229 pp. 240, 283]

As a foundation for his series, Blacker drew up a scheme based on the same projection as that used for the great military map of France. This was a modified Flamsteed projection, as described by Puissant in his *Traité de Topographie* … of 1810. An alternative name for the projection was 'Conique alterée'. Although described as being originally devised by Bonne about 1750 it was later modified by Flamsteed. Unfortunately the Directors had already taken the initiative in England and were working on a 'globular projection' devised by Aaron Arrowsmith.

A communication problem between the Directors on the one hand and Blacker and Hodgson on the other resulted in the production of 12 sheets in India in 1828 on the Blacker choice of projection. This led to definitive instructions that all material was to be transmitted to England for production. [263 (217) 231] The extent of the projection was calculated to cover the range 8° to 32° N latitude and 67° to 93° E longitude. [263 (231) 49] The parameters used, although not specifically stated, would almost certainly have been those recently determined by Lambton, with the central meridian at 76° 30' E. Individual sheets were 38 by 27.4 inches. [229 p. 295]

When Arrowsmith died there was a period of inactivity before a new map publisher was willing to take the project on. This was John Walker who designed the Atlas to consist of 177 sheets each covering 160 by 108 miles. (Note that the index to sheets for the Atlas of India [231 p. 23] only shows 146 sheets).

Lithography was, at that time, a new art. It was only established in England in 1817 by the Bohemian Rudolph Ackerman. So although Hodgson asked in 1822 for a lithographer and equipment to be sent to India their availability was scarce. The Survey Department was not an early owner of a press, instead it had to use the Government Lithographic Press when it was established in 1823. [229 p. 298]

Secrecy

In 1826 the Surveyor General had requested one of his staff to:

> ... be particularly careful that all the work ... be done ... under your immediate inspection, and that nothing ... is made public by private lithographists. Each sheet [of the Atlas] when finished should be kept by you, locked up and under your seal ... [263 (184) 421] We draw your attention to the frequent, and we believe, often unnecessary, calls for maps by our officers, civil and military, caused by the neglectful custody, or irregular appropriation, of those documents; and we desire that it may be intimated to all our servants that maps supplied from our public offices are public property, to be carefully kept.

Throughout, the Directors were very much opposed to any form of personal acquisition of copies of individual sheets. In 1828 they stated that 'All surveys made at public expense are public property, and we direct that no copies of any surveys, so made, be delivered to any persons except those appointed by Government to receive them'. [263 (90) 55] One reason for such secrecy was that the East India Company had no wish for the French or Dutch to get reliable maps of its territory. [229 p. 388]

The Map Committee

Although Everest's prime interest was in the Great Arc he did not neglect the Atlas. In fact, at one stage when Lord Auckland was becoming anxious to have it completed and formed a non-professional map committee that came up with half-baked ideas, Everest retorted strongly. The matter was sent to London for resolution and resulted in a restatement that the Surveyor General alone was responsible for map-making. [231 p. 11]

This Map Committee was formed in 1837 apparently on the suggestion of H.T. Prinsep, of the Supreme Council (not to be confused with the surveyor T. Prinsep). As it deliberated, the Committee trespassed more and more into the territory of the Surveyor General, including the appointment of local surveyors and draughtsmen.

The Governor General wanted a more active policy on the Atlas but, in the absence of Everest who was in the north-west of the country, communication was somewhat difficult. In 1838 General Sir William Morison of the Supreme Council commented that the appointment of Surveyor General and that of Superintendent of the Great Trigonometrical Survey were incompatible. Even the appointment of a Deputy Surveyor General had not greatly improved the situation since he was in direct charge of, and fully occupied with, the Revenue Surveys.

Morison went on to suggest that if the Surveyor General designate (Jervis) were to take over he hoped that the Directors would see to it that there was proper control of the field surveys.

When the Committee was formed, no attempt was made to consult Everest or James Bedford – who was Deputy Surveyor General, Bengal, and Superintendent of Revenue Surveys. Bedford was, however, directed to draft suitable regulations. Although the Committee must have known that it was treading on dangerous ground to try to interfere with an established department, especially one run by someone like Everest, it drew up various resolutions. One such resolution of 1839 went so far as to say that every future Surveyor General should reside in the Presidency and leave the work of the Great Trigonometrical Survey to a subordinate.

The same year the Committee said that all available mapping material, of whatever quality and source, should be compiled into a quarter inch map series for those areas not yet covered by the Atlas. They went further by suggesting that four extra map-compilers should be employed, but they made no recommendation as to where such scarce personnel could be found.

In 1840 Bedford was appointed to the Map Committee without even being approached but did not take long to decline the position. He had taken the step of scrutinising the various suggestions and could not support them. Thus he felt he would have been more of an embarrassment than help to the Committee.

Bedford wrote in the same vein to Everest who fully supported him. Lengthy correspondence passed between Bedford and the Committee with copies to Everest who himself protested most forcibly that the proposals were at complete variance with elementary principles. When it was all referred to the Directors they fully endorsed the stand by Everest and Bedford and even went so far as to stipulate that they would not engrave any more Atlas sheets unless they were founded on triangulation.

Altogether the Committee had given endless trouble introducing a few useful changes and a not so useful one. The Surveyor General was compelled to produce a provisional list of geographical co-ordinates derived from his primary and secondary triangulation before his final computations had been completed. This proved to be not at all satisfactory. [231 pp. 297–302]

Map production

Production of Atlas sheets was very slow. Basically the idea was that all those on topographic survey work were obliged, as soon as possible after the end of the field season, to draw their work at the reduced scale necessary for the Atlas. The results were then either sent straight to England or assembled to form a degree square and then forwarded. The first published sheets reached India in 1828 but by 1833 the total number of sheets had only reached 30 and by 1843 – the year Everest retired – the total was only 39 out of a projected 146.

All in all the Directors of the Honourable Court were expecting far too much from the Surveyor General. On the one hand there was the need for a rapid completion of the Great Trigonometrical Survey, and on the other, the need to tie all material for the Atlas to triangulation. At the same time Everest lost two of his deputies. It was an impossible situation. [231 pp. 303–4]

In 1832 some of the earliest such maps to emanate from the Hyderabad survey were heavily criticised by Everest. He described one as 'so confused as to be totally illegible'. 'If it is unintelligible to me who knows that country well, by how much more must it be so to the engraver at the India House who never was out of London'. He then went on to criticise it in depth. [263 (267) 99–101]

To furnish sufficient survey parties to continue the field work necessary for the Atlas, Everest had difficulty finding suitable staff. He envisaged a total of eight field parties, but most of his assistants were unsuitable to take charge of such parties. [231 p. 3] In 1869 the production of the sheets was at last transferred to India and a new projection was introduced instead of Blacker's polyconic version. [229 p. 296]

Appendix 4

EVEREST'S ROLE IN ADMINISTRATION

All Everest's efforts did not go into the observations and computations that go with any triangulation scheme; there were many other aspects to his labours.

Initially the Surveyor General's Department in Bengal came under the control of the Military Department of the Supreme Government. Then on 1 June 1818 it was transferred to the Public Department. Since a part of its work was for the military command there arose the necessity of separate accounting for military and civil work. Such an arrangement obviously added to the work of the Surveyor General and he (Mackenzie at that time) was not well pleased.

Office control

Everest kept close control over all aspects of the GTS, but this left little time to give attention to the topographical surveys or to the revenue surveys. He left the revenue surveys entirely to the Deputy Surveyor General. Even so, he always had a large amount of mail to deal with. This he normally replied to in great detail while at the same time urging his correspondents, particularly the field officers, to keep their letters as short as possible.

As a means of facilitating this he stipulated that:

… all letters [are] … to be sent in duplicate … on foolscap paper with half margin. One of these copies will be returned with replies or remarks opposite to the paragraph requiring them. … limit the length of their letters as far as practicable by curtailing all expressions not bearing upon the sense of their subject. [263 (578) 58]

A few months later he commented that:

The current business … has vastly increased, whilst the establishment of writers has remained the same. … The scientific surveys … have greatly added to my correspondence … I shall not be far wrong in estimating the … official writing at full double what it was in the time of my predecessor. [263 (283) 88–96]

The office of the Surveyor General must fall into arrears, and – when one man is thus called on to execute duties for which five persons ... were not deemed too many –,... the marvel is, not that such a result should take place, but that it should so long have been ... warded off. I scarcely know how to describe the drudgery ... to which I have been subject. ... Even now I am transgressing the rules laid down for me by my medical adviser, at the risk of incurring a relapse, for quiet and repose have been prescribed for me as the only chance of recovery from my late illness. [263 (342) 78–82]

Whilst Everest was obviously willing to push himself to the limits of human endurance, both in the field and with his office work, the impression is that he preferred to spend his energies on the former – and who could blame him. With so many observations and computations calling him, who would want instead to be writing interminable letters and reports to government officials?

In 1839 he explained that:

... crowds upon crowds of important documents still remain on the file in rough draft for want of time to copy and index them ... Every letter which arrives and requires an answer is detained on the shelf ... until I can afford time to attend the subject. Each paragraph is then diligently and carefully read and studied by myself, and ... the reply is drafted by me. When the reply is fairly copied it is brought to me for ... examination ... [263 (344) 279–82].

Unlike any modern office of comparable complex and extent, there was no one to whom even the drafting of replies could be entrusted, no one to delegate aspects of the administration to. The burden of holding two posts, with few, and sometimes no, deputies, led to the constant worry of tasks undone and areas of operation professionally unsupervised.

The appearance of a new Governor General, Lord Auckland, in 1836, did nothing to help Everest since it only increased his correspondence, and there were requests for lengthy progress reports each year. In addition, the absence of Everest from Calcutta and the consequent delay in correspondence did not please the new incumbent. At the time Everest produced a lengthy response that included:

... then I arrived at headquarters jaded to death by the night work of the observatory. ... Moreover, the current business and money matters of the office had on that account fallen into arrears, which it was my first duty to bring up, and lastly my parties ... did not arrive ... until the middle of June. The question then ... was not as to the propriety of drawing up ... the report, ... but as to which business was to be first attended to, the computations or the report. I decided that the former was ... [263 (342) 78–82]

This is reminiscent of the present day with enthusiastic staff becoming bogged down with the need to produce reams of reports rather than getting on with the prime objective. Particularly in the case of a major project such as the GTS, one might have reasonably expected that a brief progress report annually would have sufficed with a detailed submission at the conclusion of activities. To be side-tracked into detail before the complete picture was available would have been galling even to a person of lesser volatility than Everest.

Public relations

Relationships between the field parties and the local population could also be a source of difficulty. Quite often the man in charge on the spot was able to sort out any misunderstandings, grievances or problems. On occasion this did not suffice, and the Surveyor General/Superintendent GTS was involved.

It was policy that consent had to be obtained from the ruler of each State before survey work in that State. On occasions delays occurred and, if Everest happened to be the man on the spot, he was not sympathetic towards any excuses. The Political Agent was the man in the firing line, with the task of persuading any unco-operative rulers. In addition it was necessary to organise supplies and labour through the same channels. Delays led to uncontrollable impatience and fierce protests to Government.

Looking at the situation from the point of view of the local population, the influx of a large survey party requiring provisions must have put a considerable strain on resources. Spare elephants or camels or food for 20, 30 or, on occasions, 200 or more people cannot be conjured up at the drop of a hat in a small rural community. In 1833 Waugh wrote of his problems. 'The party ... consists of about 200 men, with about 50 head of cattle, and ... the provisioning of such an establishment for any length of time in jungles remote from cultivation is a source of ... anxiety'. [263 (286) 105]

On 4 November 1833 Everest wrote:

The whole of my establishment have been waiting in the Dhoon since the 1st November, and ... no elephants having yet arrived it is impossible to proceed with my business ... Punctuality is the very life and soul of my department ...

Everest wrote similarly the following year:

My camp came here on the 4th, [letter written on the 8th] and on the evening of that day more than half the people went without food. I thought this might have been accidental, but it has occurred almost every day since, in a greater or less degree, and yesterday ... a quantity of unwholesome ottah [coarse bread] was sent, which the people could not use, famished as they were. [263 (289) 6–8]

Obviously any survey party has its administrative problems, but often there can be a member of the party detailed to look after such circumstances. When the situation is that the leader has not only to do all negotiations and field all arguments, difficulties and revolts but also do all the field observations, then the constitution must suffer. Under the unhealthy natural conditions that prevailed in almost all areas where the GTS went, any additional debilitating elements such as overwork, excessive stress and continual worry must take its toll.

To get the long lines required for triangulation it was often necessary to cut numerous trees. In some parts of scantily populated country this was no problem, and intruding trees were felled at will. The situation was quite different, however, when populated areas were reached. Here the trees bore crops that were gathered, had owners, and were valuable as timber. Despite the offer of compensation, it was not always an easy task to persuade the local people to agree to the felling.

Even more contentious situations occurred when buildings were involved. No matter how flimsy they might be, they were homes, and no one lightly allows destruction of one's house. To arrange such clearance required patience and skilful negotiation. This would often involve the local *tahsildar* (chief officer of a sub-district) to get a price acceptable to the villager yet not excessive from the point of view of the Government. As one would expect with a character such as Everest, there were occasions when he violently disagreed with the negotiated price and refused to make a payment unless it was sanctioned from higher authority.

On the whole the field parties got on well with the local population. As anywhere else, however, there were occasions, such as the destruction of monuments, that obviously could not be tolerated. At other times various members of the team were assaulted and robbed. Such events then made it difficult to get the camp followers to go out.

Often it was advantageous to occupy high buildings, but what if it should be felt that women's quarters were being overlooked and their privacy infringed? Everest and his assistants always tried to avoid such accusations, but on one occasion a wealthy man even offered to have a replacement high point built a few yards away.

> Persuaded ... that our telescopes which invert have magic powers, and are able to turn women upside down (an indecent posture no doubt and very shocking to contemplate) it is natural enough that they should assign to us the propensity of sitting all day long, spying through the stone walls at those whom they deem so enchanting. [263 (346) 134–5]

Then there was the propensity to destroy survey station marks. Everest appealed to the Government:

> to allow more vigorous action to prevent such wanton interference. [He] ... urged that steps should be taken to persuade the more intelligent of the leaders that the operations of the G.T. Survey were not in any wise connected with the black arts, or with the finding of gold, but were purely to find out the length, breadth and thickness of the Earth ... [263 (286) 108–9]

The local population treated some stations with superstitious veneration or with suspicion and dread. As Everest wrote to Government in 1836:

> In some parts of the country, the platforms are regarded as objects of idolatrous veneration, and there are amongst the villagers persons who, I am told, make money by pointing out the centre and circle engraved on the stone, and descanting on its holiness and marvellous powers. In other parts of the country the most preposterous stories are promulgated, regarding the incantations and ceremonies gone through in achieving this magic mark and laying it in the platform, such as that young pagan children – obtained by stealth if possible, or otherwise by purchase, at any price – are offered up as sacrifices, and that their skulls bear testimony to the fact. [231 p. 86]

Staff

Considering the size of the country and the magnitude of the task of the GTS, it might be thought that Everest had a considerable number of surveyors. In fact, the GTS was very poorly staffed. In 1815 Lambton had just four assistants and by the end of 1818 this was increased to six, one of whom was George Everest. At the time of Lambton's death in 1823 Everest inherited six survey staff and when he went on sick leave in 1825 this was down to four. The situation improved noticeably after his return and in 1832 there were five military surveyors and over 20 civil assistants. For the whole period 1832–43 Everest averaged a staff of five military surveyors and 23 civil assistants. As Surveyor General he had additional separate staff, but these were engaged on topographic and revenue surveys rather than primary triangulation.

The size of party required when observing on the GTS was enormous and took considerable organisation. At Kalianpur in 1841 Everest had with him 16 assistants, 63 sepoys and similar, 2 native doctors, 353 khalasis, 100 servants, 6 elephants, 115 camels, 50 horses, 100 bullocks and cows, 25 tattoos. A total party of 535 persons and 296 animals. [263 (406) 11]

Everest had regular problems with staff who were very poorly paid requesting either an increase or resigning for lack of it. In 1833 there was a period when one officer was in the position of having no more of his own money that he could use for advances of wages to his staff and '... the operations of the party were suspended for want of funds ... The native establishment were grumbling and discontented ... not receiving their pay regularly...' [263 (286) c–f]. With little more than subsistence wages, any staff who were at all educated or able were in demand. Certainly Everest and others always had the greatest difficulty in filling vacancies. In 1835 six of his seven computers left for the prospect of some eight times greater salary elsewhere. [263 (348) 140–3]

Other requirements

In addition to being a surveyor, Everest had to be negotiator, personnel manager, transport manager, politician, improviser, medical assistant, finance manager and chief general factotum. Many have held similar posts this century, when it would have been far easier to delegate some of the work than it was in Everest's situation. It must have made him a very lonely and aloof man, far more than the Kumpass Walla he was once referred to as, yet in keeping with the nickname of Neverrest.

Appendix 5

THE FIGURE OF THE EARTH AND GEODESY

*T*he eighteenth century saw a dramatic increase in activity related to the figure of the earth. Most notable among these were the two expeditions to Peru and Lapland that finally proved Newton to be correct in his theory of the oblate shape. Then at the end of the century the major arc measurement in France assisted in the definition of the metre as a unit of length. The accuracy of the methods and instrumentation used, however, were little better than the noise level. In other words observational errors were of a similar magnitude to the quantities that were required to be evaluated if accurate parameters for the earth were to be obtained. True, there were improvements during the eighteenth century, but far more were required if reliable values were to materialise. Everest, and his predecessor Lambton, played no small part in the development of the science of geodesy. From 1799 to 1843 they executed the longest arc of meridian measured to that date, and from it Everest derived two sets of values for the earth's parameters, one in 1830 and the other in 1847. This was in addition to a set by Lambton in 1821. To appreciate the significance of this, some background is required so that the magnitude of the undertaking and value of the results can be put into perspective.

Background

Late in 1799 Captain William Lambton of HM's 33rd Regiment of Foot drew up a project for a mathematical and geographical survey that would extend right across the Southern Peninsula of India from the Coromandel coast to the Malabar coast and in due course from the southern tip to northern extreme. This would form the basis or framework for a general survey of the country. 'He proposed to execute a triangulation emanating from a measured baseline and verified from time to time by other such baselines but in order that the results might be correctly employed in determining the spheroidal coordinates of the triangulation stations it was necessary that the figure of the earth, and the length of the polar or of the equatorial axis should be accurately known. Captain Lambton was aware that it was incumbent upon him to execute a geodetic survey *pari passu* with his geographical survey as a necessary preliminary to the determination of the elements of the trigonometrical stations and other marks'. [281 p. xvi] Since, at that time, knowledge of the lands under the control of the East India Company was very scant, the initiative of Lambton was appreciated and acted upon with almost undue speed. Approval came within three months and after obtaining equipment Lambton was able to start field work within ten months of putting his idea forward – no mean achievement in a country where communications were poor, distances long and logistics difficult. Although the prime aim was to measure a meridional arc up the centre of India from Cape Comorin at 8° N to the Himalayan foothills at 30+° N, there were also plans for arcs along lines of latitude. 'Col Lambton decided to determine the figure of that portion of the earth's surface to which his operations would be restricted, by measuring the lengths of meridional arcs in successive parallels of latitude from Cape Comorin northwards and the lengths of arcs of great circles perpendicular to the meridian in the parallels of Madras and Bombay'. [281 p. xx]. The first work on these arcs was the measurement of a baseline at St Thomas Mount near Madras in April and May 1802. [228 p. 257]

The idea for a Great Trigonometrical Survey

In 1800 Lambton had written a Plan of a Mathematical and Geographical Survey [51], [263 (63)], [196] where he set out the principles of a geodetic survey and the requirements for the determination of the figure of the earth.

He indicated that it appeared from experiments with pendulums that the gravity at latitude 10° diminished suddenly. 'If so ... a degree on the meridian from that parallel to the equator must be very short compared with a degree immediately to the northward of 10°. It will not only be necessary to attend to this circumstance in the course of a mathematical survey as needing a correction, but as an object leading to something curious with regard to the figure of the earth...'

He was aware also of apparent difficulties arising from observations of precession up to that time. 'For most ... assumptions have been that ... the equatorial is to the polar diameter as 231 to 230; for by allowing any other ratio the results will make the effects of precession different from what they are observed to be; and yet the measurements which have been made on the meridian in different latitudes give the protuberance at the equator 1:312 instead of 1:230..."' [51], [263 (63)] In a technical journal he went further 'Should the earth prove to be neither an ellipsoid, nor a figure generated by any particular curve of known properties, but a figure whose meridional section is bounded by no law of curvature, then we can obtain nothing until we have an actual measurement'..." [197 p. 318].

Thus even from the beginning Lambton was conscious of the complicated situation and this was to be similarly appreciated by his successor. They both had as their aim the determination of a network of positions on which to found any detailed surveys of India but, at the same time, saw how this would lead to a considerably improved knowledge of the figure of the earth.

When he started his triangulation Lambton would measure a baseline then determine its azimuth from observations to Polaris at each end. A base extension net would join the base to the main triangles. In the principal triangles all angles were measured three or four times but without changing zero. Then at selected points the zenith sector was used for latitude observations. The connection of heights to sea level was somewhat tenuous. At Bangalore a barometer height was brought up from Madras in 1800. Then in 1802 the St Thomas base was related to low water, although this could only be considered a local datum. The base at Tanjore was connected to the beach at Nagore by triangulation. [228 pp 258–9]

Start of the Great Trigonometrical Survey

Lambton's first meridional arc was measured along the Coromandel coast during 1802-3 and he regarded the start of his Great Trigonometrical Survey as the measurement of the St Thomas' Point baseline in April and May 1802. At six stations from Cape Comorin to Bidar he took observations for latitude and azimuth to divide the arc into five sections. As a basis to work from, before his own observations would give appropriate constants, he took the length of a degree of the meridian at 13° N as 60 191 fathoms (as determined by the French academicians and by General Roy) and took Newton's value of 1/230 for the ellipticity. Subsequently from his arc of about 1° 35' measured in 1802–3 on the Coromandel coast he found 1° as 60 494 fathoms which he then used in subsequent computations across the peninsula. His results at this stage were discordant and did not even fit the assumed ellipticity value. He made the length of a degree:

| Bangalore to Putchapolliam | 11° 59' 55" N as | 60 530 fathoms |
| Bangalore to Paughur | 12 33 09 | 60 466 |

[281 p. xxiii]

These indicated that either the oblate ellipsoidal hypothesis was incorrect, since the lengths appeared to decrease instead of increase with latitude or that the work was in error. The problem he had not grappled with sufficiently was to prove to be that of deflection of the plumb line. Needless to say, he felt that deflection was the problem but was not then in a position to quantify it. He had found the effects of deflection near Bangalore which could only be accounted for by supposing there to be differing densities through the earth. 'Let the Figure of the Earth be what it will, we may venture to abandon the hypothesis of uniform density from which Sir I Newton drew his conclusions'. [228 p. 261] Although both Lambton and Everest had a grasp of the problems of deflection of the plumb line, that information seemed not to have been widely circulated. (As late as 1820 – just before Everest went to the Cape of Good Hope to investigate that very problem – Hodgson and Herbert were looking to explain a 10 second discrepancy in an area that would have been highly susceptible to deflection. [229 p. 177])

Early results

From Lambton's work of 1802–5 the early results for great circles perpendicular to the meridian gave

At 12° 32' 12" N   1° = 61 061.0 fathoms
   12 55 10   60 743.8
             and   60 751.8

[281 p. xx]

The next meridian arc was an extension southwards towards Cape Comorin then moving northwards to latitude 13° 02' 55"

Arc	Amplitude	1° in fathoms
Putchapolliam to Punnae	2° 50' 10.54"	60 472.83
Putchapolliam to Namthabad	4 06 11.28	60 487.56

[281 p. xxiii]

Lambton's figure of the earth

In 1812 Lambton received new values for the earth parameters as deduced in Europe and these gave the ellipticity as nearly 1 : 304. As a result he threw out all his earlier values and recomputed the whole of the central arc up to Gooty (lat. 15° long. c78° E). Then he extended it to Bidar (lat. 18° long. c78° E) to give an overall amplitude of nearly 10° from Cape Comorin. He decided also from these overseas values that his adopted value for the degree perpendicular to the meridian was too small by about 120 fathoms, that the mpv (most probable value) should be 60 687 fathoms and that the meridional degree in latitude 13° 34' 44" should be 60 491.4 fathoms. [281 p. xxi] In 1821, however, the Parliamentary Commission adopted as standard measure the scale of 1760 by Bird. The effect of this was, according to Kater, that the Indian standard scale required to be decreased by 0.0018% of its length while Ramsden's bar used in the survey of Great Britain had to be increased by 0.007% Since the Indian arc depended on both of these, it had to be corrected. [229 pp. 237–8] Lambton actually recomputed his work not just once but twice. From this he derived his own earth constants as:

a	b	$e = (a - b)/a$
20 918 747 feet	20 851 326 feet	1:310.28

He had combined the known results from the English, French and Swedish arcs to obtain his value. In 1808 it was written

> The value of Major Lambton's Survey, is ... but little understood. I fear there are but few among us who consider the ascertaining the lengths of three or four degrees of meridian, and as many of Longitudes, as of any importance, or who conceive that much scientific knowledge, or much labour, is necessary for accomplishing it. ... The

opinions of the Learned in Europe, however, are very different; ... Major Lambton will, if not prematurely interrupted, in a short time have ascertained the length of a greater arc of the Meridian than was done either in Lapland or Peru ...[228 p.265]

Although that statement was correct, since the Peru arc was only 3°, the complete task of the arc in India was to taken another 32 years to complete, by which time it would cover some 22°. Lambton sent his various results to learned societies in Calcutta, London and Paris and was very gratified to have a response from Delambre. Comment from the English scientific world was noticeable by its absence. The portions of the arc surveyed by Lambton and Everest joined together at Damargidda in latitude 18°.

In 1842 Everest suggested to the Government that much of Lambton's work be reobserved. The suggestion was declined but in 1849 Waugh put the same idea forward. In 1861, long after he had retired, Everest suggested to the Royal Society that the scattered records of Lambton's work should be collated and re-examined. Unfortunately the Royal Society committee set up to decide about this felt that it would do no good whatsoever. [228 pp. 265–6]. Not until the period 1866 to 1874 was all the work by Lambton to the south of Bidar finally reobserved. [263 (462) 265–307; (401) 203]

Reduction of angles

In 1819 Everest developed two new formulae for reducing observed terrestrial angles to the angles of the chords. He had a paper on the subject read at the Astronomical Society on 9 April 1824. Between 1819 and 1824 he was aware that others in Europe may have devised the same expressions, in which case he requested that his paper be suppressed and returned to him.

If the spherical angle is C and its corresponding chord equivalent $C' = C - x$, then he found the expression for x as

$$x'' = \mu.\alpha.\beta.\cot(C/2)/4R^2 - [\mu \cot C(\alpha+\beta)^2/8R^2](1+(\alpha-\beta)^2/16R^2)$$

alternatively

$$x'' = \mu.\alpha.\beta.\tan(C/2)/4R^2 - [\mu \cot C(\alpha-\beta)^2/8R^2](1+(\alpha+\beta)^2/16R^2)$$

where $\mu = 206\,264.8''$ and $\alpha, \beta =$ chord distances forming angle C'

In both cases the second terms are exceedingly small and can be omitted. In addition, $\mu/4R^2$ and $\mu/8R^2$ are constant coefficients, and equal to $1.182\,29 \times 10^{-10}$ and $5.911\,47 \times 10^{-11}$ respectively. (Everest used 3955.36 miles for R.) His worked examples agreed to the third decimal of a second with the calculated spherical excess for the triangles. The formula he used for spherical excess was

$$\log E = \log \beta + \log \alpha + \log \sin C + \text{const.} \log \overline{10.373\,755\,3} \qquad [91]$$

Everest's first set of values

By now Lambton had died and Everest was in charge of the GTS. Between October 1823 and March 1825 he completed the fifth section to the south of Takalkhera and the sixth section to the north up to Sironj. Then while on sick leave in England he computed his first set of values for the figure of the earth. He considered his most trustworthy values to go into the calculation to be those of the arc Punnae-Kalianpur which he combined with the values for the European arc Formentera-Greenwich to give earth parameters of

a	b	$(a-b)/a$
20 922 931.80 feet	20 853 374.58 feet	1 : 300.80

These became known as Everest's First Set of Constants and were used in all calculations of the Survey of India.

Deflection of the vertical

By this time Everest was very well aware of the problems of attraction because of his stay in the Cape Province but as yet he had no quantitative measures. The arcs on which he was basing his calculations were almost equally north and south of Takalkhera. Initial calculations showed that the curvature of the meridian was the opposite to what was expected, i.e. increasing towards the north. He reasoned that this could be due to a latitude error at the central station, due possibly to attraction by the tableland to the north of Takalkhera. This was a rectangular area some 120 miles east–west by 60 miles north–south, with the station some 20 miles to the south and centrally placed. Everest calculated what the change in the latitude of Takalkhera would have to be for the result to then conform to the expected value of 1 : 300. From his theory, and using a ratio for the density of the hills to the mean density of the earth of 0.6 and height of the plateau as 0.3 miles he determined that an attraction of 5.0" would be required. He then used the known information for the two sections of arc in the following manner. If x is a small correction to the amplitudes α and α'; ρ, ρ', are the radii of curvature and φ, φ', the mean latitudes of each section; then he derived the approximation that

$$x/\alpha = [(\rho - \rho')/(\rho + \rho')] + [sin(\varphi' - \varphi)sin(\varphi' + \varphi)]/200.$$

Using this relationship he calculated that the deflection at Takalkhera would be about 4.5": a very satisfactory agreement with the theoretical value. [66 pp. 28–9]

It might be asked why Everest used Kaliana as the end of his arc with regards to the computation of the earth's parameters. As Shortrede wrote in 1848, 'The fear of disturbance from this immense mountainous mass [the Himalayas] induced Colonel Everest to limit his arc to Kaliana, sixty-nine miles to the south of Banog, and fifty-seven miles south of the Dehra Dun base'. [254 p. 93] For Banog, various estimates put the deflection to be of the order of 20" and obviously something to be avoided if possible.

Everest was also aware of the possibility of assessing the problem by gravity observations but he possessed no equipment. [229 p. 254]

Appendix 5 The figure of the Earth and geodesy

Everest's second set of values

Everest's second set of constants were derived in his volume of 1847 and given as:

a	b	$(a-b)/a$
20 920 902.48 feet	20 853 642.00 feet	1 : 311.043
or 3 486 817.08 fathoms	3 475 607.00 fathoms	

[116 p. clxxix]

These values were never widely accepted. [231 p. 104]

When Everest was computing latitude, longitude and azimuth for successive stations he used formulae that were infinite series. They converged rapidly, however, and four terms were sufficient. For some work even just the first two terms were enough. His approach was as follows.

If the known point A has latitude λ and longitude L then the unknown point B has latitude $\lambda' = \lambda + \Delta\lambda$ and longitude $L' = L + \Delta L$

A, B = azimuth A–B and B–A respectively

c = distance AB

Then $\delta_1\lambda + \delta_2\lambda = \Delta\lambda$

$\delta_1 L + \delta_2 L = \Delta L$

and $\delta_1 A + \delta_2 A = \Delta A$

where $\delta_1\lambda = P\cos A.c$

$\delta_2\lambda = \delta_1 A.R \sin A.c$

$\delta_1 L = \delta_1\lambda.Q \sec\lambda\tan A$

$\delta_2 L = \delta_2\lambda.S \cot A$

$\delta_1 A = \delta_1 L \sin A$

$\delta_2 A = \delta_2 L.T$

$B = (\pi + A) + \Delta A$

P, Q, R, S, T = numerical functions of the earth parameters a and b and were tabulated in increments of 10' of arc for rapid interpolation. [272 pp. 493–9]

> In apportioning the residual error among the angles of his triangles, Colonel Everest was one of the first (if not the first) to recognise the existence of equations of condition (to be fulfilled by the true angles) other than the old one which made the sum of the angles of a triangle = 180° + the spherical excess. [271 p. 108]

Azimuths

For determining the azimuth of a Reference Object (RO) from observations on a circumpolar star Everest modified the method of using a star *at* its maximum elongation to observing *near* maximum elongation and thus allowing scope for many more observations. About an hour before maximum elongation the theodolite was set up and a reading taken to the RO.

> Sight the telescope to the star's computed altitude, move it azimuthally until the star appears in the field of vision. Tighten horizontal clamp and use slow motion to place star in upper angle of the wires if it is ascending – or vice versa. Mark chronometer time of star reaching the intersection and read off azimuth limb. Repeat the process. Then observe RO again. This would be one set on one face. Change face and repeat. Repeat whole several times. [272 pp 593–9]

Calculations were then made to correct all observations to the position of maximum elongation and the required azimuth determined.

Pendulum tests

While in England in 1829 Everest superintended experiments on new pendulums that he had been requested to take back to India. In doing so, he took the opportunity to instruct cadets from Addiscombe in the technique whilst the equipment was to hand. As he met with some problems he read a paper on the topic at the Astronomical Society on 13 March 1829.

> I was at first in hopes of being able to shew satisfactorily, that a deviation from the truth in the position of the knife edges, unless it were considerably larger than what would be likely to occur, would have but an insensible effect on the length of the pendulum; and having separated the inequalities into two particular cases, the solution of the first of these tended much to strengthen my opinion. I had not, however, proceeded far with the second case, before it became evident that the effects consequent on it were much more formidable; strange anomalies seemed to arise, to oppose my progress and make me doubt the solidity of the reasoning; and as it has cost me some trouble to reconcile the apparent inconsistencies, I have not thought that my paper would be overloaded by containing my solutions of them. [99 p. 25]

Basically what he was trying to determine were the effects of the axis of the pendulum not being correctly perpendicular to both the breadth and length of the bar. For the breadth he found it safe to neglect any non-perpendicularity but the length presented a problem. The question was 'Is the deviation serious or not?... answered this in the affirmative, it becomes me to say that I sought it not; that I stumbled upon it as a truth in my way, which it would have been unworthy of me to conceal from my scientific brethren; and that I had originally no thought or design of impugning any established theory'. [98 p. 117]

Everest, in effect, turned his mind to determining how much the greatest deviation from perpendicularity in a pendulum vibrating on a knife edge may be expected to affect the results. He stated the problem thus:

> In a pendulum vibrating on a knife-edge, we have altogether to rely on the artist who constructed it for the truth of position: no method is known either of adjustment or

of detecting the deviation from perpendicularity; yet, as it is hardly possible that a knife-edge ever can have been quite correctly placed, it is desirable to ascertain how far the greatest inequality that may be apprehended would affect the result; and it is to this inquiry in its simplest form that I mean to confine myself. [99 p. 26]

After some pages of mathematics (which are set out fully in [99]) he took details from a paper by Mr Baily and showed that an error of less than 49' in the fixing of the knife edges would account for all the discordance found by Mr Baily. [99 p. 36] In effect he found that a much more serious error would be likely to arise from this source than had been first thought possible. [99 p. 30]

Archdeacon John H. Pratt

The problems of the deflection of the plumb line and of the figure of the earth were taken up by Archdeacon John H. Pratt after Everest had retired. He was not unfamiliar with the situation in which Everest had operated since he was Archdeacon of Calcutta. In a few years from 1855 Pratt wrote a number of scientific articles on the subject.

> I believe myself,…that the earth's form is not an exact spheroid. …The spheroid (of depression 1-300th) which has been determined … is therefore, the *average* spheroid…some parts being above it and some below it … We can no longer assume that the arcs of meridian are all equal ellipses, or are ellipses at all, or that the arcs of longitude are circular. [238 p. 203]

In 1855 Pratt had used the arc from Kaliana to Damargidda and found that it most nearly coincided with an ellipse of compression 1 : 426 rather than 1 : 300. Following from this he criticised the attempt by Tennant to test the arc used by Pratt against other arcs but all that would be highlighted from that exercise would be whether or not all the arcs used belonged to one and the same spheroid. In no way would such a comparison show whether calculations were correct or not. [238 p. 205]

Pratt then concentrated on the problems of the plumb line.

> …the results of the measurement of the great arc of the meridian of India afford undoubted indications of the disturbing influence exercised upon the direction of the plumb-line by the attraction of the Himalaya Mountains and the elevated regions lying beyond them. Thus, the amplitude of the northern division of the arc included between Kaliana and Kalianpur, when determined by astronomical observations of latitude at the two extreme stations, was found to be 5° 23' 37.058"; whereas when computed geodetically, assuming the usually admitted values of the major axis and ellipticity of the earth, the value of the amplitude appeared to be 5° 23' 42.294".

Pratt indicated that the discrepancy was of the right sign but was it of the right magnitude? If it was, the expected deviation would be greater at the northern station than the southern and hence there would be a smaller astronomical amplitude. As part of his investigations Pratt found what effect a small error in amplitude (5.236" in this case) would have on the value of the ellipticity. He found it to be at least 1/25th of the whole value.

It is not, therefore, sufficient to attribute the discordance to the influence of mountain attraction; but, on the contrary, it becomes an object of the highest importance to ascertain whether the effects due to this cause are capable of accounting exactly for the observed anomaly. [234 pp. 36–7]

Pratt was about to demonstrate his ideas on isostasy and, with the situation in India, had ideal material to work on. After deriving necessary expressions he then investigated an approximate value of the intensity of mountain attraction at the two extremities and the middle of the arc – with Kaliana at 29° 30' 48", Kalianpur at 24° 07' 11" and Damargidda at 18° 03' 15", which he labelled respectively A, B and C.

On a large map of India he traced out compartments of terrain and calculated the possible influence of each compartment on the stations. His results were:

Deflection of plumb line in meridian at	A	27.853"
	B	11.968
	C	6.909
in prime vertical at	A	16.942
	B	4.763
	C	2.723

(Some of the above values were later corrected by small but insignificant amounts.)

These values were much greater than those indicated by the Indian Survey. He then modified some of his assumptions such as that for the density of the elevated region but to no avail. He concluded that 'The ellipticity which results from taking account of mountain attraction flattens the middle point of the arc by 0.0298 of a mile, or 157 feet; whereas the ellipticity, when mountain attraction is neglected, is more curved by 0.0532 of a mile, or 281 feet'. [234 pp. 37–41] Note that a correction is incorporated in the above text that is pointed out in the later paper. [238 p. 204]

When referring to Everest's volume of 1847 Pratt commented that:

Colonel Everest, ... assuming ... that the Indian Arc is curved like the average ellipse of the earth; and ignoring, ... the effect of the Himalayas, brought out this result ... that by his geodetic computation Kaliana was farther north from Kalianpur by 1-10th of a mile than by the astronomical latitudes. He attributes this important discrepancy to mountain-attraction; but does not prove that mountain-attraction will produce this exact amount of error.

Pratt's calculations not only showed that effects of this magnitude could occur but that they could be some three times larger. This he illustrated for the same arc where 36 separate stars were observed for determination of the amplitude. Of the results, 29 were only 1" from the mean; 6 less than 2" and one just greater than 2".

...and yet to get an accurate result, 36 observations were thought necessary – showing that even a deviation of 1" was considered of importance. But Himalayan attraction produces in this amplitude ... an error of more than 15", and surely cannot be passed by. [242 p. 26]

He concluded that the only way to make things tally was to assume that the form of the Indian meridian was not that of the average ellipse. Later investigations found that in the Ganges Plain, over 100 miles from the Himalayas, the deflection was in the opposite, i.e. negative, direction.

Determination of longitude

In the period under review the determination of longitude was a difficult problem. There were some five different methods adopted to solve this:

1. distances of the moon from the sun or a star,
2. transits of the moon,
3. zenith distances of the moon,
4. occultations of the stars by the moon,
5. eclipses of Jupiter's satellites.

As these methods are no longer in vogue a brief comment on each is in order.

1. Distances of the moon from the sun or a star. Lunar distances had to be combined with the zenith distance of the moon and of the sun or a star. The sequence of observations was :

> Several lunar distances with time noted. As accurate as possible.
> Zenith distances of moon and star with time noted. Not high accuracy.
> Several lunar distances.
> Zenith distance of the star.
> Zenith distance of the moon.
> Several lunar distances.

Means were taken of the various quantities and after manipulation the Greenwich mean solar time could be converted to Greenwich sidereal time and comparison with sidereal time of observation gave the longitude.

2. Transits of the moon. Although very simple and tolerably accurate only about eight observations could be taken in any one month and even then the moon had to be on the meridian so use of the method was restricted.

3. Zenith distances of the moon. This was a good method so long as observations were not far from 6 hours sidereal time. From observations of the moon across all horizontal wires of an altitude/azimuth instrument the apparent zenith distance would be well known for a certain chronometer time. From observation of the chronometer error the sidereal time could be calculated.

4. Occultations of the stars by the moon. If the disappearance, or re-appearance of the star behind the moon was noted at a particular chronometer time, and the chronometer error was determined, then the sidereal time could be found. In the solution two longitudes were assumed, one lower and one greater than the reputed value of the point.

5. *Eclipses of Jupiter's satellites*. This was only considered to be a rough method. The requirement was to note the moment of disappearance or re-appearance of the satellite. More details of each of these can be found in [272 pp. 607–23].

Intercomparison of arcs

In his Report of 1830 [96] Everest made 15 intercomparisons between major arcs. The arcs themselves were:

> Damargidda to Kalianpur
> Punnae to Kalianpur
> Punnae to Damargidda
> Bidar to Sironj
> Greenwich to Formentera
> Clifton to Dunnose
> French arc
> Peruvian arc
> Lapland arc

His results ranged from $c/a = 1/290.44$ to $1/418.23$ and $a = 3\,481\,441.3$ to $3\,487\,363.3$ [96 pp. 109–16]

He accepted the values of $a = 3\,487\,155.3$ fathoms and $c/a = 1/300.80$ for his first set of constants for the figure of the earth.

For calculation of the compression he used the theorem set out by Playfair in volume 5 of the *Edinburgh Philosophical Transactions*. [96 p. 105] From the same work it appears that he used seven-figure logarithmic tables for his computations; and directions in the field were read to 0.5 seconds although the angles were meaned out to 0.001 seconds. Triangle closures after allowance for spherical excess were seldom greater than 3.5 seconds. [96 p. 214]

For the application of spherical excess and chord corrections he used the method of Legendre. [96 p. 57]

When it came to the apportionment of errors, a variety of techniques were used. Lambton had distributed angular misclosures proportional to the size of the angle. Everest did not agree with this approach and indicated that the common method was to distribute equally to all angles if they were from the same number of observations. If they were from unequal sets of observations then distribution could be inversely proportional to the number of observations. He mentioned least squares and the fact that though it was simple for a single triangle it became complicated with overlapping figures. He tried examples with condition equations but it is not certain how widely he used the technique on his great arc. [96 p. 74]

In his subsequent magnus opus of 1847 [116] Everest intercompared 12 arcs from different parts of the world. Five of these were in India, six in Europe and one in Peru. They were:

Appendix 5 The figure of the Earth and geodesy

Arc	Observer
1 Kalianpur to Kaliana	Everest
2 Damargidda to Kalianpur	Everest
3 Damargidda to Kaliana	Everest
4 Damargidda to Punnae	Lambton
5 Punnae to Kaliana	Lambton and Everest
6 Formentera to Dunkirk	Delambre, Mechain, D'Arago and Biot
7 Jacobstadt to Hochland	Struve
8 Tarqui to Cotchesqui	Bouguer and De la Condamine
9 Dunnose to Clifton	Mudge
10 Pahtavara to Mallorn	Ofverbom, Svanberg, Holmquist and Palander
11 Jacobstadt to Dorpat	Struve
12 Dorpat to Hochland	Struve

Between these he made 42 intercomparisons for values of e^2, c, a, c/a, and the meridional quadrant length. The results ranged from:

$e^2 = 0.005\ 1188$ on 2 to 4 to $0.010\ 4108$ on 1 to 2;

$c/a = 1/191.6$ on 1 to 2 to $1/390.2$ on 2 to 4.

Despite these extremes, 31 out of the 42 results for c/a were in the range 1/275 to 1/325.

The meridian quadrant values ranged from 5 479 601 to 5 463 352 units of 6 English feet, i.e. 32 877 606 to 32 780 112 English feet but with the average value of 5 468 705 units or 32 812 230 English feet. This, in terms of the accepted value for the foot-metre conversion, gave the quadrant as 10 001 289 metres and hence immediately highlighted the inadequacies of the acceptance of a definition of the metre as 1/10 000 000 part of the quadrant. [116]

Baselines measured in India 1800–1870

Date	Place	Observer	Lat	Long	Height(ft)	Length(ft)	Method	Ref
1800	Bangalore	Lambton	12°50	77°35		39 267.706	100ft chain in coffers	[197]
1802	St Thomas Mt.	Lambton	12 57	80 16		40 006.4450	100ft chain in coffers	[198, 5, 265]
1804	Bangalore	Warren	12 57	77 42		39 793.7	100ft chain in coffers	[199, 265]
1806	Coimbatore	Lambton	10 58	77 43		32 301.2769	100ft chain in coffers	[200, 5, 265]
1808	Tanjore	Lambton	10 44	79 08	0	21 705.00	100ft chain on ground	[265]
1809	Palamcottah	Lambton	8 47	77 43	204	30 507.8	100ft chain on ground	[5, 265]
1811	Gooty	Lambton	15 03	77 40	1182	32 607.6000	100ft chain on ground	[201,5]
1812	Guntur	Lambton	16 17	80 31	118	26 407.4374	100ft chain on ground	[265]
1814	Kumta	Penning	14 28	74 25		21 609.1248	100ft chain on ground	[265]
1815	Bidar	Lambton	18 03	77 41	1957	30 806.176	100ft chain	[96, 203, 265]
1822	Takalkhera	Lambton	21 07	77 42	1250	37 912.5613	100ft chain on ground	[265]
1824	Sironj	Everest	24 07	77 52	1574	38 411.8991	100ft chain in coffers	[96, 5]
1832	Calcutta	Everest	22 40	88 25	15	33 959.9174	Compensating bars	
1835	Dehra Dun	Everest	30 18	77 58	1980	39 183.8734	Compensating bars	[116]
1835	Dehra Dun	Waugh/Renny	30 18	77 58	1980	39 184.0321	Compensating bars	[116]
1838	Sironj	Everest	24 07	77 51	1529	38 413.3675	Compensating bars	[116]
1841	Bidar	Waugh	17 55	77 37	2030	41 578.5361	Compensating bars	[116, 265]
1848	Sonakhoda	Waugh	26 17	88 17	223	36 685.7946		
1854	Chach	Waugh	33 55	72 29	1015	41 345.4219		
1855	Karachi	Waugh	24 56	67 13	46	38 624.3215		
1863	Vizagapatam	Walker	18 00	83 01	311	34 778.3908		
1868	Bangalore	Hennessey	13 03	77 40	3318	36 083.6258		
1869	Comorin	Basevi	8 15	77 45	135	8912.5279		

Appendix 5 The figure of the Earth and geodesy

The figure of the earth from Indian arc measures.

Date	Observer	Arcs	(a)	(b)	(f)	
1818	Lambton	Trivandeporum–Paudree Punnae–Damargidda French, English, Extended Swedish	20921114.4		304	English feet
1821[*]	Lambton	French, English, Swedish Damargidda–Sironj Punnae–Damargidda Trivandeporum–Paudree	20918747	20851326	310.28	
1830	Everest	Punnae–Kalianpur Formentera–Greenwich	20922931.80	20853374.58	300.8017	Indian feet
1847	Everest	Formentera–Dunkirk Jacobstadt–Hochland Tarqui–Cotchesqui Dunnose–Clifton Pahtavara–Mallorn Jacobstadt–Dorpat	20920902.48	20853642.00	311.043	
1860	Clarke	The arcs of Everest 1857 + other comparisons	20926217	20855221	294.754	
1939	Bomford	Everest 1830	20922840.95	20853284.03		Eng. ft. 1865–94
	Bomford	Everest 1830	20922859	20853302		Eng. ft. 1926

[*] 1818 values adjusted for new standard length values

The length of 1° as derived from Indian arcs.

Observer	Date	Arc lengths from	to	Mid. Lat.	Amplitude	Length(ft)	1° in feet
				° ′ ″	° ′ ″		
Lambton	1802–3	Trivandeporum	Paudree	12 32 20.80	1 34 56.43	574327.96	362960.78
Lambton	1818	Punnae	Putchapolliam	9 34 43.66	2 50 10.54	1029100.50	362837.01
		Putchapolliam	Namthabad	13 02 54.57	4 06 11.28	1489131.18	362925.37
		Namthabad	Damargidda	16 34 41.87	2 57 23.32	1073428.20	363076.70
		Punnae	Damargidda	13 06 31.16	9 53 45.14	3591659.88	362945.26
Lambton	1821	Trivandeporum	Paudree	12 32 20.80	1 34 56.43	574368	362988
		Punnae	Putchapolliam	9 34 43.66	2 50 10.54	1029173	362864
		Putchapolliam	Namthabad	13 02 54.57	4 06 11.28	1489198	362941
		Namthabad	Damargidda	16 34 41.87	2 57 23.32	1073409	363071
		Punnae	Damargidda	13 06 31.16	9 53 45.14	3591780	362957.39
Everest	1830	Punnae	Damargidda	13 06 24.29	9 53 43.57	3591779.04	362973.33
		Damargidda	Takalkhera	19 34 34.01	3 02 35.87	1105499.52	363257.33
		Takalkhera	Kalianpur	22 36 31.90	3 01 19.91	1097320.02	363086.80
		Punnae	Kalianpur	16 08 22.18	15 57 39.34	5794598.58	363048.98
Everest	1847	Kalianpur	Kaliana	26 49 00.50	5 23 37.06	1961157.12	363606.31
		Damargidda	Kalianpur	21 05 13.96	6 03 55.97	2202926.22	363186.72
		Kaliana	Damargidda	23 47 02.50	11 27 33.03	4164083.34	363384.22
		Damargidda	Punnae	13 06 24.29	9 53 43.57	3591779.04	362973.30
		Punnae	Kaliana	18 50 10.35	21 21 16.60	7755862.38	363193.80
Clarke	1861	Punnae	Putchapolliam	9 34 36.65	2 50 11.14	1029173.7	362841.50
		Putchapolliam	Dodagoonta	11 59 47.22	2 00 09.89	727386.3	363194.26
		Dodagoonta	Namthabad	14 02 52.86	2 06 01.40	761813.4	362701.12
		Namthabad	Damargidda	16 34 34.62	2 57 21.73	1073410.9	363125.10
		Damargidda	Takalkhera	19 34 33.41	3 02 36.24	1105539.8	363258.13
		Takalkhera	Kalianpur	22 36 31.40	3 01 19.73	1097364.9	363107.69
		Kalianpur	Kaliana	26 48 59.79	5 23 37.06	1961138.0	363602.77

REFERENCES

Abbreviations

As Soc	Asiatic Society
BL	British Library
EIC	Honorable East India Company
ESR	Empire Survey Review
FIG	International Federation of Surveyors
GTS	Great Trigonometrical Survey
IAG	International Association of Geodesy
IOL	India Office Library
Mem RAS	Memoir of the RAS
MNRAS	Monthly Notices of the RAS
Phil Trans	Philosophical Transactions, Royal Society
RAS	Royal Astronomical Society
RGO	Royal Greenwich Observatory
RGS	Royal Geographical Society
RI (Roy Inst)	Royal Institution of Great Britain
SoI	Survey of India

1 Airy, G.B. 1839. Letters to Herschel. Jan/Feb Royal Society HS.1.75–7

2 Airy, G.B. 1839. Letter to Herschel re Duke of Sussex episode. 3 Oct. Royal Society HS.1.82

3 Airy, G.B. 1839. Letters to J. Melvill. EIC. RGO 6-417, Sect. 25–6. fols. 324, 326

4 Airy, G.B. 1839. Letter to Professor Whewell. RGO 6-417, Sect. 25–6. fol. 329

5 Airy, G.B. 1845. Figure of the Earth. Encyclopaedia Metropolitana

6 Altherr, W. & Gruen, A. 1990. The new 1:50 000 map of Mount Everest. Bicentenary Conference RGS Nov.

7 Anderson, R.G.W. 1982. Land Surveys in India. Science Museum pp. 61–3

8 Angus-Leppan, P. 1982. The height of Mount Everest. *Survey Review* Vol. 26 No. 206 Oct. pp. 365–85

9 Anon, 1791. St Alphege's Church Register transcript. 27 Jan.

10 Anon, 1806–20. Acts & Proc. of Corpus Christi College, Oxford

11 Anon, 1827. Royal Society document detailing election of Everest as a Fellow

12 Anon, 1838. 'Jervis' document with all 38 signatures. RGO 6-417, Sect. 25–6, fol. 332

13 Anon, 1851. Records of the Royal Military Academy, Woolwich
14 Anon, 1856. Announcement of name and height of highest peak. *Proc. Asiatic Soc. Bengal* Aug. pp. 437–9
15 Anon, 1856. Note on name and height of highest peak. *Proc. Asiatic Soc. Bengal* Sept. pp. 455–6
16 Anon, 1857. Presentation of Patron's Gold Medal to Waugh. *Geog. Jl.* Vol. 27 pp. lxxxix–xciii
17 Anon, 1867. Obituary of George Everest. Annual Report Royal Asiatic Soc. Bengal Vol. III pp. xvi–xviii
18 Anon, 1868. Obituary of George Everest. *Proceedings Royal Society* Vol. 16 pp. xi–xiv
19 Anon, 1868. *Jl. Asiatic Soc. Bengal.* pp. 25–6
20 Anon, Obituary of George Everest. *Encyclopaedia Britannica.* 11th edn.
21 Anon, 1903. *Dictionary of National Biography.* Entries for Colebrooke, Dicey, Everest and the Duke of Sussex
22 Anon, 1906. *Dictionary of Indian Biography,* entry for Everest
23 Anon, 1931. The name of Mount Everest. *Nature.* No. 3209 Vol. 127 2 May p. 686
24 Anon, 1967. Generous tributes to British Surveyors General. *Br. Review.* 11 Nov.
25 Anon, 1974. Historic watch stolen. *Kentish Independent.* 28 Feb.
26 Anon, 1977. Around the villages. Claybrooke. *Hinckley & District Focus.* Sept. pp. 19–22.
27 Anon, 1983. The Manor of Gwernvale, Crickhowell heritage.
28 Anon, 1984? Peaks gain, lose height. *Nepal Times*
29 Anon, 1990. Mountain man's hidden past. *Brighton Evening Argus.* 29 June
30 Anon, 1990. Honour of the highest order. *Brighton Evening Argus.* 4 July
31 Anon, 1990. Dehra Doon Base-line. Everest Bicentenary Souvenir. Dehra Dun.
32 Babbage, G. 1829. Account of the Great Congress of Philosophers at Berlin 18 Sept 1828. *Edin. Jl. Sc.* Vol. X Oct-Apr pp. 225–34
33 Bhattacharji, J.C. 1961. The Indian foot-metre ratio. *ESR* No. 119 pp. 13–18
34 Biddle, C. 1968. The bicentenary of the Survey of India. *Chartered Surveyor* Vol. 100 No. 10 pp. 500–505
35 Black, C.E.D. 1891. Memorandum on Indian Surveys 1875–90 412 pp. *IOL* L/MIL/17/11/19
36 Boase, F. 1892. Obituary of George Everest. *Modern English Biography.* 3 volumes p.1007.
37 Boeck 1893. Durch Indien im verschlossenen Land Nepal
38 Bomford, G. 1939. The readjustment of the Indian triangulation. SoI Professional paper 28
39 Bomford, G. 1971. *Geodesy.* 3rd edn. Clarendon Press
40 Boole, M.E. 1878. Home-side of a scientific mind. *The University Magazine.* Jan–Apr
41 Boole, M.E. 1903. The building of the Idol. *Brotherhood.* June 1903–Feb 1904
42 Boole, M.E. 1904. *The preparation of a Child for Science.* Clarendon Press
43 Boole, M.E. 1905. The religious philosopher as a social harmoniser. *East and West* (Bombay) Oct–Nov
44 Boole, M.E. 1905. The naming of Mount Everest. *East and West* (Bombay) March
45 Boole, M.E. 1905. Scientific research by natives of India. *The Indian Spectator.* 25 February
46 Boole, M.E. 1909. Indian thought and Western science. *Ceylon National Review*, June
47 Boole, M.E. 1910. The forging of passion into power.
48 Boole, M.E. 1911. A child's idyll. *East and West* (Bombay) May–July
49 Boole, M.E. 1923. At the foot of the Cotswolds

References

50 Brookes, W.R. 1988. Note on height of Mount Everest. *Canadian Surveyor* Summer p. 175

51 Burrard, S.G. 1888. Operations under Col Lambton. [265 (XII)] Appendix

52 Burrard, S.G. 1901. The attraction of the Himalaya mountains upon the plumb line in India. SoI, Professional paper 5

53 Burrard, S.G. 1903. On the values of longitude employed on maps of the Survey of India. Prof paper No. 7, Sect. 1 pp. 1–7

54 Burrard, S.G. 1904. The Everest controversy. *Nature.* 10 Nov. pp. 42–6

55 Burrard, S.G. & Hayden, H.H. 1907. A sketch of the geography and geology of the Himalayan Mountains and Tibet, Part I

56 Burrard, S.G. 1931. Mount Everest and its Tibetan names. SoI Professional paper 26.

57 Burrard, S.G. 1934. The place of Mount Everest in history. *ESR* No. 14 Sept pp. 450–64

58 Butler, P.R. 1947. Basevi of the pendulums. *Blackwood's Magazine*, Apr

59 Butterfield, A.D. 1906. *A History of the Determination of the Figure of the Earth from Arc Measurements.* Davis. Paris. pp. 113–39

60 Cameron, A. 1990. The Shape of the Earth and the Arcs of Meridian of De laCaille and Maclear. *Mon. Notices Astro. Soc. South Africa*, Vol. 49 p.169.

61 Chadha, S.M. 1990. Sir George Everest birth centenary. Everest bicentenary Souvenir. Dehra Dun.

62 Chadha, S.M. 1990. Survey of India through the ages. Proc. Bicent. Conf. RGS

63 Chen, J.-Y. 1980. A new determination of the height of the world's highest peak. *Ost. ZfV und Photo.* Heft I pp. 1–19

64 Clarke, A.R. 1861. On the figure of the Earth. Mem RAS Vol. XXIX pp. 25–57

65 Clarke, A.R. 1866. Comparison of the standards of length of England, France, Belgium, Prussia, India and Australia. HMSO.

66 Clarke, A.R. 1880. *Geodesy.* Oxford University Press. pp. 27–31

67 Cobham, E.M. 1931. *Mary Everest Boole. Collected Works.* C.W. Daniel Co. London

68 Colebrooke, H.T. 1816. On the heights of the Himalaya Mountains. *Asiatic Researches.* Vol. 12 pp. 251–85

69 Cook, A. 1990. The achievements of Sir George Everest in Geodesy. *Proc. Bicent. Conf. RGS*

70 Corpus Christi, 1806-20. Acts & Proceedings of Corpus Christi College. B/4/1/3 Note inside front cover and pp. 42–3

71 Cowell, E.B. 1860. Letter re election of Everest as Honorary member Asiatic Society Bengal

72 Dansheng, G. 1979. The height of Quomolangma Feng. IAG General Assembly. Canberra. Dec

73 Dawson, R. 1803. Memo on Survey Training

74 Desio, A. 1988. Which is the highest mountain in the world? Report to Academia dei Lincei

75 Duhan History of GTS. ed. Medley. *Indian Engineering.* Vol. II pp. 285, 398. Vol. III pp. 94, 305, 402

76 Dyson, A.M. 1913 Lutterworth. John Wycliffe's Town.

77 EIC, 1806. Register of EIC Cadets. IOL L/MIL/9/282

78 EIC, 1821. Lambton's work on triangulation. IOL L/MIL/5/386 (96)

79 Edmonstone, N.B. 1827. Memo to EIC re Trigonometrical Survey of India. IOL L/MIL/5/407 Collection 263. fol. 191–200

80 Edney, M.H. 1989. The Ordnance Survey and British Surveys in India. Sheetlines. *Jl. Charles Close Soc.* No. 26 Dec. pp. 3–8

81 Edney, M.H. 1990. Systematic surveys and mapping policy in British India 1757–1830. *Proc. Bicen. Conf. RGS.*

82 Edney, M.H. 1997. *Mapping an Empire.* Univ. of Chicago Press.

83 Edwardes, H.B. 1873. Life of Sir Henry Lawrence. 3rd edn. Original in 2 vols c.1837

84 Everest, G. 1817. Letter to Major Craigie [263 (171) 15–16]

85 Everest, G. 1821. Various letters, relating particularly to Cape of Good Hope IOL E/1/145, fols. 186–210. Later pub. as [87]

86 Everest, G. 1822. Letter to Priv. Sec. Gov. Gen. [263 (171) 256]

87 Everest, G. 1822. On the triangulation of the Cape of Good Hope. *Mem. Astro. Soc.* Vol. I Paper XXI pp. 255–70 Read 10 May 1822

88 Everest, G. 1823. Letters to Morton, Captain Barnett and Sir C. Metcalfe [263 (171) 7–10, 70–1, 75]

89 Everest, G. 1823–4. Letters to S.G. India [263 (171) 3–5; 11–12; 116–38; 174–9; 155–6]

90 Everest, G. 1826–30. Various correspondence including that on the Irish Survey and GTS. IOL L/MIL/5/402 Collection 205, fols. 295–465 = BL Add.MS 14380 (72–94) = Nottingham University PwJf 2767/1–7 = [93] and [94]

91 Everest, G. 1826. On the corrections requisite for the triangles which occur in geodetic operations. *Mem. Astro. Soc.* Vol. II Paper III pp. 37–44 Read 9 Apr 1824

92 Everest, G. *et al.,* 1828–9. Miscellaneous correspondence to EIC. IOL L/MIL/5/413 (312)

93 Everest, G. 1829. Report on Irish Survey. Nottingham University. PwJf 2767/1–7 = IOL L/MIL/5/402 Collection 205 fol. 295–317 = [90].

94 Everest, G. 1829. Report on the GTS. IOL L/MIL/5/402 Coll.205 fols. 358–406 = [90]

95 Everest, G. 1830. Letter by Everest to Salmond. IOL E/1/171, fols. 574–6

96 Everest, G. 1830. *An Account of the Measurement of an Arc of the Meridian between Parallels of 18° 3' and 24° 7'.* Parbury Allen & Co. London.

97 Everest, G. 1830. Letter to Babbage. BL Add. Ms. 37185 ff. 198

98 Everest, G. 1831. On the errors likely to arise …Read 1829 See [99] MNRAS Vol. I pp. 117–20

99 Everest, G. 1831. Remarks respecting the errors likely to arise in the determination of the length of the pendulum from the false position of the fixed axes. *Mem. Astro. Soc.* 1831 Vol. IV Paper III pp. 25–37 Read 13 Mar 1829 See [98]

100 Everest, G. 1831–4. Miscellaneous correspondence. Nottingham University. PwJf 2836

101 Everest, G. c.1834. Map of part of Great Arc. Nottingham University PwJf 2938

102 Everest, G. 1835. Formulae for calculating azimuth in triangulation operations. *Asiatic Researches.* Vol. 18 pp. 93–106

103 Everest, G. 1835. On the compensation measuring apparatus of the GTS of India. *Asiatic Researches.* Vol. 18 pp. 189–215

104 Everest, G. 1837. Operations of Trigonometrical Survey. Jan 1823–37 Great Arc Series Bidar-Sironj-Dehra. Vol. 7 parts i & ii

105 Everest, G. 1838. Memorial to EIC re a CB. RGO 6–417 Sect. 27–31, fol. 367

106 Everest, G. 1839. On compensating bars. *Trans. As. Soc. Bengal*

107 Everest, G. 1839. Letters to HRH the Duke of Sussex. Wm. Pickering, London

108 Everest, G. 1839. Some account of the progress of the Trigonometrical Survey now carrying on in India. MNRAS Vol. IV pp. 206–10 Communicated by Jervis 14 June

109 Everest, G. 1842. On the astronomical circles of the Grand Trigonometrical Survey of India… Mem. RAS. Vol. XII pp. 141–52 Read 11 Dec 1840

References

110 Everest, G. 1843. Description of a method of dividing one circle B by copying from another A previously divided. MNRAS Vol.V pp. 66–8 Read 11 Dec 1840 NB [110] is contained in [109]

111 Everest, G. 1843. Letter to Airy from India. RGO 6–417 Sect. 25–6 fol. 336

112 Everest, G. 1843? Dehra Dun base. Prof Paper on Indian Engineering. Vol. 4 p. 303

113 Everest, G. 1844. Brief note on [114] British Association Notices pp. 2–3

114 Everest, G. 1845. Address 'On geodetic operations in India' British Association p. 25

115 Everest, G. 1845. Memorial to EIC re a CB. RGO 6-417 Sect. 27–31, fol. 369

116 Everest, G. 1847. *An Account of the Measurement of Two Sections of the Meridional Arc of India* …London. 2 volumes

117 Everest, G. 1847. Letter to G.B. Airy. RGO 6-417 Sect. 27–31. fol. 404

118 Everest, G. 1848. Contribution in the *Edinburgh Review* Vol. 87 p. 372

119 Everest, G. 1851–62 Miscellaneous letters. RGS Archives

120 Everest, G. 1859. Rectification of logarithmic errors in the measurement of two sections of the Meridional Arc of India. *Proc. Royal Society* Vol. IX Jan 27 pp. 620–6

121 Everest, G. 1861. Letter to Dr Shaw re perambulator 28 Jan. RGS Archives.

122 Everest, G. 1861 Letter to Sec. Royal Society re Lambton's triangulation results. Royal Society MC. 6.130

123 Everest, G. 1865. Letter to Col. Walker 15 March. SoI Records

124 Everest, G. 1865. Letter to Sabine. 16 March Royal Society

125 Everest, G. 1867. Will of G. Everest

126 Everest, G. 1860. On instruments & observations for longitude for land travellers. *Geog. Jl.* Vol. 30 pp. 315–24

127 Everest, G. Correspondence of Charles Babbage. BL. Add. 37182-99

128 Everest, G. Letter to Lord Ellenborough. PRO 30/12/20/10

129 Everest, G. Calcutta Longitudinal Series. GTS Vol. 8

130 Everest, L.F. 1934. Memo on Sir George Everest by his son. Soc. Genealogists IND/G 63

131 Everest, R. 1830–2. Contribution in *Gleanings in Sc*. III 125. As. Soc. Bengal.

132 Everest, T. 1820. Will of Tristram Everest

133 Faux, R. 1986. Making a mountain out of naming Everest. *The Times*. 31 August

134 Fellowes, P.F.M. *et al*., 1933. *First over Everest. The Houston Mount Everest Expedition 1933*, Bodley Head.

135 Fisher, R., Lloyd, G.C.W. & Wonnacott, R.T. 1986. The Cape of Good Hope arc of meridian. FIG Toronto. Paper 106

136 Fowler, T. 1893. History of Corpus Christi College Oxford

137 Freshfield, D.W. 1884–6. The great peaks of the Himalayas. *Alpine Jl*. Vol. XII pp. 438–60

138 Freshfield, D.W. 1885. Notes on Col. Tanner's Report. *RGS Proc*. Vol. 7 pp. 753–6

139 Freshfield, D.W. 1886. Further notes on Mont Everest. *RGS Proc*. Vol. 8 pp. 176–88

140 Freshfield, D.W. 1902-3. Mount Everest or Jomokankar. *Alpine Jl*. Vol. 21 pp. 33–5

141 Freshfield, D.W. 1903. The highest mountain in the world. *Geog. Jl*. Vol. 21 June. pp. 294–8

142 Freshfield, D.W. 1904. Notes on Tibet. *Geog. Jl*. Jan. p. 89

143 Freshfield, D.W. 1904. Notes on Tibet. *Geog. Jl*. Mar. pp. 361–3

144 Freshfield, D.W. 1922. Mount Everest v Chomolungma. *Alpine Jl*. Vol. 34 pp. 300–3

145 Gauss, C.F. 1903. *Collected Works*. Heliotrop. Werke. Vol. IX pp. 459–84

146 Gill, D. 1913. *A History of the Royal Observatory, Cape of Good Hope*. London.

147 Gore, J.H. 1891. *Geodesy*. Heinemann. pp. 178–83

148 Graham, K. 1990. Maclear's sector station al Klyp Fontein, Aurora, private paper.

149 Graham, W.W. 1884. Reply to Hooker. *RGS Proc*. Vol. 6 pp. 139–40, 425, 447

150 Graham, W.W. 1884. Ascents in the Himalayas. *RGS Proc*. Vol. 6 pp. 68–9

151 Grant, R. 1852. *History of Physical Astronomy* Bohn, London. Reprint, Johnson Reprint Corp., London, 1966

152 Graves, R.P. 1882. *Life of Sir Wm. Rowan Hamilton*. 3 volumes. Contains letters from George Everest. Hodges Figgis, Dublin

153 Gray, H. 1990. Personal correspondence

154 Green, E. 1886. *Pedigree of the family Wing 1486–1886*. Mitchell & Hughes London

155 Greer, J.D. 1990. Photographing Everest. *Photo. Eng & Remote Sensing* 56(1) Jan. pp. 110–15

156 Guggisberg, F.G. 1900. *The Shop: the Story of the Royal Military Academy*. London.

157 Gulatee, B.L. 1950. Mount Everest, its height and name. SoI Technical paper 4

158 Gulatee, B.L. 1954. Ht of Mount Everest 1952-4. SoI Technical paper 8

159 Hanks, P. & Hodges, F. 1988. *A Dictionary of Surnames*. Oxford University Press

160 Heaney, G.F. 1957. The story of the Survey of India. *Geog. Mag*. Vol. 30 No. 4 pp. 182–90

161 Heaney, G.F. 1967. Sir George Everest. *Geog. Jl*. Vol. 133 June. pp. 209-11

162 Heaney, G.F. 1968. Rennell and the surveyors of India. *Geog. Jl*. Vol. 134 Sept. pp. 318–27

163 Hedin, S. 1926. *Mount Everest*. FU Brodhaus, Leipzig

164 Herschel, J.F.W. 1839. Letter to Airy. 9 Oct. Royal Society HS.1.83

165 Herschel, J.F.W. 1840? Letter re Jervis. Royal Society HS.3.283

166 Herschel, J.F.W. 1840. Letter to Melvill. 22 Jan. Royal Society HS.13.34

167 Herschel, J.F.W. 1848. Remarks on Everest's 'Account of the measurement … [158] Vol. VIII pp. 116–18

168 Hodgson. B.H. 1848. Memo on the Seven Cosis, with sketch map. *Jl. As. Soc. Bengal* Dec pp. 646–9

169 Hodgson, B.H. 1856. Native name of Mount Everest. Letter *Jl. As. Soc. Bengal* Vol. XXV pp. 467–70

170 Hodson, V.C.P. 1922–47. *List of Officers of Bengal Army 1758-1834* London 4 volumes

171 Holdich, T. 1905. Mount Everest. *The Standard*. 24 Jan.

172 Hooker, J. 1854. *Himalayan Journals*

173 Howard-Bury, C.K. 1922. *Mount Everest: the Reconnaissance 1921*. Arnold

174 Hunter, J. de G. 1918. The Earth's axes and triangulation. SoI Professional paper 16

175 Hunter, J. de G. 1924. Height of Mount Everest and other peaks. SoI Geodetic Report 1

176 Hunter, J. de G. 1928. The figure of the earth. 15th Indian Science Cong. Calcutta

177 Hunter, J. de G. 1929. Geodesy. SoI Departmental paper 12 pp. 287–99

178 Hunter, J. de G. 1953. Heights and names of Mount Everest. Occ. Notes RAS Vol. 3 No. 15 pp. 1–13

179 Hunter, J. de G. 1955. Various determinations over a century of the height of Mount Everest. *Geog. Jl*. Vol. 121 March pp. 21–6

180 Hyde, J.F. Correspondence with EIC. IOL L/MIL/5/393 (150)

181 Hyman, A. 1982. *Charles Babbage: Pioneer of the Computer*. Oxford University Press

References 295

182 Insley, J. 1989. *Everest: The Man, the Mountain and the Theodolite*. S. Asia Library Group. Jan

183 Insley, J. 1990. Instruments of inspiration. George Everest in England 1825–1830. *Proc. Everest Bicen. Conf. RGS.*

184 Jelly, J.S.O. 1991. Personal correspondence

185 Jervis, T.B. 1837. *Geog Jl*. Vol. 7 pp. 127–43

186 Jervis, T.B. *et al*. 1837–9. Miscellaneous correspondence with EIC. IOL L/MIL/5/413 (306)

187 Jervis, T.B. 1838. Address to Geog. Sect. Br. Assn. 26 Aug. Contained in [107] = BM Addnl. MS 14380(5)

188 Jervis, T.B. 1840. As [108] *Bombay Geog. Soc. Jl*. pp. 157–89

189 Jervis, W. 1898. *Thomas Best Jervis: A Centenary Tribute*. Elliott Stock, London

190 Jones, T. 1898. History of Brecknockshire. p. 400

191 Kochhar, R.K. 1990. Astronomy in British India: science in the service of the State. Everest Bicentenary Souvenir. Dehra Dun.

192 Kochhar, R.K. 1990. *Modern Astronomy in India, 1651-1960*. Vistas in Astronomy. Pergamon Press

193 Kochhar, R.K. 1991. Astronomy in British India: Science in the service of the State. *Current Science* Vol. 60 No. 2, 25 January, pp. 124-9

194 Krakaver, J. 1988? After all is said and done, will Everest still be number one? *Smithsonian Jl*. pp. 176–98

195 LaCaille, N.-L. 1763. Journal Historique du Voyage fait au Cap de Bonne-Esperance. Paris.

196 Lambton, W. 1800. A Plan of a Mathematical and Geographical survey... IOL P/254/52 pp. 746–57

197 Lambton, W. 1801. An account of a method of extending a geographical survey across the Peninsula of India. *Asiatic Researches* (Bengal) Vol. VII pp. 312–37

198 Lambton, W. 1805. An account of the measurement of an arc of the meridian on the Coast of Coromandel. *Asiatic Researches* (Bengal) Vol. VIII pp. 137–93

199 Lambton, W. 1808. An account of trigonometrical operations crossing the peninsula of India. *Asiatic Researches* (Bengal) Vol. X pp. 290–384

200 Lambton, W. 1816. An account of the measurement of an arc of the meridian ... 8° 9' 33" and 10° 59' 48"... *Asiatic Researches* Vol. XII pp. 1–101

201 Lambton, W. 1816. An account of the measurement of an arc of the meridian... 10° 59' 49" to 15° 6' 0" N ... *Asiatic Researches* Vol. XII pp. 286–356

202 Lambton, W. 1818. An abstract of the results deduced from the measurement of an arc on the meridian extending from latitude 8° 9' 38.4" to latitude 18° 3' 23.6" N being an amplitude of 9° 53' 45.2" *Phil. Trans* Vol. 108 pp. 486–517

203 Lambton, W. 1820. An account of the measurement of an arc of the meridian ... 15° 6' 2" to 18° 3' 45" N... *Asiatic Researches* Vol. XIII pp. 1–127

204 Lambton, W. 1822. Letter to G. Everest 18 Sept = [263 (171) 257–8]

205 Lambton, W. 1823. Corrections applied to the Great Trigonometrical Arc extending from latitude 8° 9' 38.39" to latitude 18° 3' 23.64" to reduce it to the Parliamentary Standard. *Phil. Trans*. Vol. 113 pp. 27–33

206 MacHale, D. 1985. *George Boole: His Life and Work*. Boole Press, Dublin.

207 Maclear, T. 1840. On the position of LaCaille's stations at the Cape of Good Hope. Mem RAS Vol. XI. pp. 91–137

208 Maclear, T. 1866. *Verification and extension of LaCaille's Arc of Meridian at the Cape of Good Hope*. London. 2 volumes, Lords Commissioners of the Admiralty.

209 Macleod, M.N. 1931. Review of 'Mount Everest and its Tibetan names' by Burrard [56] *ESR* Vol. 1 July, pp. 36–8

210 Markham, C.R. 1871. Memoir on the Survey of India.

211 Martin, B.E. 1972. Parsons and Prisons. Published privately

212 Martin, C.G.C. 1990. George Everest and the triangulation of the Cape of Good Hope. *Proc. Bicent. Conf. RGS*

213 Martyn, J. 1973. What George Everest did. *Himalayan Jl.* Vol. 33 pp. 1–6

214 Mason, K. 1934. The official height of Mount Everest. *Himalayan Jl.* Vol. 6 pp. 154–7

215 Mason, K. 1955. *Abode of Snow*. Rupert Hart-Davis.

216 Mason, K. 1958. Map of Mount Everest. *Himalayan Jl.* Vol. 21 pp. 157–8

217 Melvill, J. 1839. Letter to Airy. RGO 6-417, Sect. 25–6 fol. 325.

218 Murchison, R. 1867. Obituary of Everest. *Geog. Jl.* Vol. 37 pp. cxv-cxviii Also in *Proc. RGS* Vol. 11 pp. 185–8

219 Nagar, V.K. 1990. The development of the Survey of India from time of Sir George Everest to modern date. Everest Bicent. Souvenir. Dehra Dun, also *Proc. Bicent. Conf. RGS*

220 Nichols, J. 1807. *History and Antiquities of Leicestershire*. Little Claybrook & Great Claybrook. pp. 103-5

221 Odell, N.E. 1925. Identity of Everest. *Alpine Jl.* Vol. 37 May

222 Odell, N.E. 1935. The supposed Tibetan or Nepalese name of Mount Everest. *Alpine Jl.* Vol. 47 pp. 127–9

223 Pangtey, S.S. 1990. Proposal for the development and conservation of the Park Estate. Everest bicentenary Conf. Dehra Dun

224 Papworth, K.M. 1986. Ms notes. RGS Archives, Everest file

225 Parks, Fanny. 1849. *Wanderings of a Pilgrim*. 2 volumes. Oxford University Press, London. Reprint Oxford University Press, Karachi, 1975

226 Pevsner, N. 1966. *The Buildings of England: Warwickshire*. Penguin Books.

227 Phillimore, R.H. 1945. *Historical Records of The Survey of India*. Vol. I 18th century

228 Phillimore, R.H. 1950. *Historical Records of The Survey of India*. Vol. II 1800–1815

229 Phillimore, R.H. 1954. *Historical Records of The Survey of India*. Vol. III 1815–1830

230 Phillimore, R.H. c.1956. Rough notes on the Park Estate. IND G/69 Soc. of Genealogists

231 Phillimore, R.H. 1958. *Historical Records of The Survey of India*. Vol. IV 1830–1843

232 Phillimore, R.H. 1968. *Historical Records of The Survey of India*. Vol. V 1844–1861

233 Phillimore, R.H. Miscellaneous notes. Soc. Genealogists. IND G/63-9

234 Pratt, J.H. 1855. On the attraction of the Himalaya mountains. MNRAS XVI pp. 36–41

235 Pratt, J.H. 1855. On the attraction of the Himalaya Mountains and of the elevated regions beyond them, upon the plumb line in India. *Phil. Trans.* Vol. 145 pp. 53–104

236 Pratt, J.H. 1856. On the effect of local attraction upon the plumb line at stations on the English Arc of meridian between Dunnose and Burleigh Moor...*Phil. Trans.* Vol. 146 pp. 31–52

237 Pratt, J.H. 1857. On the deflection of the plumb line in India. (Abstract of [239]) *Proc. Royal Soc.* Vol. IX pp. 493–7

238 Pratt, J.H. 1858. The Great Indian Arc and the Figure of the Earth. *Jl. As. Soc. Bengal*. Vol. XXVII pp. 201–13

239 Pratt, J.H. 1859. On the deflection of the plumb line in India. *Phil. Trans.* Vol. 149 pp. 745–78

References

240 Pratt, J.H. 1859. On the influence of the Ocean on the plumb line in India. *Phil. Trans.* Vol. 149 pp. 779–96

241 Pratt, J.H. 1859. On the influence of mountain attraction … *Jl. As. Soc. Bengal*. Vol. 28 pp. 310–16

242 Pratt, J.G. & Tennant, J.F. 1859. Letters re the Indian arc of meridian. *Jl. As. Soc. Bengal*. Vol. 28 pp. 17–27

243 Prinsep, J. 1832. Progress of the Indian Trigonometrical Survey. *Jl. As. Soc. Bengal*. Vol. I pp. 71–3

244 Prinsep, J. 1833. Determination of the constant of expansion of the standard 10 ft bar…*Jl. As. Soc. Bengal*. Vol. II pp. 131–43

245 Prinsep, J. 1833. Table for ascertaining the heights of mountains from the boiling point of water. *Jl. As. Soc. Bengal* Vol. II pp. 194–201

246 Procházka, E. 1987. *Uvod do dejin zememerictvi*. Part VI. CVUT, Prague

247 Roy, W. 1787. An account of the Mode proposed to be followed in determining the relative Situation of the Royal Observatories of Greenwich and Paris. *Phil. Trans*. Vol. 77 pp. 188–226

248 Roy Inst. 1851–96. Minutes books of the RI. 1838–1851

249 Ruttledge, H. 1934. *Everest 1933*. Hodder & Stoughton

250 Sandes, E.W.C. 1933. *The Military Engineer in India*. Inst. Royal Engineers. 2 vols.

251 Schlagintweit. 1855. IOL L/MIL/5/423 (400)

252 Schlagintweit. 1862. Results of a scientific mission to India and High Asia.

253 Sheffield, A.F.W. 1991. Personal correspondence

254 Shortrede, R. 1849. On the latitude of Dera by observation, and on the disturbing attraction of the Himalayas. Mem RAS Vol. XVII pp. 93–105

255 Smith, J.R. 1990. Everest, Man and Mountain. *Land & Minerals Surv*. July. pp. 321–8

256 Smith, J. R. 1990. George Everest – Surveyor General Kishewar Hind. *Survey Review*.   July

257 Smith, J.R. 1990. Sir George Everest and the Surveying of India. *Prof. Surveyor*. July/Aug. pp. 14–20

258 Smith, J.R. 1990. Col Sir George Everest CB, FRS. Everest bicentenary conference Dehra Dun. Oct. and Everest bicentenary conference. RGS. November

259 Stanley, W.F. 1901. *Surveying and Levelling Instruments*. Stanley, London

260 Strange, A. 1883. Report on the GTS [265 (IX)]

261 Stubbs, F.W. 1877. *History of the Organisation, equipment, and war services of the Regiment of Bengal Artillery*

262 Stubbs, F.W. 1892. List of Officers who have served in the Regiment of the Bengal Artillery… London

263 Surv.of India. 1801–40. Volumes of Records held at National Archives, Delhi. Referenced as [263 (vol) page]

264 Surv.of India. 1851. Auxiliary Tables. 6th edn. 1936

265 Surv.of India. 1870–1910. *Account of the Operations of the Great Trigonometrical Survey of India*. Dehra Dun. 19 volumes referenced as [265 (vol) page]

266 Tanner, H.C.B. 1883. Report on Indian Survey. Comments on height

267 Tanner, H.C.B. 1884–6. The great peaks of the Himalaya. *Alpine Jl*. Vol. XII pp. 438–48 Part of [106]

268 Tennant, J.F. 1857. An Examination of the Figure of the Indian Meridian as deduced by Archdeacon Pratt from the two Northern India Arcs … RAS Notices. 9 Jan.

269 Tennant, J.F. 1857. Abstract of a report on the determination of the longitude of Karachi. MNRAS Vol. XVII pp. 56–8

270 Tennant, J.F. 1857. On the effect of local attraction in modifying the apparent form of the earth... MNRAS Vol. XVII pp. 236–41

271 Tennant, J.F. 1866. Everest obituary. MNRAS Vol. 27 pp. 104–8

272 Thuillier, H.L. & Smyth, R. 1851. *Manual of Surveying for India*. 3rd edn. 1875 Thacker Spink, Calcutta.

273 Troughton, E. 1809. An account of a method of dividing astronomical and other instruments... *Phil. Trans.* Vol. 99 pp. 105–45

274 Troughton, E. 1822. An account of the Repeating Circle and the Altitude and Azimuth instrument ... Mem RAS Vol. I pp. 33–54

275 Turner, R. & Goulden, S.L. *Great Engineers and Pioneers in Technology*. Vol. I St Martin's Press, New York

276 Various. 1833. Memos on production of Atlas of India. Nott. Univ. Pw Jf 2861

277 Vigne, G.T. 1844. Travels in Kashmir ... and the Himalaya. G.T. Vigne, London

278 Visagie, J.C. 1971. William Fredrik Hertzog, 1792–1847 MSc. Thesis University of Cape Town

279 Waddell, L.H. 1899. Among the Himalayas

280 Walker, J.T. 1865. On the methods of determining heights in the Trigonometrical Survey of India. MNRAS Vol. XXXIII pp. 103–14

281 Walker, J.T. 1870. Standardisation of measured baselines. Account of operations of GTS [265 (I)]

282 Walker, J.T. 1886. Notes on Mont Everest. *Proc. RGS* Vol. 8 pp. 88–94

283 Walker, J.T. 1895. India's contribution to Geodesy. *Phil. Trans.* Vol. A186 pp. 745–816

284 Warner, B. 1979. *Astronomers at the Royal Observatory at the Cape of Good Hope*. Balkema.

285 Warner, B. & N. 1984. *Maclear and Herschel letters and diaries at the Cape of Good Hope*.

286 Waugh, A. 1850. (Parliamentary) Report on the Great Trigonometrical Survey of India. London

287 Waugh, A. 1857. Letter to Thuillier with annexes *Jl. As. Soc. Bengal* Vol. XXVI pp. 297–312

288 Waugh, A. 1858. Mounts Everest and Deodanga - 4 reports by eminent surveyors *Proc. RGS* Vol. II pp. 102–15

289 Waugh, A.S. & Hodgson, B.H. 1857. The Himalaya and Mount Everest. *Proc. RGS* Vol. 2 1855–57 pp. 345–350

290 Webb, R.S. 1929. MSS notes on the History of Geodesy. Surveys & Mapping, Cape Town

291 White, 1846. Leicester Directory. Claybrook Magna & Parva pp. 383–5

292 Wood, H. 1904. Report on the identification and nomenclature of Himalayan peaks. SoI

293 Yolland, W. 1847. An account of the measurement of the Lough Foyle base in Ireland

294 Younghusband, F. 1926. *The Epic of Mount Everest*. Edward Arnold

INDEX

Academy of Science, Paris 27, 54, 55
Airy, George (1801–1892) 32, 33, 107, 109, 113, 115, 119, 120, 132, 193
alt-azimuth instrument 60
Amherst, Lord (formerly William Pitt) (1773–1857) 53
angle, horizontal by repetition 67
　reduction formulae 277
Arc, Great 69, 101, 115, 126, 132, 136, 147, 190, 235, 263
　　description of 18
　　intercomparisons 284, 285
　　Lapland 55, 273
　　length of, LaCaille 32, 34
　　longitude 18
　　measurement, proposal for 16
　　meridional 18, 274
　　meridional, Coromandel coast 275
　　Peru 31, 55, 123, 273, 277
Argand, Aimé (1755–1803) 238
Armstrong, John W. (b. 1812) 101, 191, 192, 193, 217
Arrowsmith, Aaron (1750–1823) 262, 263
Asiatic Society of Bengal 69, 74, 212, 214
Astronomical Society (Royal) 116, 237, 277, 280
Atlas of India 261–5
　of Bengal 261
Aubert, Alexander (1730–1805) 76, 79
　scale 76, 242
Auckland, Lord George (1784–1849) 263, 268
azimuth 279

Babbage, Charles (1792–1871) 8, 159, 161, 162

Bachelors' Hall 149, 151, 153, 154
Baily, Francis (1774–1844) 54, 107, 109, 115, 119, 120, 281
Baker, Godfrey (1786–1850) 12
Bangalore 17, 18, 248, 275
Banog 87, 93, 94, 98, 147, 278
barometer 20, 231
　height by 275
　mountain 231
　pump, Everest 117, 232
Barrow, Henry (1790–1870) 59, 65, 79, 94, 99, 100, 150, 237, 250–7
　at Kaliana 98, 253
　behaviour 252
　criticisms of 255
　Fellow of RAS 257
　left service 256
　request for leave 254
　return to Calcutta 256
　　England 257
bars, comparisons 78, 82
　compensating 60, 64, 68, 69, 73, 78, 118, 125, 233, 235
　　expansions 236
　　theory 236
　standard 68, 73, 76, 77, 242
　　theory of comparison 242, 243
baseline 206
　Ardenelle 15
　Bangalore 17, 18, 286
　Bidar 18, 49, 99, 100, 102, 124, 125, 127, 129, 259, 286
　by wooden rods 29
　by chain 38, 44, 53, 62
　Calcutta 69, 71, 74, 79, 286

Coimbatore 18
Dehra Dun 80, 86, 89, 92, 93, 99, 278, 286
Gooty 18, 286
Jellinghi 15
LaCaille 28, 29
Lough Foyle 60, 65
Madras 17
Pachapalaiyam = Coimbatore 18, 286
Porto Novo 15
Romney Marsh 77, 232
Sironj 44, 48, 49, 86, 98, 99, 101, 102, 118, 180, 233, 286
St Thomas' Mount 274, 275, 286
Takalkhera 38, 44, 286
Tanjore 18, 275, 286
results tabulated 286
Basevi, Adelaide (1796–1885) 174
Beaufort, F. (1774–1857) 107, 108, 109, 113
Bedford, Sir James (1788–1871) 264
Bell, Sir Charles (1870–1945) 222, 223
Bensley, Sir William 10
Bentinck, Lord William (1774–1839) 62, 63, 72, 86, 87
Bidar 20, 39, 47, 99, 100, 124, 277
Blacker, Valentine (1778–1826) 42, 47, 49, 50, 51, 62, 70, 258, 262, 263
Boeck, Dr 221
Boileau, Alexander (1807–1862) 77, 80, 88, 100, 190, 258
Bombay 35, 39, 40
Bomford, Guy (1899–1996) 202
Bontein family 140, 169
  John (1809–1878) 178
Boole, George (1815–1864) 8, 176
  Mary Everest (1832–1916) 3, 7, 168, 176
    comment on records 3
    letter to Dr Bose 162
    writings 155–64
      A child's idyll 163
      At the foot of the Cotswolds 163
      Home-side of a Scientific Mind 160
      Indian thought and Western Science 162
      Scientific research by natives of India 161
      The Building of an Idol 164
      The forging of passion into power 163
      The Naming of Mount Everest 156–60
      The Preparation of a Child for Science 161
      The religious philosopher as a social harmoniser 164

Bose, Dr (1858–1937) 161, 162
Boss, Emil 208, 209
Bouguer, Pierre (1698–1758) 31, 123
Bradley, James (1692–1762) 33
British Association 106, 107, 110
Bruce, General 222
buildings, destruction of 270
Burrard, Sir Sidney (1860–1943) 144, 193, 202, 213, 214, 216, 218, 220, 223
  height of mountain 200
Burrow, Reuben (1747–1792) 15

Calcutta 14, 50, 52, 68, 69, 79, 88
Call, Thomas (1748–1788) 61
Cape Comorin 18, 274, 276
Cary, William (1759–1825) 44, 76, 78
  brass scale 44, 45, 49
Casement Col. W. 71, 74, 77, 79, 149, 252
Cavendish, Henry (1731–1810) 29
chain 232
  Birge 76
  Gilbert 76
  Lambton 232
  Ramsden 76
  standard 44
chains, comparison of 76
Chandra Das, Sarah 222, 224
Chaur, Mount 91
Chimborazo, Mount 31, 32, 189
Chippens Bank 169
chronometer 233
  gold 183, 233
Chunar 14, 15, 20
circle, astronomical 96, 98
  repair 254
  tests of 254
  azimuth 94
  division by hand 251, 257
  reading 124
  repeating 234
circumferentor 234
Clarke, Alexander (1828–1914) 39, 78
Clive, Lord Edward (1754–1839) 16, 114
Colby, Thomas (1784–1852) 60, 61, 78, 107, 108, 110, 111, 180, 235
Cole, Susannah (b. 1718) 5, 165
Colebrooke, Emma (b. 1799) 177
  Robert Hyde (1762–1808) 61, 70
  William Macbean G. (1787–1870) 11, 12, 17, 177
Colonso, Bishop, John W. (1814–1883) 160
compass 234

Index

Kater 235
computing section 69, 127
Condamine, Charles de la (1701–1774) 24, 31, 123
coordinates, geographical, formulae 279
   provisional list 265
Cornwall 63, 65
corporal punishment 21
Court of Directors 9, 53, 55, 57, 58, 59, 62, 102, 105, 106, 109, 112, 117, 129, 131, 262, 265
Crawford, Charles (1760–1836) 61, 211
Curzon, Lord (1859–1925) 189

Dalrymple, Alexander (1736–1808) 16
Damargidda 18, 124, 277, 281, 282
Dart, Joseph Henry (1817–1887) 34, 59
datum at Kalianpur 196, 197
De Morgan, Augustus (1806–1871) 161, 162, 163
degree, early results 276
   length of 29, 31, 275
      tabulated 288
   meridian by bamboo rods 15
Dehra Dun 85, 121, 125, 131, 132, 147, 150, 193, 253, 258
   Headquarters 71, 78, 258
Delambre, Jean-B. (1749–1822) 28, 55, 114, 277
Desio, Professor 207
Dhaulagiri 189, 198, 208
Dodagoonta 17, 18, 79
Drummond, Thomas (1797–1840) 65, 77, 240

earth, ellipticity 275, 282
   figure of 35, 56, 114, 273–88
      Lambton assumption 17
      Everest 1st 273, 277, 278, 284
      Everest 2nd 273, 279, 285
      Lambton 35, 276
      tabulated 287
East India Company 33, 41, 55, 58, 61, 63, 102, 105, 109, 110, 119, 130, 233, 249, 250
   agent for Government 9
   Everest recommended to 10
   Officers from Woolwich 8
   Robert Everest chaplain to 7
   Royal Charter 8, 9
   scant knowledge 274
Edmonstone N.B. 55, 56, 58
Everest, Alfred Wing (1856–1928) 134, 136, 169

Benigna Edith (1859–1860) 134, 140, 170
Charles (1802–1803) 168
Cyril Feilding (1887–1916) 169
Emma Colebrooke (1849–1852) 134, 144, 169
Emma, Lady (1823–1889) née Wing 134, 136
   death and cremation 144
   family 135
Ethel Gertrude (1855–1916) 134, 161, 169
George (1790–1866) 62
   administration 267–71
   appointment as Surveyor General 61
   appreciation 130
   arrived in India 11
   baptism 6
   belief in Golden Mean 134
   bicentenary of birth 227
   birth 6
   CB and knighthood 102, 133, 138
   certificate from Woolwich 10
   children 134
   citation from Sir J. Herschel 135
   Claybrook Hall 134
   Clubs 133
   Coat of Arms 139, 182
   death 135, 143
   dominant figure 213
   donations 138
   dual role 67
   Duke of Sussex letters 107, 111
   fever 12, 25, 43, 46, 50, 94
   first joined Lambton 18, 20
   freemason 112, 133
   FRS 54
   gentleman cadet 8, 19
   gold chronometer 144
   handover to Waugh 123
   health concerns 133
   Java survey 11
   justifying fitness 46
   Lambton's assistant 19
   marriage 134, 182
   Manager at the Royal Institution 136
   mathematical ability and potential 19
   Member of Royal Asiatic Society 137
      Royal Institution 136
   memo on GTS 11
   memoir to EIC 53
   military education 8
   move to Mussoorie 147
   Neverrest 4, 271

302   Index

portraits 140
qualities 39, 225
religious beliefs 157
report, to Blacker 51
  1830 20, 54, 59, 63, 284
  1847 131, 135, 284
retirement 132, 182
river clearance 12
sailed for Cape of Good Hope 26
  England 52, 53
  India 10, 65
salary 58, 95
sick leave 47, 50
successor designate 105
Superintendent of GTS 41, 180
support from India House 55
syllabus studied 9
takes over 41
telegraph line survey 14
testimonial from RAS 135
theodolite 54
visitor to Greenwich Observatory 137
Westbourne Street 134
wish to retire 129
George John (1835–1908) 168
George Wilfrid (1890–1913) 169
John (1718–1769) 5, 165
John (1788–1820) 7, 167
Lancelot Feilding (1853–1935) 7, 121, 134, 136, 144, 169
Lucetta Mary (1787–1857) 7, 140, 167, 179
Mont 215
Mount 132
  accepted height values 201
  announcement of height 197
  Chinese expedition 207
  computation of height 192, 193
  'discovery' of 193
  early results for height 193, 194
  first expedition 222
  height 189–209
    accuracy 196, 197
    by GPS 207, 208
    on WGS 84
  Italian-Chinese expedition 208
  modern observations 206
  name 198, 211–24
  not prominent 199
  over-fly 204
  photo-survey 204
  rename by Chinese 223
  spheroidal height 202

summary of heights 209
  names 224
  what is measured 195
Robert (1798–1874) 7, 84, 136, 147, 167
Thomas Roupell (1800–1855) 7, 159, 160, 163
Tristram (1656?–1721) 5, 165
Tristram (1688–1752) 5, 165
William Tristram (1747–1825) 6, 139, 165, 179
  will 7
Winifrid Crew (1851–1910) 134, 169, 176
excess, spherical 75, 284
expansion, coefficient of 76, 79, 151

Fallows, Fearon (1789–1831) 27, 33
Faraday, Michael (1791–1867) 107, 136
Feilding, William George (1784–1868) 50, 63, 178
Fergusson R.B. (1790–1825) 14
Fisher, Anne 5
Foster, Henry 54
Freshfield D.W. 201, 218, 219, 220, 221, 222

Garstin, John (1756–1820) 61
Gauss, Carl F. (1777–1855) 129, 237
geodesy 17, 46, 55, 117, 273–88,
geoid separation 196, 197, 206,
geoidal height 197, 200, 202,
Gilbert, Davies 54
Gompertz, Benjamin (1779–1865) 54
GPS 207, 208
gravity value 55, 274
Great Trigonometrical Survey 12, 16, 19, 42, 53, 114, 118, 131, 233, 265
  idea for 274
  computations 59
  start 275
Gulatee B.L. 194, 196, 197, 200, 201, 202, 206
  summary of heights 201
Gwalior State 89, 96, 101, 226
Gwernvale 6, 139

Hack, Mary 5
Hahnemann, Samuel (1755–1843) 7, 8, 163, 164
Hamilton, Sir William Rowan (1805–1865) 60, 63, 107, 180
Hastings, Lord Francis (formerly Lord Moira) (1754–1826) 19
Hathipaon 87, 93, 147, 148, 154
headquarters, move of 79

Hedin, Sven (1865–1952) 212, 213, 218
heliotrope 23, 60, 82, 83, 84, 91, 237
    invented by Gauss 237
Hennessey, John B.N. (1829–1910) 132, 192, 193, 199, 217
Herschel, Sir John F.W. (1792–1871) 8, 33, 54, 107, 109, 115, 118, 119, 120, 135, 136, 159, 161, 162
Himalayas 89, 91
Hinton, Howard Everest (1912–1977) 176
Hodgson Brian H. 212, 214, 215, 216, 218, 219
    mistaken 217
Hodgson, John (1800–1894) 45, 62, 70, 91, 93, 231, 249, 258, 262, 263, 275
Hooker, Sir Joseph 219
Howard-Bury, Col. 193, 213, 222
Humbolt, Baron von (1769–1859) 63
Hunter, J.De Graaff (1882–1967) 195, 196, 200, 201, 202, 204
Hussain, Saiyid Mir Mohsin (d.1864) 70, 95, 98, 123, 125, 131, 150, 232, 234, 241, 247, 252, 255, 258–9
    at baselines 259
    return to Calcutta 259
    work on circles 258
Hutton, Dr Charles (1737–1823) 9, 32
Hyderabad 15, 20, 22, 25, 35, 40, 43, 44, 46, 100, 124, 265
hygrometer 237

Ichamati river 12, 179
India, Atlas of 42, 106, 261–5
    maps of 4
    projection 262
    secrecy 263
Indian, Father of Geodesy 39
    peninsula, width of 17
instrumentation, new 58
Irish Survey 60, 65, 118
isostasy 282
Ivory, Sir James (1765–1842) 9, 107, 116

Jacob, William (1813–1862) 102, 125
Java Survey 11, 179
Jervis, Thomas Best (1796–1857) 105–21, 264
    address to British Association 106
    resignation 129
    retirement 121
    survey of South Konkan 117
Jones, William (1812–1871) 102, 113, 115

Kaliana 120, 123, 124, 136, 253, 254, 278, 281, 282
    astronomical station 96, 123
Kalianpur 88, 89, 123, 124, 125, 259, 271, 281, 282
    astronomical station 96, 97, 123
    datum 196, 197
    latitude 48
Kalinjar, siege of 11, 179
Kallonas, Nicholas 82
Kanchenjunga 189, 191, 198, 208
Kater, Henry (1777–1835) 17, 54, 59, 234, 238, 239, 240, 276
Keelan, Henry (1817–1887) 82, 83, 85
Kidd (Kyd), Alexander (1754–1826) 61
Konkan Survey 105, 117, 261
Kumpass Walla 4, 271

LaCaille, l'Abbé de (1713–1762) 27
    arc of 28, 33, 180
    Journal 27, 28, 29, 33
    observatory 28, 31
Lady Barlow 12
Lambton, William (c.1753–1823) 29, 37, 41, 47, 54, 55, 56, 57, 76, 77, 99, 107, 109, 110, 111, 114, 118, 125, 127, 190, 234, 237, 239, 242, 248, 250
    Academy of Sciences 114
    angle misclosures 284
    death 39, 40, 90, 180
    Diploma from Academy 55
    disposal of property 40
    Everest appointment 20
        as assistant 14
        letters from 27
    Fellow of Royal Society 114
    last report 107
    observing in rainy season 23, 37
    original arc 58, 273, 274
    permission for arc 16, 17
    refraction 190
    Superintendent of GTS 20
lamp or light 23, 37, 46, 238
    Argand 95, 181, 238, 257
    blue 37, 38, 82, 83, 84, 86, 239
        manufacture 88, 239
    burning of 83, 85, 97
    Drummond 60, 240
    reverberatory 37, 60, 73, 238, 257
    vase 37, 180, 239
Lane, Charles (b. 1817) 101, 191, 192, 194, 214

Index

Lawrence, Henry (1806–1857) 47, 130
level 240
lithography 263
Logan, George (1809–1854) 94
Logarithm Lodge 149, 151, 153
longitude, degree of at Cawksally 15
   determination of 283
   series, Bombay 35, 90, 102, 127, 180
      Calcutta 69, 74, 102, 246
      North connecting 126
      North East 90, 132, 190, 192, 194
Lords Cricket Ground 65, 74, 180
Low, Robert (1791–1863) 48

Mackenzie, Colin (1753–1821) 12, 18, 62, 262
   Surveyor General 20
Maclear, Thomas (1794–1879) 28, 33
Madras 17, 26, 34, 70
map(s)
   Committee 263, 264
   D'Anville 213, 214
   early 213
   Jesuit Fathers' version 213, 214
   production 265
Marlow, Royal Military College 8, 9, 47, 179
Matabhanga river 1, 179
Mathematical Instrument Maker 71, 98, 249, 250, 255, 259
Melvill J.C. 111, 119
meridian
   arc station Damargidda 39
   series, Amua 90, 101, 126, 127
      Budhon 89, 100, 126
      Calcutta 190
      Karara 102, 126, 127
      Ranghir 90, 101, 126
      South Parasnath 75, 89, 100
Minto, Lord 11
mistake, perpetuation of 32
Moira, Lord Francis (later Lord Hastings) (1754–1826) 12, 19
Montgomerie J. 198, 216
Morison, Sir William (1781–1851) 264
Morrison, Charles 92, 147, 149, 150
Morton, Dr John (1796–1866) 40, 41
Mudge, William (1762–1820) 10
Murphy, Charles (b. 1798) 60, 65, 77, 82, 101
Mussoorie 80, 95, 131, 148, 150, 152, 200, 241, 254

Nanda Devi 189
Napoleonic Wars 11

Natha Singh 222
Nepal, opening of 206
Nicholson, James (1819–1869) 191, 192, 193, 194

Observatory, Bombay 35, 39
   Cape 27, 28, 33
   Damargidda 124
   Dublin 60
   Greenwich Royal 5, 51, 59, 250
   Madras 17, 35, 39, 198
Olliver, Joseph (b. 1785) 22, 25, 26, 45, 46, 48, 69, 72, 80, 85, 101, 125, 246
Ordnance Survey 77, 78, 108

Pachitan Bay, survey of 12
Park Estate 71, 80, 87, 147–54, 254, 258
   boundary pillars 151
   disposal of 152, 181
   future 154
   measurement of 149
   office 150
   purchase by Everest 148, 181
   remains 153
Park House 149, 151
Parks, Fanny 150, 181
passport, Tibetan 222, 223
Paudree 17
Peak b, h, γ 191, 192
   K2 198
   XV 132, 190, 191, 192, 193, 194, 196, 197, 201, 211
      coordinates 198, 202
pendulum, apparatus 59, 60, 116, 240
   experiments 109, 151, 274, 280
   Kater variety 240
Penning, Joshua de (1784–1645) 40, 41, 44, 45, 47, 69, 94, 128, 253, 258
perambulator 14, 20, 84, 234, 240
   Everest design 240, 241
Peyton, John (b. 1804) 94, 113, 150, 151
Phillimore R.H. 3, 17, 100, 108, 152, 153, 155, 193, 226
Picard, Jean (1620–1682) 55
plane table 241
plumb line deflection 31, 55, 275, 281, 282
Pratt, John H. (1809–1871) 281, 282
Prinsep, Henry Thoby (1792–1878) 264
   James (1799–1840) 74, 79
problems, staff 45
   survey 46
projection, Bonne 262

Flamsteed 262
Puissant, Louis (1769–1843) *Traité de Topographie* 262
Traité de Géodésie 190
punishment 73
Punnae 18, 136

Raffles, Sir Stamford (1781–1826) 11
Ram Chundra 161, 162, 163
Ramsden, Jesse (1735–1800) 60, 114, 247, 256
  bar 76
ray trace method 83, 84, 94
refraction 37, 42, 189, 190, 191, 192, 193, 194, 199, 211, 237
  coefficient of 194, 200
Reichenbach 63
Rennell, James (1742–1830) 17, 39, 61, 110, 111, 261, 262
Renny-Tailyour, Thomas (1812–1885) 90, 100, 101, 113, 115, 116, 123, 124, 125, 126, 129, 131
resection 33
revenue surveys 267
Reynolds, Charles (1756–1819) 62
Richardson, William 51, 59, 250
Rock, Joseph 204
Rossenrode, William (1792–1852) 22, 24, 49, 80, 84, 85, 88, 89, 96, 97
Roxburgh Castle 65
Roy, William (1726–1790) 16, 232
Royal Asiatic Society 214, 215
Royal Astronomical Society 124
Royal Geographical Society 140, 197, 198, 214, 218
Royal Society 54, 55, 106, 107, 108, 113, 116, 138, 277
Ruge, Dr S. Mont or Mount 220
Ryall, Dr John 7
  Mary (1809–1895) 163, 168

Sabine, Edward (1788–1883) 107, 138
Schehallien 31
Schlagintweits, the 216, 217, 218, 219, 220, 221
Scott William H. (1812–1873) 217
sextant 241
Shaw, Dr 134, 135, 241
Shortrede, Robert (1801–1868) 75, 78, 90, 91, 102, 126, 127, 226, 243, 278
Sickdhar, Radhanath (1813–1870) 127, 128, 131, 132, 191, 193, 211

member of RAS 132
Simms, William (1793–1860) 60, 77, 180, 238, 243
Sironj 44, 47, 48, 69, 74, 88, 89, 93, 97, 99, 100, 102, 125, 278
Smith, Lucetta Mary (b. 1809) 6, 10, 165
Smyth R. 132
Snellius, Willibrord (1580–1626) 15
Solo river survey 12, 177, 179
South, Sir James (1786–1867) 54, 136
spheroid, Everest 196, 197
  international 206
staffing problems 271
standard, comparison of 78, 92
  Parliamentary 78, 79
  scale, Indian 109, 276
star observations 124
station location 29, 81
Struve, Friedrich (1793–1864) 78
Superintendent of GTS 56, 61, 67, 68, 129, 131, 254
survey, destruction of marks 270
  party size 270, 271
Surveyor General of India 56, 61, 67, 68, 95, 97, 103, 113, 119, 127, 129, 131, 249, 254
  prime aim of 261
Sussex, Duke of (1773–1843) 46, 107, 108, 111, 112, 119, 120, 181
Sydenham, Thomas (1780–1816) 15

tables, logarithmic 47, 243
  presented to Royal Society 243
  traverse 243
tahsildar 270
Takalkhera 35, 44, 47, 48, 278
Tandy, Sir Edward 223
tape measure by Chesterman 69, 181
Tate W.A. (1795–1871) 107, 110
Taylor, Sir Geoffrey Ingram (1886–1975) 176
  Thomas 51, 59, 65, 86, 87
telegraph line survey 14, 51
Tennant, James F. (1829–1915) 191, 216, 217, 281
Thatcher, Robert (1812–1899) 152
theodolite, Cary/great/36 inch 17, 38, 51, 60, 62, 94, 114, 245, 246
  18 inch 61, 234, 247
  24 inch 60
  36 inch, damage to 60, 61, 62, 245, 250
  3 ft 86, 89, 95, 246
  Everest 54, 181, 243, 244
  observations 75, 180, 238

thermometer, wet and dry 42
Thuillier, Sir Henry L. (1813–1906) 132, 198, 211, 212, 215, 217
tidal observations 191
Topping, Michael (c.1747–1796) 15
towers, observation 35, 36, 68, 70, 73, 82, 94, 98, 101, 126
triangulation 15
Trivandeporum 17
Troughton & Simms 77, 98, 180, 243, 244, 246, 254
Troughton, Edward (1753–1836) 60, 79, 116, 253
Tulbagh, Governor 28, 33

units, conversion of xii

Vernet, Du, James (1803–1872) 126, 127
vertical angle observation 39, 42
vertical, deflection of 197, 200, 202, 278
Vigne, Godfrey (1801–1863) 91, 92
Voysey, Henry (d. 1824) 21, 22, 23, 25, 38, 42, 43, 44, 45, 50, 238

Waddell, Col. 220, 222, 224
Walker, J.T. (1826–1896) 132, 199, 216, 217, 218, 220
  John, Atlas design 263
Wallerstein, Prof. George 207
Walpole, Henry (1787–1854) 62
Warren, John (1769–1830) 17, 18
Washburn, Bradford 209
  mapping of Everest area 226
Waugh, Andrew (1810–1878) 39, 90, 93, 95, 98, 101, 102, 106, 113, 115, 116, 123, 124, 125, 126, 129, 130, 131, 132, 189, 191, 192, 193, 194, 195, 197, 198, 202, 211, 212, 214, 218, 219, 220
  appointed Surveyor General 129, 130
  at Bidar base 102, 125, 259
  at Damargidda 124
  at Dehra Dun baseline 93
  at Sironj baseline 99
  Himalayan heights 189
  mountain computations 132
  reobserving Lambton work 277
  resource problems 269
  simultaneous observations 123
Webb, Ronald Stretton (1892–1976) 28, 32
Wellesley, Arthur (later Duke of Wellington) (1768–1852) 16
Western, James (1812–1871) 59, 65, 70, 73, 75, 89, 100
Wheeler, E.O. 204
Whewell, W. 107, 109, 115, 119, 120
Whish, William Simpson (1787–1853) 148
Wickwar, Everest, T. R. rector to 7
Wilcox, Richard (1802–1848) 80, 86, 252
Wilkins, Charles (1749?–1836) 54
Williams, Monier (1777–1823) 62
Wing, Arthur (1828–1873) 174
  Emma = Lady Everest (1823–1889) 176
  Henry Tryan (1851–1886) 136
  Henry Vincent (b. 1834) 174
  John (1643–1726) 171
  Mary Ann (1798–1880) 143, 174
  Thomas (1796–1850) 143, 174
  Thomas Twining (1826–1904) 174
  Tycho (1696–1750) 171, 174
  Vincent (1619–1668) 170
  Vincent (1840–1885) 176
Wood, Henry 221
  expedition to mountain 202
  Mark (1750–1829) 61
Woolwich, Royal Military Academy 8, 10, 47, 179
workshop in Calcutta 251
Worthington & Allan chain 76

Yellapooram 25, 43
Younghusband, Francis expedition to mountain 202

zenith circle 61
  sector 17, 18, 29, 30, 33, 38, 44, 46, 247
  by Dolland 248